Water and Wastewater Management

Volume-2

Water and Wastewater Management

Volume-2

D. R. Khanna
R. Bhutiani
Gagan Matta
Vikas Singh

2024

Daya Publishing House®
A Division of
Astral International Pvt. Ltd.
New Delhi – 110 002

First Published, 2010

Reprinted, 2024

ISBN: 978-81-7035-748-3 (HB)

Published by : **Daya Publishing House®**
A Division of
Astral International Pvt. Ltd.
– ISO 9001:2015 Certified Company –
4736/23, Ansari Road, Darya Ganj,
New Delhi-110 002
Ph. 011-43549197, 23278134
E-mail: info@astralint.com
Website: www.astralint.com

Printed at : **Replika Press Pvt. Ltd.**

Preface

Indian rivers are known to be sacred for us because a large number of populations depend directly or indirectly on it for their daily use. Water resources are of critical importance to both natural ecosystems and human development. But industrialization, urbanization and rapid increase in population has resulted in the degradation of surface as well as ground water quality. Various industrial processes coupled with domestic activities often produce harmful discharges which ultimately found their way to the water resources. Water pollution not only affects inhabited plants and animals, individuals, species and populations but also to the natural biotic communities as well.

This book presents an overview of wastewater pollution and its abatement techniques which are being used in the present world. This book will also result in the preservation of the precious data of various significant water resources which could be helpful for present and future both. Through this book we have tried to determine the present status of water resources, their conservation strategies *etc.* We are grateful to the eminent scientists, researchers, scholars *etc.*, for contributing their research work for the benefits of humankind. We acknowledge for their great support and hope best for this book in fulfilling the purpose of making awareness among social community.

Editors

Contents

— Volume 2 —

Water and Wastewater Management in 2 Vols (2011) *Pages* **343–347**
Editors: **D.R. Khanna, R. Bhutiani & Gagan Matta**
Published by: **DAYA PUBLISHING HOUSE, NEW DELHI**

Chapter 54

Water Quality Management: Strategies for Conservation of Bhopal Waters

Jaya Garg and H.K. Garg
Oriental Institute of Science and Technology, Bhopal – 462 021
Department of Zoology and Aquaculture, Govt. P.G. College, Sehore – 466 001

ABSTRACT

The city of Bhopal is crowned with lakes and reservoirs, which are major sources of potable water, recreational activities and aquaculture. These reservoirs are under great environmental stress due to agricultural run offs, human encroachments, siltation and growth of aquatic needs. In order to maintain Upper lake as a healthy water- resource for drinking purposes, lower lake as a recreational ground for boating, water games, tourism and surface transport and Shahpura reservoir as a pisciculture station, an attempt has been made to formulate certain measures, that can cure the already spoiled lakes and check the water sheds from getting eutrophicated further.

Keywords: *Lake, Conservation, Restoration, Bhopal waters, Eutrophication.*

Introduction

Man made eutrophication of freshwater emerged in sixties as one of the most relevant causes of water quality deterioration. Excessive enrichment of inland waters, coupled with dramatic urbanization, has led to formidable algal blooms in many parts of the world. These algal blooms are, not only believed to impair the over all quality of water but, also known to produce toxins, which when accumulate in fish, can along with food chain, be transferred to man.

Since 1980's Italy has progressively introduced legislation limiting phosphate content from 26 per cent (1982) to just 4 per cent (1988) in waters (Water and Environment International, 1992). In

South Sweden, the agricultural areas especially of Lake Rings fen ere given so much attention that in 1985, Government restricted the use of fertilizers, storage and distribution of manure and operation of sewage treatment systems (Forsberg, 1987). Countries like USA, Sweden and Israel have taken several concrete measures to restore the already eutrophied lakes but unfortunately very little information are available on such programmes in India and that too on Bhopal waters. In the present communication, a few restoration measures have been explored after thoroughly consulting a number of successful examples from abroad and from the data compiled in the laboratory during last ten years. The purpose is directed to plough right through the problem practically rather than giving mere deductions.

Exploration Zone

Bhopal, the capital of Madhya Pradesh, is situated amidst Vindhyachal ranges at an altitude of 495 m above MSL. The city has a pride possession of lakes and reservoirs. Upper lake has been the lone source of drinking water to 12 lac population of the city. The reservoir is said to have been a part of much larger sheet of water that existed in the 11th century. The present stretch of 32 kms^2 has been resulted by impounding a seasonal rivulet Kolans.

The lake is dammed at two places, the eastern end near Kamla Park and the western bank, which is surrounded by agricultural fields. Forested slopes of Shamla hills adorn the southern fringe of the lake, contributing raw sewage and agricultural runoffs. During festivals, nearly 2000 idols are immersed every year that adds to 75 tons of silt into lake. Consequently the sinuosity of lake has receded over the years. Lower lake, located in a densely populated area of the town, was created in 1749 by constructing an earthen dam across the Upper Lake. Human interventions like bathing, washing of clothes and dumping of garbage have impinged radical changes in the composition of lake water. Cultivation of lotus with a primary aim to add beauty has been causing reduction in the depth of the lake. Further, the southern arm of this lake is completely occupied by huge vegetation comprising *Hydrilla, Lemna* and *Azolla* in the name of *Jal Upvan.*

Shahpura lake is an elliptical receptacle situated in thickly populated southern part of Bhopal. Created due to stone quarrying activity adjacent to village Shahpura, the lake was filled up by storm water and developed into the present form. Due to inadequate catchment area, most of the impounded water contains sewage inflow, which imparts colour, odour and taste to it. Despite immense strain experienced due to extrinsic as well as intrinsic factors, the lake is serving as a pisciculture station for the Department of Fisheries, Govt. of M.P. To save the water shed from extinction, its management has been taken over by E.P.C.O.

Thus very little attention has so far been given to these beautiful reservoirs, which offer drinking water supply and wonderful opportunities for water sports, tourism and surface transport.

Recommendations

In order to save Bhopal waters from hypereutrophy as well as propagation of obnoxious algal species, it is necessary to minimize harmful fluxes from land to water. Diversion of sewage is likely to be one of the most effective ways of checking eutrophication as nutrients input is largely derived from this source (Edmondson, 1972 and Bjork, 1972). In U.S.A., restoration of lakes has been successful in Lake Washington and Shangava Lake using wastewater (sewage) treatment (Goldman and Home, 1983). In Sweden, similar efforts were made by lowering detergents, phosphorus contents and nutrient load.

Studies conducted by Garg and Garg 1997, 1998, 1999, 2000 and 2001 on Bhopal waters under microcosmic conditions, also revealed that control of phosphorus had a check on propagation of algal growth. In controlled microcosm, when no nutrient was added, observed that algae found it arduous to multiply obviously due to deficiency of nutrients mainly phosphorus. In treated microcosms, they flourished well, till the time; the added nutrients already present in natural waters were available. In the later days, very low abundance was determined in microcosms. Thus in order to check the entry of nutrients, in upper lake, the sewage should be immediately diverted to Biogas based sewage treatment in the northern and southern side of the lake (long pipes for sewage transport are not recommended as they are undesirable and uneconomical). In lower lake, washer men colony should be shifted away from the bank to prevent detergent in flow, which accrues hardness and phosphorus inputs to water. Shahpura Lake is being polluted by *Panchsheel Nullah.* The suspended nitrates and phosphates of the Nullah should be prevented by putting a solid masonry barrier across it so as to head up the water and then the water should be taken into sever line by passing through a grit chamber.

Some scientists believe that nitrogen control is needed to maximize the benefits of phosphorus reduction in freshwaters. Phosphorus stimulates algal growth, which, in turn, consumes nitrogen. When phosphorus is removed, the algal growth stops and the unconsumed nitrogen is left as such where it is available to create even larger blooms. To overcome this situation, use of wetlands is expedient, which serve as a means of reducing the nutrient load to adjacent bodies of water (Kadlec 1987 and Kusler *et al.*, 1994). It is a combination of saturated soil, plants and microorganisms that provide aerobic and anaerobic conditions and encourages nutrient removal (especially nitrogen) from the overlying flood water (Johnston, 1991). These buffer zones act as sink-preventing nitrate built up (nitrification) and in turn recycle it to nitrogen gas (denitrification) by the same mechanism. Thus, in order to remove excessive nitrogen from water, wetlands should be created in the vicinity of Bhopal lakes.

Encroachments on the water margins of lakes should be taken off and dumping grounds on catchment areas should be completely banished. Human and animal intervention should be minimized in the fringe areas by making fences and a live barrier in the form of a highway.

Introduction of grass carp *Ctenopharyngodon idella* in 90 water systems of Sweden proved quite successful. Judicious stocking of selected fish types like *Ctenopharyngodon idella* (grass carp), *Hypophthalmicthys montrix* (Silver Carp), *Cyprinis carpio* (Common Carp) may effectively be used as biological agents to control plankton in the ecosystem of lakes (Bhatiya *et al.*, 1973 and Juanico *et al.*, 1994). This indirectly curbs the growing eutrophy if proper type is selected. If planktiferous fish are massive in number, they consume zooplankton consumer fish varieties are bred in Shahpura Lake or selective zooplankton community is allowed to flourish, lake water becomes comparatively cleaner. The task already taken up by fisheries department, Govt. of Madhya Pradesh is praiseworthy, still better varieties are to be bred and techniques improved. Besides, aquatic macrophytes are also reported to control eutrophication as they shade algal growth and assimilate nutrients released from both external and internal sources. This situation of limiting nutrients by macrophytes in open lakes has been proposed by Goel *et al.* (1985), Petr (1987) and Sen *et al.* (1992). Two species of Chlorophyceae *viz., Chlorella* and *Scendesmus*, are also found to reduce nitrogen and phosphorus load from sewage water (Tam and Wong, 1989) and hence can be proved effective in preventing eutrophication. In upper and lower lakes, useful plants like lotus and trapa should be sown to check the entry of nutrients, which would also serve as an additional source of revenue.

The lake waters can also be provided with copper treatment, as it is economical as well as effective to kill the floating masses. The chemical is non-toxic to fish and zooplankton and is rapidly lost to sediments. Besides, nitrate can also be reduced by carbon additions as performed experimentally in aquifer microcosms by Obenhuber and Lawrence (1991).

Bubbling and hypolimnetic aeration removes oxygen deficiency (Forsberg, 1987). When nutrients diversion is not possible, excess nutrients and low oxygen in bottom waters may be altered by physical or chemical means as flushing and aeration. Trees should be planted around the three water formations to check soil erosion, siltation and flooding. Deweeding campaign should be launched for manual removal of weeds, which can be used later as compost. This step is especially needed in Shahpura lake whose 75 per cent of lake area is covered by weeds. In upper lake de-silting should be done at least at the shores especially from where the water enters the lake and turns towards Bhadbhada while in Shahpura reservoir, drain entering it should be cemented at both banks, up to the stream of barrier.

Process of sediments removal should be started early and huge amount of organic matter that comes out can be used for agricultural purposes in the form of fertilizers. Sign boards with slogans depicting the importance and cleanliness of water should be put all around to create/awareness. A central lake authority should be set up for periodical monitoring of lakes and implementation of schemes for lake restoration. Economic development should be promoted by bringing back aesthetic beauty and vitality.

In this direction, Ministry of Environment and Forests, Govt. of India, has drafted a scheme for conservation and management of twin lakes of Bhopal (the upper and the lower) on the guidelines laid down at Ramsar Convention held in Egypt and a Grant worth of Rs. 335 crores has been sanctioned by O.E.C.F. Japan. The project has taken off with an aim to deepen and widen the spill channels, restoration of island, installation of fountains and improvement in the quality of water. However, to ensure the availability of potable water in increased quantity and satisfactory quality from upper lake and to preserve the aesthetics of lower lake of Shahpura reservoir, it is of vital importance to curb the entry of agricultural run offs, divert the inlets of sewage of impose a ban on anthropogenic activities in and around the lake basin.

References

Bhatia, H.L., Sastry, C.A. and Nigam, R.K., 1973. Control of aquatic weeds trom water bodies using grass carp. *Ind. J. Environ. Hlth.*, 15: 92–99

Bjork, S., 1972. Ecosystem studies in connection with the restoration of lakes. *Verhandlungen der Internalen FurTheoretische und Angewandte Limnologie*, 18: 379–387.

Edmondson, W.T., 1972. Nutrients and phytoplankton in lake Washington. In: *Nutrients and Controversy*, (Ed.) G.E. Likens. Limnol. Oceanogr. Spec. Symp., 1: 172–193.

Forsberg, C., 1987. Evaluation of lake restoration in Sweden. *Schweiz Z Hydrol.*, 42: 260–274.

Goel, P.K., Trivedi, R.K. and Vaidya, R.R., 1985. Accumulation of nutrient from wastewater by water hyacinth *Eichhornia crassipes*. *Geobios*, 12: 115–119.

Goldman, C. R. and Home, A.I., (Eds.) 1983. *Limnology*. McGraw Hil Inc., Tokya, pp. 464.

Garg, H.K. and Garg, J., 1997. Bhopal waters: Heading towards extinction. In: *Paper Presented in Nat. Res. Sem. on New Economic Policy, Opportunities and Challenges for Sustainable Development of Fisheries and Conservation of Natural Resources*, p. 33.

Garg, H.K. and Garg, J. Impact of magnesium acetate on algaerin experimental hydro-systems. In: *Proc. Nat. Seminar on Interdisciplinary Area of Sci. and Technology: Present and Future*, p. 24.

Garg, H.K. and Garg, J., 1999. Effect of inorganic phosphorus on algae excised trom inland waters. *Ind. J. Zool. Spectrum*, 9(1): 9–13.

Garg, J. and Garg, H.K., 2000. Impact of phosphorus on hydrobiology of Bhopal waters. *J. Ecotoxicol. Environ. Monit.* (In press).

Garg, J. and Garg, H.K., 2001. Nutrient loading and its consequences on lake ecosystem. *Tropical Ecology* (In press).

Johnston, A., 1991. Sediment and nutrient retention by freshwater wetlands: Effects on surface water quality. *Rev. Environ. Cont.*, 21(5–6): 491–565.

Juanico, M., Azov, Y., Teltsch, B. and Shelef, G., 1994. Effluents addition to a shallow freshwater reservoir used for irrigation. II. Short term effects on the filter clogging capacity of water. *Wat. Res.*

Kadlec, J.A., 1987. Nutriet dynamics. In: *Aquatic Plants for Water Treatment and Resource Recovery*, (Eds.) K.R. Reddy and W.H. Smith. Mangolia Publ. Inc., Ornaldo, pp. 393.

Kusler, J.A., Mitsch, W.J. and Larson, J.S., 1994. Wetlands. *Scientific American Current Contents*, 37: 3.

Obenhuber, D.C. and Lawerence, R., 1991. Reduction of nitrate in aquifer microcosms by carbon additions. *J. Environ. Aual.*, 20(1): 255–258.

Peter, T., 1987. Fish, fisheries, aquatic macrophytes and water quality in inland waters. *Water Qual. Bull.*, 12(3): 103–129.

Sen, N.S., Hora, P. and Mishra, V.K., 1992. Physico-chemistry, nutrient budget and the factors influencing primary production in a tropical lake. *J. Freshwat. Biol.*, 4(2): 141–153.

Tam, N.F.Y. and Wong, Y.S., 1989. Wastewater nutrient removal by *Chlorella pyrenoidosa* and *Scenesdesmus* species. *Environ. Poll.*, 58: 19–34.

Water Environment International, 1992. *Excerpts.* Vol. 2(14), 6. Internal. Trade Publ. Ltd., Redhill, England.

Water and Wastewater Management in 2 Vols (2011) *Pages 348–367*
Editors: D.R. Khanna, R. Bhutiani & Gagan Matta
Published by: DAYA PUBLISHING HOUSE, NEW DELHI

Chapter 55

Removal of Nitrate from Drinking Water: A Review

Samina Hasan and S.P. Pande
National Environmental Engineering Research Institute, Nagpur – 40 020

ABSTRACT

The concentration of nitrates in groundwater is observed to be increasing in the world in general and Asian countries in particular. Groundwater is still a main source of drinking water especially in rural areas of Asian countries. Nitrates are known to cause methaemoglobinemia in babies and intestinal cancer due to formation of nitrosamines in general population. Several treatment processes for the removal of nitrates from drinking water have been studied. These processes comprise; vegetative de-nitrification, reverse osmosis, electrodialysis, catalytic reduction, chemical reduction, ion exchange and biological denitrification. Combinations of two methods like combined ion exchange-biological denitrification have also been studied. Every method have merits and demerits, however, the methods based on ion exchange for the removal of nitrates from drinking water and regeneration of resin by biological denitrification appear to have edge over other methods. This communication presents a review on existing denitrification processes and spells out needs for future research.

Keywords: Nitrate, Denitrification, Drinking water, Pollution.

Introduction

Nitrogen exists in the environment in several oxidation states, with ammonia, nitrate and diatomic nitrogen gas being the main forms. Nitrates constitute the final stage in the oxidation of nitrogen compounds and are; therefore a measure of the original quantity of organic matter with which water is associated. The nitrate contained in pure well waters derived from extensive catchments is largely the result of biological activity in the surface layers of soil enhanced by cultivation and the application of manures. The nitrate of stream of rivers may, however receive substantial increment from well-nitrified sewage effluents. Excessive use of nitrogenous fertilizers in agriculture has been one of the primary sources of high nitrates in groundwater (De Roo, 1980; Scheper *et al.*, 1984). The increase in levels of nitrate nitrogen can also arise in intensively cultivated area under horticultural lots and under animal feed lots (Wilson *et al.*, 1999). Presence of high level of nitrates has been reported in several countries (Pande and Hasan, 1979; Fried, 1991; Spalding and Exner, 1991, 1993 and Nixon, 1992).

Sources of Nitrate

The primary source of all nitrates is atmospheric nitrogen gas, which is converted to organic nitrogen by some plant species by a process called nitrogen fixation. On the death of the plants, the organic compounds are decomposed by microorganisms to inorganic ammonium salts (ammonia fixation), which in turn are converted to nitrates by a process called nitrification. The intermediate product nitrite is generally short lived and seldom accumulates in significant quantities in any natural environment in environment that are depleted in oxygen, some microorganisms can use nitrate in place of gaseous oxygen to carry out their metabolic processes. The products of this reaction are nitrogen gas and/or nitrous oxide. This process is called denitrification (Hounslow, 1995). Nitrate occurs in almost all natural waters and its concentration can go up to hundreds of mg/l except when contamination is present, they seldom exceed 20 mg/l. However 10 mg/l as NO_3^- N or greater may be regarded as probable indication of contamination from fertilizers, municipal wastewater, feedlots, septic systems and sometimes the cultivation of grasslands.

Nitrogen chemistry in natural water is complex because nitrogen exists in different oxidation states such as NO_3^- (+5), NO_2^-(+3) and ammonium NH_4^- (-3). The NO_3^- (+5) form is expected under oxidized conditions and NO_2^- is found under moderately reduced conditions. In addition, nitric oxide NO (g), nitrous oxide N_2O (g), and nitrogen N_2 (g) are also expected in moderately reducing conditions (Stumn and Morgan, 1996). Among several N species, NO_3^- is the most stable and is commonly found in groundwater associated with intensive agriculture. Nitrate itself is nontoxic. Doses up to 9 gm/day have been used to treat phosphatic kidney stones in humans without any adverse effect. Contamination of drinking water with nitrate presents a health hazard because NO_3 can be reduced to NO_2 in the gastrointestinal tract and causes 'Methaemoglobinaemia' alternatively called as 'Blue Baby Syndrome', a some times fatal disease to which infants are particularly susceptible and also affects adults deficient in glucose phosphate dehydrogenase. In addition nitrate and nitrite have a potential to form carcinogenic N-nitroso compounds, which is a postulated cause of stomach cancer (Majumdar and Gupta, 2000). Thus it is important to maintain low levels of nitrate in drinking water. WHO and U.S.E.P.A. have suggested the guideline value of 10 mg/l as NO_3^- N (Lunkad, 1994).

Nitrate contamination of our ground and surface water resources is becoming an ever increasing problem. In central Himalayan snow and ice, NO_3^- content is about 0.5 mg/l (Lunkad, 1994) while world average river water contains 1.0 mg/l NO_3^- and the ultimate sink of terrestrial waters, the oceans, on average, have 0.67 mg/l NO_2 (Mason and Moore 1985). In some parts of Europe, nitrate concentrations in groundwater reached serious levels 20 years ago and have continued to increase (Barrenstein *et al.*, 1986). In 1970, nitrate levels in 60 public supply groundwater sources in England and water intermittently exceeded guideline value, in 1980 this number increased to about 90; and in 1987 about 142 (Hourse of Lords, 1989). Surveys conducted in India have shown excessive level of nitrates in groundwater (Pande *et al.*, 1979; Gopal *et al.*, 1977; Gupta 1981 and Handa *et al.*, 1982).

Methods of Nitrate Removal

Nitrate is highly soluble in water and there are very few known insoluble compounds of nitrate. The removal of nitrate is therefore difficult using conventional flocculation and sedimentation methods. In view of this, technologies based on physico-chemical principles and biological denitrification has been used for the removal of nitrate from drinking water. Some of the prominent methods used for denitrification of drinking water are described below.

Reverse Osmosis

Reverse Osmosis (RG) is a pressure driven process, which is accomplished by passage of water through a membrane against the natural osmotic pressure to accomplish separation of water and ions. Membranes commonly used are made up of cellulose acetate (0.10 to 0.15 mm thick), while membranes made up of polyamides and composite membranes are also available.

In the reverse osmosis process, a thin membrane separates two salt solutions. Water from the side of lower salt concentration flows through the membrane to the solution of high concentration attempting to equalize the salt content, while the membrane allowing water flow blocks passage of salt ions. If pressure is applied to side of higher salt content, flow of water can be prevented. If pressure is increased the water flow is reversed and passes from salt water to the freshwater; in this manner the salt are separated from the solution. Reverse osmosis operating pressures very between 350 and 1500 psi with a typical range of 600 to 800 psi. Because of high rejection of salt ions, the necessary reduction of the nitrate concentration of well water may be achieved by treating only part of the total amount of water and bypassing the rest. Reverse osmosis, however, cannot separate nitrate selectively. Reverse osmosis generally results in reduction of mineral content in water.

Rautenbach *et al.*, 1986 commented that the economics of RO process requires high nitrate rejection of the membranes, high water flux at low driving forces and a high recovery rate. They informed that in most of the cases of high nitrate in groundwater in Germany, the dissolved solids are below 600 mg/l and therefore, low pressure modules can be used for nitrate removal. Based on a pilot plant studies they reported that (1) all reverse osmosis membrane are not suitable for nitrate separation (2) feed treatment and recovery rate have to be adjusted individually to the feed water quality and composition, specially the levels of Ca, Sr, Ba, Si, HCO_3 and SO_4 and (3) permeate must be conditioned by process like CQ stripping, lime treatment to overcome CO_2 and consequently avoid passing of CO_2 through the membranes. They finally commented that the removal of nitrates by RO system is economically viable. Well water having nitrate concentration of 94 mg/l was reduced to 18 mg/l by RO process.

The concentrate resulting from the reverse osmosis system had a pH value of approximately 7.0 so that the wastewater could be directly discharged into the sewers without any further treatment.

Electro Dialysis

In the electro dialysis (ED) process, ions are separated from the water by attracting them through selective ionpermeable membranes using an electrical potential. The unit consists of membranes interposed in the path of a direct current generated by end electrodes. Electro dialysis employs two different kinds of membranes- anion and cation selective membranes-, which are stacked alternatively from the diluate and retentate compartments. The feed is concentrated in the retentate compartment and depleted in the diluate compartment as a consequence of the selectivity of the membranes (Hammer, 1991). An electro dialysis system requires a supply of pressurized water (50-75 psi), a membrane stack and a direct current power source. Nitrate removal efficiency in ED is about equal to that of reverse osmosis and the process requires lesser acid dosages than RO and higher water recovery rates.

Inorganic and organic both types of membranes are used for electro dialysis process. White *et al.* (1991) used inorganic membranes for the electro dialysis (barium silicate, phosphate composite membrane). Adhikari *et al.* (1991) used electro dialysis for desalination of brackish water and 86.7-97.8 per cent reduction in nitrate concentration resulted in water containing nitrate in the range of 26.0-144.0 mg/l and feed water IDS in the range of 6000-36000 mg/l. A selective nitrate removal process was developed by Miquel and Oldani (1991), which was effective in removing nitrate concentration from 50 mg NO_3/l to 25 mg NO_3/l without the addition of any chemical.

Nitrate has been removed by electro dialysis method using a nitrate specific anion exchange membrane (Indusekhar *et al.*, 1991). Amine groups were incorporated in the membrane prepared from chloromethylated poly sulphone. The membrane exhibited a larger nitrate flux in comparison to chloride flux.

A pilot dialysis unit 'Micro Acilyzer. G5' with monovalent permselective membrane was used by Saddi (1998) for the removal of nitrate and chloride from groundwater. It was found that ED reduced both the nitrate and chloride level below the WHO guidelines at 77 per cent recovery without any acid dose. The energy consumption per Kwh/g salt/m^3 product water was 0.48. The membrane selectivity depended strongly on the individual ion concentrations of the feed water. Linear flow velocity increase needed a longer process path length (membrane length) in order to reach the same product water quality.

Segall and Clifford (1992) presented theoretical fluid transport and chemical reaction equations for one and two dimensional fluid flow induced by an electric field. Laboratory columns and two dimensional model studies examine electroosmosis (EO) as a method of contaminant removal. Nitrates dissolved in EO column influent water more through a column and are reduced at the cathode as follows:

$$NO_3^- + 2H^+ + 2e^- \rightarrow NO_2^- + H_2 + O \qquad E_o = +0.84 \text{ volts}$$

$$NO_2^- + 8H^+ + 6e^- \rightarrow NH_4 + 2H_2O \qquad E_o = +0.89 \text{ volts}$$

$$2NO_3^- + 12H^+ + 10e^- \rightarrow N_2\,(g) + 6H_2O \qquad E_o = +1.24 \text{ volts}$$

At high pH, little N_2 is expected, but ammonia gas does evolve from the collected column effluent.

$$NH_4^+ + OH^- \rightarrow NH_3 + H_2O$$

Catalytic Reduction

The catalytic reduction of nitrate represents an effective, selective and almost residue free process for the selective removal of nitrate from water. Nitrate and its intermediates like nitrite are selectively reduced to nitrogen on Pd-based bimetallic supported catalysts.

Horold *et al.* (1993) tested bimetallic catalysts to remove nitrate from drinking water. A catalyst with lead (5 per cent) copper (1.25 per cent) impregnated on Al_2O_3 was observed to completely remove nitrate from water having initial nitrate concentration of 100 mg NO_3^-/l. The reaction was completed in 50 minutes. The nitrate removal activity of the catalyst was 3.13 mg NO_3^-/min/g of catalyst, which was 30 times greater than that of microbial denitrification. The process operated effectively in the pH range of 6–8.

Reddy *et al.* (2000) designed and developed an apparatus for testing the catalytic reduction process under different redox potentials and evaluated the effectiveness of different catalysts in removal of nitrate from groundwater.

Three catalysts Palladium (Pd), Platinum (Pt) and Rhodium (Rh) on 5-10 per cent carbon were tested in this study. One litre of groundwater was amended with 0.5 g catalysts and reacted at different redox potentials (340 to –400 mv) and reaction times (1-6 h). During the catalytic reduction process the pH was maintained around 6.5 by bubbling 5 per cent carbon dioxide (CO_2). Initial nitrate concentrations ranged between 32 and 41 mg/l. Among the three catalysts, Rh was observed to be most effective the removing NO_3 in groundwater samples. Results with Pt catalyst showed very slow reduction *e.g.* at –400 mv and 6 and 14 hours reaction times nitrate level decreased only by 25 per cent and 44 per cent respectively. Results suggest that Rh catalyst at –400 mv and 6 hours reaction time can decrease NO_3^- concentration from 40 to 11.9 mg/l (68 per cent). Nitrate was not detected during the nitrate reduction process. The re-oxidation of formerly reduced sample to 390 mv resulted in no increase in the concentration of NO_3^- and its reduction rate significantly increased on application of a small flow of current. Application of 4.6-6.1 v to 250 mv and 6 hours reaction time resulted in 80 per cent (40 to 7.9 mg/l) decrease in NO_3 concentration. These results suggested that the Rh catalyst was effective in lowering NO_3 concentrations in groundwater (Horold *et al.*, 1993). The reaction is as under.

$$2NO_3^-(aq.) + 5H_2(g) \rightarrow 7N_2(g) + 2OH^-(aq.) + 4H_2O$$

The proposed process has a potential for field application because Rh catalyst can be coated onto a fiberglass mesh and the desired redox potential can be acquired with a photo volatic cell. This process could become potentially inexpensive method as the sunlight could be used as the energy source to activate the catalyst.

Chemical Reduction

Nitrate can be removed by its chemical reduction, to nitrogen. Murphy (1991) described a chemical denitrification process by use of powdered AI, which is a powerful reductant, can decompose water, NO_3 and SO_4 and has an optimum pH 10.25 for nitrate reduction. Al powder reduces nitrate to ammonia, nitrogen and nitrite. The removal of nitrate ion by Al may first occur by adsorption onto the particles. The reactions probably proceed to nitrite as under.

$$3NO_3^- + 2Al + 3H_2O \rightarrow 3NO_2^- + 2Al(OH)_3$$

and from nitrite to either ammonia

$$NO_2^-\ 2Al + 5H_2O \rightarrow NH_3 + 2Al(OH)_3 + OH^-$$

or Nitrogen

$$2NO_2^- + 2Al + 4H_2O \rightarrow N_2 + 2Al(OH)_3 + 2OH^-$$

This process can be effectively used in water treatment plants using lime for water softening. The pH is usually raised to or above 9.1 using lime, thus very little additional cost for pH change will be required. Al also reacts with water according to the following reaction.

$$2Al + 6H_2O \rightarrow 2Al\ (OH)_3 + 3H_2$$

The loss of reductant decomposition of water can be minimized to less than 2 per cent between pH 9.1 and 9.3 and 1.16 g of Al is required to reduce 1 g of nitrate. Nitrate reduction also occurs with iron under basic pH conditions. This reaction requires Fe : NO_3 ratio of about 15 : 1 in presence of Cu catalyst and the process proceeds according to the following reaction (Sova, 1986).

$$NO_3^- + 8\,Fe\,(OH)_2 + 6\,H_2O \rightarrow NH_3 + 8Fe(OH)_3 + OH^-$$

Iron is used by many workers for the successful removal of nitrate from water. Chemical reduction of nitrate by iron in hot sulphuric acid is quantitative but ammonia is not the principal product (Pepin-Lahllur, 1930). Szabo *et al.* (1951) described NO_3^- reduction quantitatively by boiling ferrous hydroxide suspension in the presence of catalysts such as colloidal silver or copper hydroxide.

Young (1964) described reduction of nitrate with Ferrous Hydroxide at room temperature at pH 7.0-10.3. The initial nitrate concentration was 100 mg/l (as N) and colloidal copper hydroxide employed as a catalyst was proved to be optimum as 25 mg/l (as Cu) and was used at that level in subsequent experiment. The best result obtained was 90 per cent reduction of NO_3 in four hours at pH 10.3. The reduction was slower and less complete at lower pH.

Young (1964) used powdered and coarse iron for the reduction of NO_3^- in water. Powdered iron (2 g/l) was added in water at various pH levels, which results in 94 and 93 per cent reduction of nitrate in five hours at pH 3.0 and 2.5 respectively. Coarse iron at 2 g/l yielded 57 per cent reduction in NO_3^-. Ammonia was the principal end product in each case. The balanced equation for this mechanism is,

$$NO_3^- + 4Fe^o + 6H_2O \rightarrow 4Fe^{+2} + NH_3 + 9OH^-$$

The rate of reaction for the reduction of nitrate decreased when coarse iron was used. This is because of decreased surface area with increased particle size.

Cheng *et al.*, 1997 described a method for the reduction of 12.5 mM nitrate solution to less than 0.1 mM by zero valent iron under aerobic conditions at room temperature and normal pressure. At pH 5.0 with a 0.05M sodium acetate/acetic acid buffer, nitrate was rapidly reduced to ammonia is a pseudo first order reaction.

Ottley *et al.* (1997) described the significance of trace metal catalysed reduction of nitrate in deep groundwater where organic carbon concentration were very low, laboratory experiments were conducted in which nitrate was reduced chemically by Fe (II) to ammonium at pH 8 and at 20°C in the presence of copper (II). Solid phase Cu rather than dissolved copper was directly involved in the nitrate reduction reactions catalysed by goethite.

Zawaideh *et al.* (1998) demonstrated the effects of pH and addition of an organic buffer (HEPES) (4-(2hydroxyethyl)-1-piperazine ethane sulfonic acid) on nitrate transformation in zero valent iron systems, Nitrate nitrogen was removed by 94 per cent when 0.01 M of HEPES was added to a non shaking batch reactor containing 20 mg/l nitrate-nitrogen and 41 per cent (WN) of Fe^o. Shaking was proved to be effective in removal of nitrate. Using the response surface technology, it was found that nitrate removal was closely related to pH. At low pH (*e.g.* pH < 2), the nitrate removal was fast and efficient (95 per cent to 100 per cent) this was because of formation of Fe^+ at low pH. At high pH (*e.g.* > 11), the transformation of nitrate was fast and efficient only for low concentrations nitrate in Fe^o–H_2O system. At high pH (*e.g.* of 11), some solid compounds such as $Fe(OH)_2$, $Fe(OH)_3$ or $FeCO_3$ are formed which would eventually form a surface layer on Fe^o surface that would inhibit further decomposition of Fe^o (Stumn *et al.*, 1996). Therefore the nitrate removal will not increase with initial nitrate concentration. At normal pH range (pH=6 to 8), nitrate removal was usually lower than 50 per cent without buffer treatment. Mass transport of substrate to the iron surface proved to be important in achieving high removal efficiency.

Siantar *et al.* (1996) used zero valent iron on hydrogen/palladium catalyst (H_2/Pd-Alumina) for the treatment of nitrate-contaminated water.

Flis (1991) reported that iron may reduce nitrate to nitrogen, nitrite or ammonia depending on the reaction conditions.

$$10Fe^{o} + 6NO_3^- + 3H_2O \rightarrow 5Fe_2O_3 + 60H^- + 3N_2$$

$$Fe^{o} + NO_3^- + 2H^+ \rightarrow Fe_2^+ + H_2O + NO_2$$

$$NO^- + 6H_2O + 8e^- \rightarrow NH_3 + 9OH^-$$

Nitrate can also be reduced to nitrogen using H_2 and Pd on copper under water treatment conditions (Sell *et al.*, 1993)

$$5H_2 + 2NO_3^- \xrightarrow{Pd/Cd} N_2 + 4H_2O + 2OH$$

Chew and Zhang 1998 indicated that electrokinetics/iron wall process can be used to remediate nitrate contaminated groundwater. In the iron wall oxidative dissolution of Fe^{o} takes place resulting in the formation of Fe^{2+} ions.

($Fe^{o} \rightarrow Fe^{2+} + 2e^-$) having a reduction potential of 0.44v (Sawyer *et al.*, 1994)

$$2NO_3^- + 12H^+ + 10e^- \rightarrow N_2\,(g) + 4H_2O,\ E_o = 1.24V,$$

The combined reaction is

$$5Fe^{o} + 2NO_3^- + 12\,H^+ \rightarrow 5Fe_2^+ + N_2(g) + 6H_2O, E_{rxn} = 1.68v.$$

Nitrates are anions, which will move to the anode in the elctrokinetics. When only electrokinetics was used at various constant voltages, 25 to 37 per cent of nitrate nitrogen was transformed. Amount of nitrate-nitrogen transformed improved when a Fe^{o} wall (20 g or about 8-10 per cent by weight) was placed near the anode. For test runs at various constant voltages, the amount of nitrate-nitrogen ranged from 54 to 87 per cent. By switching to constant currents, the amount of nitrate-nitrogen transformed was about 84 to 88 per cent. The major transformation products were ammonia nitrogen and nitrogen gas. Nitrite nitrogen was less than 1 per cent in all-experimental runs. Two localized pH conditions exist in the system, a low pH region near the anode and a high pH region near the cathode. Placing of an iron wall near the anode increase the pH in that area as time increases. Movement of the acid front did not flush across the cathode. High nitrate transformation efficiencies can be achieved with pH in the range of 4 to 10.

Ion Exchange

Ion exchange is a unit process in which ions of a given species are displaced from an insoluble exchange material by ions of different species from solution. In practice an ion exchange material is placed in a bed and the contaminant to be treated is passed through it. When ion exchange capacity of the bed is depleted, the feed is stopped and a regenerating solution (concentrated Solution of NaCl or $NaHCO_3$) is passed through bed. Ion exchange materials both natural and synthetic are adsorbents which carry charged ionic groups. To maintain electroneutrality, each ionic site must have associated with it an ion of opposite charge (counter ion). The success of ion exchange depends on the ability of the counter ion to be replaced or exchanged for another ion of the same charge. When all counter ions have been replaced, the ion exchange material is exhausted. Acid or cationic resins will exchange cations such as Ca^{2+} or Mg^{2+} and base or anionic resin will exchange OH^- or Cl^- anions (Reeves, 1972).

Guter (1981) tested a pilot-scale ion- exchange study conducted at McFarland, California for the removal of nitrate from groundwater containing nitrate-nitrogen in the range of 16-23 mg/l. He was able to remove nitrate almost completely.

Lin (1997) used Dowex- Sar as a strong base, anionic ion exchange resin of the Cl type. Regeneration was attained with a salt solution containing 10 per cent w/w salt in about 10 hours. The pH and temperature effect on the equilibrium exchange capacities for nitrate removal was seen to be rather small. A higher initial concentration of nitrate was observed to enhance the ion exchange process.

Basic bismuth nitrate has low solubility in water. Utilizing this property several bismuth compounds were examined by Fritsche (1993) for their ability to remove, nitrate from water. Yellow Bismuth hydroxide, prepared by the reaction of bismuth salt solution with excessive sodium hydroxide proved to be an effective material for anion exchange. Besides nitrate, it also removes phosphates, sulfates and chlorides. The selectivity of the method, which is based on ion exchange with hydroxyl ion (OH^-) as counter ion depends on the pH value. Regeneration can be achieved by sodium hydroxide.

Hoek *et al.* (1988) used a strong base, macro porous anion exchange resin Amberlite IRA 996 for nitrate removal from groundwater. Amberlite appeared to be more nitrate selective than sulfate selective in treating high nitrate concentrations in potable water. High nitrate concentration of 18 mg NO_3^- N/l was reduced to 5-6 mg NO_3^- N/l after the treatment with this resin at a flow rate of 35 Bed Volume (BV)/hour.

Liang *et al.* (1999) studied the removal of nitrate from contaminated groundwater using three types of strong base anion resin. In type I resin three methyl groups make up the functional group, in type II an ethanol group replaces one of the methyl group of resin I and third type of resins were nitrate/sulfate selective (NSS) resins and similar to type I resin containing tri methyl amine functionality but ethyl, propyl-o-butyl groups substitutes. They observed that type I resin provided longer run before regeneration than the type II or NSS resins. The removal of nitrate was almost 100 per cent up to 400 bed volumes in type I resin, up to 300 bed volumes in type II resin and up to 250 bed volumes in NSS resin. It was also observed that in type I resin bicarbonate interferes less with nitrate removal than it did in type II resin. Less regenerant salt was also required for regeneration producing less waste in type I resin than in type II or NSS resin.

Hoell and Feuerstein (1985) developed a process called CARIX process to provide simultaneous removal of nitrate, sulfate and hardness. The CARIX process is based on the combined use of a weak acidic cation exchanger in the hydrogen ion form and a strongly basic anion-exchanger in the bicarbonate (HCO_3^-) form. Both resins are regenerated simultaneously with carbon dioxide in a non-polluting fashion.

The nitrate ion exchange process involves multicomponent ion exchange with sulfate as the major competitor with nitrate for ion exchange sites. Conventional strong base anion exchange resins prefer divalent hydrophilic sulfate over monovalent hydrophobic nitrate. This sulfate preference needs to early nitrate breakthrough and causes nitrate peaking when nitrate breakthrough is exceeded. For high sulfate water greater than 140 mg/l, the nitrate selective TEA and TBA (Tri Ethyl Amine and Tri Butyl Amine) resins generally perform better than the conventional TMA (Tri Methyl Amine) resins for nitrate removal.

Biological Denitrification

The process refers to the dissimilatory reduction by essentially aerobic bacteria of one or both of the ionic nitrogen oxides (NO_3^- and NO_2^-) to the gaseous (NO and N_2O) which may themselves be further reduced to dinitrogen (N_z). Denitrification is catalysed by either heterotrophic or autotrophic bacteria that derive their energy requirements by the oxidation of organic and inorganic material, respectively (Hiscock *et al.*, 1991).

The oxidation half reaction of carbohydrate is:

$$CH_2O + H_2O \rightarrow CO_2(g) + 4H^+ + 4e^-$$

and that of pyrite is

$$1/14\ FeS_2 + 4/7\ H_2O \rightleftharpoons SO_4^{2-} + 1/14\ Fe^{2+} + 8/7\ H^+ + e^-$$

Heterotrophic denitrifying microorganisms can use a variety of organic carbon sources; however most of the work regarding denitrification of water involves the use of methanol or acetic acid.

$$5CH_3OH + 6NO_3^- \rightarrow 3N_2 + 5CO_2 + 7H_2O + 6OH^-$$ (Sherrard, 1988)

$$5C_2H_5OH + 12NO_3 \rightarrow 6N_2 + 10CO_2 + 9H_zO + 12OH$$ (Richard *et al.*, 1980)

$$5CH_3COOH + 8NO_3^- \rightarrow 4N_2 + 10CO_2 + 6H_2O + 8OH^-$$ (Frick and Richard, 1985)

Methane and Carbon mono oxide can also be used as substrate for denitrification.

$$5CH_4 + 8NO_3^- + 8H^+ \rightarrow 5CO_2 + 4N_2 + 14H_2O$$ (Brrenstein *et al.*, 1986)

$$2NO_3^- + 5CO + H_2O \rightarrow N_2 + 2OH^- + 5CO_2$$ (Sherrard, 1988)

Denitrification can also be accomplished by autotrophic bacteria, which can use hydrogen or various reduced-sulphur compounds as energy sources.

The following stoichiometric relationship for hydrogen and sulfur has been reported.

$$2NO_3^- + 5H_2 \rightarrow N_2 + 4H_2O + 2OH^-$$ (Kurt *et al.*, 1987)

$$5S_2O_3^- + 8NO_3^- + H_2O + 2OH^-$$ (Kurt *et al.*, 1987)

$$5S^- + 8NO_3^- + 8H^+ \rightarrow 5SO_4^- + 4N_2 + 4H_2O$$ (Barrenstein *et al.*, 1986)

Biological denitrification is commonly used in municipal and industrial wastewater treatment. However, it may be used to treat drinking water if bacterial contamination of the treated water can be avoided. Denitrification takes place under anionic conditions. Nitrate reduction to nitrogen gas occurs through a series of steps as follows (Knowles, 1982).

$$NO_3^- \rightarrow NO_2^- \rightarrow NO \rightarrow N_2O \rightarrow N_2$$

Autotrophic Denitrification

Flere and Zhang (1999) evaluated sulfur and limestone autotrophic denitrification (SLAD) processes with four laboratory scale fixed bed reactors. A maximum denitrification rate of 384g NO_3^-–N/(m^3 day) was achieved at a loading rate between 600 and 700g NO_3^-–N/(m^3 day). The effluent nitrite concentration started to rise gradually once the loading rate was above 600g NO_3^-–N/

(m^3 day). A loading rate between 175 and 225 g of $NO_3^-N/(m^3$ day) achieved the maximum nitrate-N removal efficiency (approximately 95 per cent). A SLAD column requires backwashing after 6 months of operation when the influent is synthetic groundwater but will foul and requires backwashing within 1-2 months when the influent is real groundwater.

Schippers *et al.* (1987) described denitrification using sulfur limestone filtration. Limestone granules were added to sulfur to maintain the pH because pH was reduced by the hydrogen ions, which was released in this process. Filtration was carried out at 0.5 m/h, similar to the rate used in slow sand filtration. The influent was degassed to remove nitrogen and oxygen and 0.3 mg PO_4^{-3}/l phosphate was added. The pilot plant consisted of a vaccum degasser, a slow sulphur/limestone filtration unit, a cascade tray and an infiltration pond. The concentration of NO_3^- N was reduced from 20 to 3-5 mg/l at a filtration rate of 0.25 m/h.

Denitrification with sulphur/limestone was also achieved by Blecon *et al.* (1983) and Hoek *et al.* (1992). Some authors have evaluated reduced sulfur compounds such as suphfide and thiosulphate for the denitrification of water and domestic and industrial wastewater (Claus, 1985; La Motta Diaz, 1985; Martin, 1982 and Batchelor, 1978). Sulphate was the byproduct of denitrification using sulfur compounds.

Lewandowski *et al.* (1987) evaluated autotrophic denitrification using calcium alginate beads containing elemental sulphur, calcium carbonate and *Thiobacillus denitrificans* suspended in a bench scale completely mixed batch reactor. Nitrate concentration was reduced from 27 mg/l to 6 mg/l in seven hours.

Nitrate has been removed in fixed film reactors using various support media composition, hydrogen generation, degassing technique and post-treatment methods. Various methods have been used to remove microorganisms and dissolved organic carbon from denitrified water (Gros and Ginocchio, 1982 and Ginocchio, 1984).

Kurt *et al.* (1987) used hydrogen for autotrophic denitrification in a bench scale fluidized bed reactor. Nitrate and hydrogen were assumed to be limiting substrates in the first denitrification step, and nitrite and hydrogen were assumed limiting in the second step. The nitrification rate was shown to be more strongly dependent on the nitrate concentration than on hydrogen. A residence time of 4.5 hours was required from complete denitrification of water containing 25 mg/l NO_3^-–N at the optimum pH of 7.5.

Gros *et al.* (1986) and Gros and Treutler (1986) reported on the performance of a commercial scale biological denitrification plant utilizing hydrogen, in Minchegladbach, West Germany and the process was given the trade name 'Denitropur. The plant design incorporates indirect hydrogen saturation, phosphate addition, four-packed bed reactor in series, post aeration, flocculent addition, filtration and UV disinfection. Carbon dioxide was added as an inorganic carbon source and to buffer against an alkaline pH shift. The sludge production was less (approximately 0.2 kg/per kg N removed). Residence time of one to two hours was required to remove 50 mg/l nitrate.

In electrode biofilm reactor, denitrifying organisms are immobilized on the cathode utilizing H_2 produced by the electrolysis of water and subsequently reduce nitrate to nitrogen gas.

$$2NO_3^- + 2H^- + 5H_2 \rightarrow N_2 + 6H_2O$$

carbon electrode was used as anode, so that the CO_2 formation seems to occur instead of O_2 formation (Kinoshita, 1988).

$$C + 2H_2O \rightarrow CO_2 + 2H_2 \text{ (emf} = 0.207\text{v)}$$

Sakakibara *et al.* (1994) used electrochemical and biological reactor for denitrification. Denitrifying microorganisms were immobilized with a sodium alginate gel on a cathode electrode as the anode. Biological reductions of nitrate through the use of H_2 at the cathode and formations of inorganic carbons at the anode was observed. Concentrations of nitrite and ammonium in effluent were less than 1 mg-N/l effluent concentration decreased with time and components of produced gas were H_2, N_2 and CO_2, nitrate .

Heterotrophic Denitrification

Heterotrophic biological denitrification is a well-established process for wastewater treatment. This process has not been used in full scale in U.S. however there are several full scale plants being operated in Europe (Dahab *et al.*, 1991 and Gayle *et al.*, 1989).

In heterotrophic biological denitrification, facultative microorganisms are contacted with the water supply containing nitrates and an added carbon source in an anoxic environment. Under the conditions the bacteria utilize nitrates as a terminal electron acceptor in lieu of molecular oxygen. In the process nitrate is reduced to nitrogen gas, which is harmless. Carbon source is necessary since it supplies the energy required by the microorganisms for respiration and synthesis (Dahab and Woodbury, 1998).

The removal of nitrate nitrogen from groundwater by using two rotating biological contactors (RBC) in series was investigated by Bandpi and Elliott (1996). The first RBC reactor was operated under anoxic conditions to remove nitrate-nitrogen. A fraction of effluent of the anoxic RBC was fed to a bench scale aerobic RBC to study the degradation of residual organic carbon and oxidation of nitrite nitrogen. Ethanol was used as a carbon source. The first reactor achieved 90 per cent nitrate removal efficiency at a loading rate of 76mg/nth. The optimum ethanol to nitrate-nitrogen ratio was found to be 2.35. Effluent removal of residual carbon and re-oxidation of nitrite was also achieved by aerobic biological process. Thus high nitrate removal with negligible residual carbon could be achieved by employing an anoxic denitrification followed by an aerobic process.

Bandpi and Elliott (1998) investigated a pilot scale rotating biological contactor (RBC) for the removal of nitrate-nitrogen from groundwater using three different carbon sources: methanol, ethanol and acetic acid. Optimum carbon sources to influent nitrate nitrogen ratio were established by varying the influent concentration of carbon sources. The optimum ratio of methanol, ethanol and acetic acid to nitrate-nitrogen ratios were found to be 2.9, 2.35 and 4.3 respectively. The nitrate-N removal efficiency averaged 93, 91 and 98 for methanol, ethanol and acetic acid at a loading rate of 76 mg/m^2h. The results of this study showed that the acetic acid was the most efficient carbon source for removal of nitrate-nitrogen.

Hamzah (1996) used an anoxic static bed column to investigate the influence of type of carbon source and nitrate loading on the nitrate contaminated drinking water. Results showed that the system operated at 0.45kg NO_3^-–N/m^3 day loading of nitrate to treat waters having a nitrate concentration of 500 mg/l. On stoichiometric basis, ethanol as a substrate for bioremediation gave better results as compared with methanol and acetic acid.

A study was carried out by Christensson (1994) in two anoxic chemostates to see the performance of ethanol and methanol as carbon source for denitrification. An efficient denitrification with ethanol was established in a short time, while denitrification with methanol required a substantial adoption time and never showed the same ability as denitrification with ethanol.

Akunna *et al.* (1993) used batch tests to determine the potentials of digested sludge to reduce nitrate in the presence of five different carbon sources; Glucose, Glycerol, Acetic acid, Lactic acid and Methanol. Ammonium accumulation was found in glucose and glycerol media. Dissimilatory reduction to ammonium accounted for up to 50 per cent of reduced nitrate and nitrite. Ammonification was higher than denitrification when glucose and glycerol were present in the media. Nitrate/Nitrite reduction in acetic and lactic acid media was essentially a denitrification study. Up to 100 per cent reduced nitrate and nitrite in the culture media with these acids were denitrified at average rates between 27 and 23 mg NO_3^-–N/g ML VSS. Soares *et al.* (1988) conducted denitrification studies on groundwater containing 22.6 mg/l NO_3^-–NN using a down flow sand column with sucrose as the carbon source. Complete nitrate removal was achieved at CN ratio of 2. Soares *et al.* (1991) studied microbiological denitrification in a sandy matrix by means of laboratory sand columns operated at continuous and pulse feed regimes. Formate was used as the carbon source. The results suggest that pulse application of carbon source is preferable to a continuous supply regime because continuous supply of carbon source leads to almost complete clogging of the column.

Liessens *et al.* (1993) studied heterotrophic denitrification in a fluidized bed reactor for surface water at low temperature. Methanol was used as the reductant with a nitrate removal efficiency of 9.0 Kg NO_3^-–/m^3 reactor day at 3.5 DC, the system, has shown superior performance compared with conventional fixed bedbiofilm reactors. With an influent concentration of 75 mg/l NO_3^- complete nitrate removal was achieved at an empty bed contact time of 15 minutes. Residual methanol was easily removed by the existing downstream drinking water treatment processes.

Various processes have been adapted and used for the denitrification of drinking water. Drinking water denitrification was studied on a membrane bioreactor pilot plant by Delanghe *et al.* (1994). The nitrate removal yields remained constant at 99 per cent. The specific denitrification activities averaged 0.16 kg N-NO_3, kg^{-1} MLSSd-1 (Mixed Liquor Suspended Solid) at 20 DC and pH 8. The permeation flux was about 0.5 $m^3.m^{-2}.d^{-1}$ throughout the study and did not vary with an increase in the suspended solid concentration. The specific denitrification activity decreased by a factor 1.9 with a temperature decrease of 10°C. The optimal pH was found to be 8.0-8.5. The ethanol was used as the carbon source and its consumption was 1.4 $gC.g^{-1}N$-NO_3. The pilot plant study showed that the shear stress on the ultra filtration membrane surfaces did not affect the specific denitrification activity of suspended cultures.

Dahab and Kalagiri (1996) studied a cyclically operated fixed film bio denitrification process to remove nitrates from drinking water. They also investigated the ability of two-stage system to remove nitrate and residual organics from treated water as compared to single stage units. The removal of nitrate was observed to be 98 and 95 per cent in single stage and two stage denitrification reactors respectively at hydraulic retention time (HRT) of 1.0 and 0.5 hrs. They also observed that the two stage cyclic operation was more effective in maintaining low nitrite concentration than the single stage reactor.

Combined Ion Exchange Biological Denitrification

Ion exchange is a physical-chemical process where by means of an anion exchange resin nitrate is exchanged for chloride or bicarbonate. The process has a problem of regeneration of the resin, which requires a large excess of salt producing a voluminous brine with high nitrate, sulphate and chloride concentrations (Gauntlett, 1975; Deguin, 1982; Richard and Leprince, 1982 and Hoek *et al.*, 1988). Brine disposal is very difficult and requires careful consideration and adds to the operating cost. Biological denitrification is a process by which nitrate is converted to nitrogen gas denitrifying bacteria.

A direct contact is created between groundwater, which is generally free of microorganism and bacteria. In the case of heterotrophic denitrification, a carbon source has to be added to the groundwater. Both cause a serious risk of the bacteriological contamination of the groundwater and extensive post treatment is necessary for the removal of microorganisms and organic substrate (Sorg, 1979; Barlog, 1980; Richard and Leprince, 1982 and Sontheimer *et al.*, 1982). Also the production of nitrite, an, intermediate product of denitrification is a serious risk.

Therefore by combining ion exchange and biological denitrification process into one process, most problems connected with the separate techniques can be avoided (Hoek, 1985; Hoek and Klapwijk, 1985, 1986). The basic concept of this process is to use sludge blanket reactor (SBR) to biologically denitrify the spent regenerant brine, which is then filtered, compensated with NaCl and reused. Bench scale test suggested that the salt consumption could be reduced by 50 per cent and the salt discharge by 90 per cent (Liu and Clifford 1996).

Hoek and Klapwijk 1987 described a combined ion- exchange (IX) and biological denitrification process in which nitrate is removed from groundwater by ion exchange and for the regeneration of nitrate loaded resin a denitrification reactor is used. In contrast with traditional denitrification process there is no direct contact between groundwater and denitrifying bacteria. Also brine production and regeneration salt requirements are minimal as compared with conventional regeneration of ion exchange resins. Ion exchange resin Duolite A 165 was used for the removal of nitrate and an up flow sludge blanket reactor (USBR) was used for the regenerant denitrification which was able to denitrify highly saline solution containing 25-30 g/l $NaHCO_3$ and 10-15g NaCl/l. The process is suitable for treating water with high sulphate concentrations. The waste brine produced was 13-20 per cent of the amount produced in ion exchange columns. A sand filter was recommended between the USB denitrification reactor and the ion exchange column to remove the suspended solids from the regenerant before they reached the resin along with resin disinfection with per acetic added during disinfection.

Hoek *et al.*, 1988 evaluated the combined process using groundwater with sulphate–selective and nitrate selective resins. Regeneration with $NaHCO_3$ was possible but the regeneration efficiency was very low. Regeneration salt requirement and brine production was minimized by using closed circuit regenerator using biological denitrification reactor to remove nitrate from the regenerant. The combined process resulted in about 95 per cent reduction of waste brine in comparison to the conventional ion exchange.

A bench scale ion exchange process with batch biological denitrification of the spent regenerant brine in a sequencing batch reactor (SBR) was developed by Clifford and Liu 1993. It consisted of chloride ion exchange and 0.5N (3 per cent) NaCl regeneration followed by batch denitrification and reuse of the spent brine. Complete (>99 per cent) denitrification of spent 0.5 NaCl brine was achieved in 20 hours using a non optimum methanol to nitrate-nitrogen ratio® of 2.2. At the optimum R value of 2.7, the time for greater than 95 per cent denitrification was 8 hours. The research indicated that the combination procedure resulted in 50 per cent reduction of regenerant consumption and 90 per cent reduction in the mass of waste salt discharged. Compared with USBR, the SBR ad several advantages that were pertinent to small systems: simplicity, flexibility of operation, stability of effluent quality and built in equalization. Although the higher rate USBR requires less reactor volume, when equalization and brine were taken into account, the space occupied by the process was not much different.

Fonseca *et al.* (2000) developed a novel ion exchange membrane bioreactor able to prevent secondary pollution of biologically treated drinking water and specifically tested for water denitrification. This system combines ion selective membrane dialysis and biological conversion. The ion selective membrane facilitate the extraction of pollutant from the water to the biological compartment, hinders the transfer of organic and inorganic nutrients and confines the microbial culture involved in the conversion process within the bioreactor. In the study hereby a system was used to investigate the removal of nitrate from a synthetic groundwater containing 50 mg/l of nitrate-No The treated water obtained was free of inorganic nutrients and ethanol, the carbon source was selected for the biological process, and the surface denitrification rate achieved was 7g-N $m^{-2}day^{-1}$. This system proved to be effective in producing a treated water effluent that does not require the extensive post treatment associated with conventional biological treatment.

Vegetative Denitrification

Vegetation can prevent herbicides, pesticides and fertilizers from contaminating surface and groundwater. Tree buffers can protect groundwater from different contaminants at the contaminant site. At an agricultural test site on a Iowa farm, a 3-year-old popular crop planted by University of Lowa research team reduced NO_3-N levels in leachate from fertilized fields. The trees were planted between the corn field and the stream along the stream bank. The average nitrate content of groundwater leaving the com field was 150 mg/l, which is more than 300 per cent greater than EPA's permissible limit of 45 mg/l for nitrate in drinking water. Level was 8 mg/l in the groundwater between the field and the stream. Poplar trees were chosen because the trees take up soluble inorganic nitrogen through their roots, converting them into protein and nitrogen gas.

The occurrence of saturated and non-saturated soils in Riparian buffer strips provides ideal condition for denitrification. The rate of denitrification depends on the amount of organic carbon, the degree of soil saturation, the activity of denitrificant bacteria, the temperature and pH of the system (Engler and Patrick, 1974; Reddy *et al.*, 1978 and Schippers *et al.*, 1993). Todd *et al.* (1983) observed that a forested buffer removed 82 per cent of water borne nitrogen from agricultural runoff. Fennessy and Cronk (1997) reported that the denitrification is the predominant process for the nitrate removal in riparian zones. Vegetation in Riparian buffers serve as a source of carbon for microbial denitrification. They also reported that sedimentation; soil adsorption and microbial transformation are mechanisms by which nitrogen, phosphorous and solids are removed from agricultural runoff. Nitrate removal is favoured in forested areas with subsurface flow and is less in grassed areas with surface flow.

Brown (1971), studied removal of nitrate from subsurface agricultural drainage using algal system consisting of algae, harvesting and disposal as a possible means of removing nitrate. The findings showed that about 15-90 per cent of the nitrate was assimilated by shallow algal culture.

Summary and Conclusion

Studies on the development of processes for the removal of nitrates from drinking water are either based on physico-chemical or biological denitrification system. Merits and demerits of a process are to be considered for finding its potential for full-scale application. In ion exchange process the water is passed through beds of ion exchange resin beads, which absorbs nitrate in exchange for another anion. Counter flow regeneration is normally used for nitrate removal by ion exchange process. In this process the spent regenerant is disposed to agricultural field. Nitrate selective ion exchange resins are now available. The high nitrate capacities of these membranes prevent the displacement of nitrate by sulphate and also lower cost for regenerant chemical and waste disposal.

Reverse osmosis results in the production of treated water stream and waste disposal concentrated waste stream. The waste volumes in this case are larger as compared with ion exchange and operating costs are also higher.

Electrodialysis treats the water by selective removal of nitrate ions through a semi permeable membrane. The removal efficiency for nitrate is almost equal to that of reverse osmosis. Nitrate selective membranes are now available which makes the process more effective. In this process however pre treatment of water is required. The wastes are also very concentrated and create disposal problems.

Biological denitrification process occurs naturally in the presence of sufficient carbon source and anoxic conditions. The method may be used for treating drinking water by avoiding bacterial contamination of the treated water. Both autotrophic and heterotrophic processes have been used for denitrification. However heterotrophic process is faster than autotrophic. Autotrophic denitrification increases the sulfate concentration in water and operated at low carbon level results in nitrite production. These reasons impair the efficiency of the process. With fluidized sand bed excess biomass is removed without shutting down the process. The waste volumes are small and due to low level of dissolved solids the waste can be disposed to agricultural land. The process of denitrification using ion exchange for nitrate removal from drinking water and treatment of regenerant by biological denitrification shows that the process may find more adaptation in future. The method of converting-nitrate into insoluble salt of some metal ion also has potential for future research. Some preliminary studies have been conducted on denitrification using zero valent iron. This process also appears to be a potential process for future research. Nature provides self-purification of water. The roots of plants are capable of absorbing nitrate through bacterial reaction and thereby reducing nitrate from water. Several studies have been conducted on the removal of nitrate by allowing the water to flow through the subsurface under plantation. This process has high potential for large-scale treatment of drinking water.

Acknowledgement

The authors are grateful to Dr. R.N. Singh, Director, NEERI, Nagpur for his kind permission to publish this chapter.

References

Adhikari, S.K., Narayanan, P.K., Thampy, S.K., Dave, N.J., Chauhan, D.K. and Indu Sekhar, V.K., 1991. Desaliantion of brackish water of higher salinity by electordialysis. *Desaliantion and Water Reuse*, 4: 189–200.

Akunna, J.C., Bizeau, C. and Moletta, 1993. Nitrate and nitrite reductions with anaerobic sludge using various carbon sources: Glucose, glycerol, acetic acid, lactic acid and methanol. *Wat Res.*, 27: 1303–1312.

Arthur, W. Hounslow, 1995. *Water Quality: Data, Analysis and Interpretation*. CRC Lewis Publishers.

Bandpi, A M. and Elliott, D.J., 1996. Nitrate removal from groundwater using an anoxic aerobic rotating biological contactor. *Water Sci. Technol.*, 34: 323–330.

Bandpi, A.M. and Elliott, D.J., 1998. Groundwater denitrification with alternative carbon sources. *Water Sci. Technol.*, 38: 237–243.

Barlog, F., 1980. Nitrate in trinkwasserursachen and problemlosungen. *Chem. Resch.* 33: 3–16.

Barrenstein, A., Kramer, U. and Obermann, P., 1986. Underground treatment of nitrate rich groundwater by infiltration with treated wastewater or methane rich natural gas. *DVGW-Schriftenreihe, Wasser, Frankfort, West Germany*, 106: 99–116.

Batrchelor, B. and Lawrence, A.W., 1978. A kinetic model for autotrophic denitrification using elemental sulfur. *Wat. Res.*, 12(12): 1075–1084.

Cheng, I.F., Muftikian, Fernando Q. and Korte, N., 1997. Reduction of nitrate to ammonia by zero valent iron.*Chemosphere,* 35(11): 2689–2695.

Chew Chin, F. and Zhang, T.C., 1998. *In situ* remediation of nitrate contaminated groundwater by electrokinetics iron wall process. *Wat. Sci. Tech.,* 38: 135–142.

Christensson, M., Lie, E. and Welander, T., 1994. A comparison between ethanol and methanol as carbon sources for denitrification. *Wat. Sci. Technol.,* 30: 83–90.

Claus, G. and Kutzner, H.J., 1985. Autotrophic denitrification *Thiobacillus denitrificans. Appl. Microbiology and Biotechnology,* 22: 289–296.

Clifford, D. and Liu, Y., 1993. Ion exchange for nitrate removal. *JAWWA,* 85(4): 135–143.

Dahab, M.F., Guter, G.A. and Rogalla, F., 1991. Experience with nitrate treatment in the United States and Europe. In: *Proceedings Annual Conference American Water Works Association,* Philadelphia, PA June, 23–27.

Dahab, M.F. and Kalagiri, J., 1996. Nitrate removal from water using cyclically operated fixed film bio-denitrification reactors. *Wat Sci. Technol.,* 34: 331–338.

Dahab, M.F. and Woodbury, B.L., 1998. Biological treatment options for nitrate removal from drinking water. *AWWA Inorganic Contaminants Workshop.* San Antonio, TX, February 22–24.

De Roo, H.C., 1980. Nitrate fluctuation in groundwater as influenced by use of fertilizer. Connecticut Agr. Exp. Station, University of Connecticut, New Haven. *CT Bull.,* 779.

Deguin, A., 1982. Elimination des nitrates par echange d'ions dans les eaux potables; Mise en equations du procede. *Trib Cebedeau.* 35: 35–51.

Delanghe, B., Nakamura, F., Myoga, H., Magara, Y. and Guibal, E., 1994. Drinking water denitrification in a membrane bioreactor. *Water Sci. T echnol.,* 30: 157–160.

Engler, R.M. and Patrick, W.H.Jr., 1974. Nitrate, removal from floodwater overlying flooded soils and sediments. *J. Environ. Qual.,* 4: 409.

Fennessy, M.S. and Cronk, J.K., 1997. The effectiveness and restoration potential of riparian ecotones for the management of non-point source pollution, particularly nitrate. *Critical Reviews in Environmental Sci. and Technol.,* 27(4): 285–317.

Flere, J.M. and Zhang, T.C., 1999. Nitrate removal with sulfur-limestone autotrophic denitrification processes. *J. Env. Engineering,* 125: 721–729.

Flis, J., 1991. Stress corrosion cracking of structural steels in nitrate solutions. In: *Corrosion of Metals and Hydrogen-Related Phenomena,* (Ed.) J. Flis. Materials Science Monographs, 59, Elsevier Amsterdam, NL, p. 57–94.

Fonseca, A.D. and Crespa, J., 2000. Drinking water denitrification using a noval ion exchange membrane bioreactor. *Env. Sci. Technol.,* 34: 1557–1562.

Frick, B.R. and Richard, Y., 1985. Experience with biological denitrification in a full Scale drinking water treatment plant. *Vom Wasser,* 64: 145–154.

Fried, J.J., 1991. Nitrates and their control in EEC aquatic environment. In: *Nitrate Contamination: Exposure, Consequence and Control,* (Eds.) I. Bogardi and R.D. Kuzelca. Springer-Verlag, Berlin, Germany, p. 3–11.

Fritsche, U., 1993. Removal of nitrate and other anions from water by yellow Bismuth Hydroxide. *J. Environ. Sci. Health,* A 28(9): 1903–1913.

Gauntlett, R.B., 1975. Nitrate removal from water by ion exchange. *Water Treatment and Examination,* 24: 172–193.

Gayle, B.P., Boardman, G.D., Sherreard, J.H. and Benoit, R.E., 1989. Biological denitrification of water. *J. Envir. Engrg. ASCE,* 115(5): 930–943.

Ginocchio, J., 1984. Nitrate levels in drinking water are becoming too high. *Water Services, Switzerland.* 88: 143–147.

Gopal, R., Bhargava, T.N. and Bhatti, O.P., 1977. Nitrates in well waters of Rajasthan. *Curr. Agric.,* 1(3): 57.

Gros, H. and Treutler, K., 1986. Biological denitrification process with hydrogen oxidizing bacteria for drinking water treatment. *Aqua,* London, England, 5: 288–290.

Gros, H., Schnoor, G. and Rutton, P., 1986. Nitrate removal from groundwater by autotrophic microorganisms. *Water Supply,* 4: 11–21.

Gros, H. and Ginocchio, J., 1982. Biological nitrification and denitrification on submerged mixing elements. *Gas Wasser Abwasser, Zurich, Switzerland,* 67: 312–321.

Gupta, S.C., 1981. Evaluation of quality of well waters in Udaipur District. *Indian J. Environ. Hlth,* 23(3): 195–202.

Guter, G.A., 1982. *Removal of Nitrate from Contaminated Water Supplies for Public Use.* Report EPA– 600/S2 82–042, U.S. Environmental Protection Agency (EP A), Cincinnati, Ohio.

Guter, G.A., 1981. *Removal of Nitrate from Contaminated Water Supplies for Public Use.* EPA–600/S2–81/029, April, Environmental Protection Agency Cincinnati, Ohio.

Hammer, M.J., 1991. *Water and Wastewater Technology,* 2nd edn.

Hamzah Ziad Abu Gharah, 1996. Biological denitrification of high nitrate water: Influence of type of carbon source and nitrate loading. *J. Env. Sci. and Hlth.,* A31(7): 1651–1668.

Handa, B.K., Goel, D.K., Kumar, A. and Sondhi, T.N., 1982. Pollution of groundwaters by nitrates in Uttar Pradesh. *IAWPC Tech. Annual,* 9: 95–103.

Hill, M.J., Hawksworth, G. and Tattersall, G., 1973. Bacteria, nitrosamines and cancer of stomach. *Br. J. Cancer,* 28: 562–567.

Hiscock, K.M., Llyod, J.W. and Lerner, D.N., 1991. Review of natural and artificial denitrification of groundwater. *Wat. Res.,* 25(9): 1099– 1111.

Hoek, J.P. Van der and Klapwijk, A., 1987. Nitrate removal from groundwater. *Wat. Res.,* 21(8): 989–997.

Hoek, J.P.van der and Klapwijk, A., 1985. Biological, physical, chemical methods for nitrate removal from groundwater. *PT Procestechniek,* 40: 22–25.

Hoek, J.P. van der and Klapwijk, A., 1986. Feasibility study to biological, physical chemical nitrate removal from groundwater. *H_2O,* 19: 53–58.

Hoek, J.P. van der, 1985. Biological, physical-chemical nitrate removal from groundwater: An environmental and financial attractive alternative. In: *Proceedings of the Congress Nitrates in Water,* SITE, 22–24 October.

Hoek, J.P. van der, Kappelhof, J.W.N.M. and Hijnen, W.A.M., 1992. Biological nitrate removal from groundwater by sulphur lime stone denitrificaton. *J. Chemical Technol. and Biotechnol.*, 54(2): 197–200.

Hoek, J.P. van der, Hoek, W.F. van der and Klapwijk, A., 1998. Nitrate removal from groundwater: Use of nitrate selective resin and low concentrated regenerate. *Water, Air and Soil Pollution*, 37: 41–43.

Hoell, W. and Feuerstein, W., 1985. Combined hardness and nitrate/sulphate removal from groundwater by the carix ion exchange process. *Water Supply*, 3: 199–209.

Horold, S., Tacke, C. and Vorlop, K.D., 1993. Catalytical removal of nitrate and nitrite from drinking water. I. Screening for hydrogenation catalysis and influence of reaction conditions on activity and selectivity. *Environmental Technology*, 14: 931–939.

Hounslow, A.W., 1995. *Water Quality Data, Analysis and Interpretation*. CRC Lewis Publishers.

House of Lords, 1989. *Nitrate in Water*. 16th Report, Session 1988–1989 of the Select Committee on the European Communities. HMSO, London.

Indushekhar, V.K., Trivedi, G.S. and Shah, B.G., 1991. Removal of nitrate by electrodialysis desalination and water re-use, Volume 4. In: *Proc. of Twelth International Symposium*, Institute of Chemical Engineers, Rugby, U.K., 15–18 April.

Kinoshita, K., 1988. *Carbon-Electro-chemical and Physico-chemical Properties*. John, Wiley and Sons, New York.

Knowles, R., 1981. Denitrification. *Microbiological Rev.*, 46(1): 43–70.

Kurt, M., Dunn, I.J. and Bourne, J.R., 1987. Biological denitrification of drinking water using autotrophic organisms with hydrogen in fluidized bed biofilm reactor. *Biotechnology Bioengineering*, 29: 493–501.

La Motta, Diaz E. and Andrade Salgado, A., 1985. Autotrophic denitrification using sulphides. *Plitechnica, Quito, Ecuador*, 10: 65–76.

Lewandowski, I., Bakke, R. and Characklis, W.G., 1987. Nitrification and autotrophic denitrification in calcium alginate beads. *Water Sci. Technol.*, 19: 175–182.

Liang, S., Mann, M.A., Guter, G.A., Kim, P.H.S. and Hardean, D.L., 1999. *JAWWA*, 91(2): 79–90.

Liessens, J., Germonpre, R., Beernaert, S. and Verstraete, W., 1993. Removing nitrate with a methyotrophic fluidized bed: Technology and operating performance. *JAWWA*, 85: 144–154.

Lin, S.H. and Wu, C.L., 1997. Nitrate removal from aqueous solution by ion exchange. *Environ. Sci. Health*, A32(5): 1575–1589.

Lunkad, S.K., 1994. Rising nitrate levels in groundwater and increasing N-fertilizer *consumption. Bhu-Jal News*, p. 4–10.

Majumdar, D. and Gupta, N., 2002. Nitrate pollution to groundwater and associated human health disorders. *Indian J. Environ. Hlth.*, 42: 28–39.

Marquardt, D.K., 1987. Reverse osmosis for removing nitrate from water. *Aqua.*, 1: 39–44.

Martin, G., 1982. *Biological Denitrification of Water*. Patent Application 79/10154 France.

Mason, B. and Moore, C.B., 1985. *Principles of Geochemistry*. Wiley Eastern Ltd., p. 350.

Miquel, A.F. and Oldani, M.A., 1991. Newly developed process for nitrate removal from drinking water. In: *Nitrate Contamination: Expouse, Consequence and Control,* (Eds.) I. Bogardiand and R.F. Kuzelka. Springer Verlag, Berlin, Germany, pp. 385–394.

Murphy, P., 1991. Chemical removal of nitrate from Water. *Nature,* 350: 223–225.

Nixon, N., 1992. English water utility tackles nitrate removal. *Water Engg. Mgmt.,* 139: 27–28.

Ottley, C.J., Davison, W. and Edmonds, W.M., 1997. Chemical catalysis of nitrate reduces by iron (II). *Geochimical et. Cosmochimica Acta,* 61: 1819–1828.

Pandey, S.P., Narayanswamy, V.S. and Hasan, M.Z., 1979. Quality of well waters of Nagpur with special regard to Nitrates and Nitrites. *Indian J. Environ. Hlth.,* 21: 35–46.

Pepin-Lahellur, J., 1930. Reduction of metallic nitrates and nitrated organic compounds in acid medium. *Ann. Acad. Brasil Sci.,* 2: 31–32.

Reddy, K., Patrick, W.H. and Phillips, R., 1978. The role of nitrate diffusion in determining the order and rate of denitrification in flooded soil. I. Experimented results. *Soil Sci. Soc. Am. J.,* 42: 268.

Reevs, T.G., 1972. Nitrogen removal: A literature review. *JWPCF,* 44(10): 1895–1908.

Richard, Y. and Leprince, A., 1982. Pollution per les nitrates: traitments disponibles. *Trib Cebedeau,* 35: 21–33.

Richard, Y., 1980. Denitrification of water for human consumption. *Progress in Water Tech.,* Oxford, England, 12: 173–191.

Routenbach, R. and Kopp, W., 1986. Separation of nitrate from well water by membrane process (reverse osmosis/electrodialysis reversal). *Aqua,* 5: 279–282.

Sakakibara, Y., Araki, K., Kuroda, M., Tanaka, T. and Watanabe, T., 1994. Denitrification and neutralization with an electrochemical and biological reactor. *Water Sci Technol.,* 30: 151–155.

Schepers, J.S., Frank, K.D. and Watts, D.G., 1984. Influence of irrigation and nitrogen fertilization on groundwater quality. In: *Proceeding Int. Union of Geodesy and Geophysics,* Hamburg, West Germay, p. 21–32.

Schippers, L.A., Copper, A.B., Harfoot, C.G. and Dyck, W.J., 1993. Regulators of denitrification in an organic riparian soil. *Soil. Biol. Biochem.,* 25: 925.

Schippers, J.C., Kriuth, J.C., Mulder, F.G. and van Lieshout, J.W., 1987. Removal of nitrate by slow sulfur (limestone filtration). *Aqua,* 5: 274–280.

Segall, D.A. and Bruell, C.J., 1992. Electroosmotic contaminant: Removal process. *J. Env. Engr.,* 118: 1.

Sell, M.B., Feist, H.R., Behling, R.D., Kneifel, K. and Peinemann, K.V., 1993. *Proceedings of the Membrane Technology Conference.* American Water Works Association, Baltimore, MD. pp. 1–4.

Siantar, D.P., Schreir, C.G., Chou, C.S. and Reinhard, M., 1996. Treatment of 1,2 dibromo-3-chloropropane and nitrate-contaminated water with zero valent iron or hydrogen/palladium catalysts. *Wat Res.,* 30: 2315–2322.

Soares, M.J.M., Belkin, S. and Abeliovich, A., 1988. Biological groundwater denitrification: Laboratory studies. *Water Sci. Technol.,* 20(3): 189–195.

Soares, M.J.M., Braester, C., Belkin, S. and Abeliovich, A., 1991. Denitrification in laboratory sand columns: Carbon regime, gas accumulation and hydraulic properties. *Wat. Res.,* 25: 325–332.

Sontheimer, H., Cornel, P., Fettig, J. and Rohmann, O., 1982. Grundwasservem.lfeinigung-Bedrohung fur die offentliche Wasserversorgung 2. *GWF–Wass. Abwass,* 123: 521–530.

Sorg, T.J., 1979. Nitrate removal from drinking water. *EPA Seminar on Nitrates in Groundwater.* Kansas City, Mo, 3–4 October.

Sova, R.J., 1986. The chemical removal of nitrate from water supplies using ferrous sulphates and pickle liquor. *MS Thesis,* Univ of Nebraska at Lincoln, Lincoln, Nebr.

Spalding, R.F. and Exner, M.E., 1991. Nitrate contamination in the contiguous United States. In: *Nitrate Contamination: Exposure, Consequences and Control,* (Eds.) Bogardi and R.D. Kuzelca. Springer-Verlag, Germany, pp. 13–49.

Spalding, R.F. and Exner, M.E., 1993. Occurrence of nitrate in groundwater: A review. *J. Envir. Quality,* 22: 392–402.

Stumn, W. and Morgan, J., 1996. *Aquatic Chemistry: Chemical Equilibrium and Rates in Natural Waters,* 3rd edn. Wiley, New York, p. 1022.

Szabo, A.G. and Bartha, L.G., 1951. Silver catalysts in the reduction of nitrates by ferrous hydroxide. *Acta Chem Hung.,* 1: 116.

Todd, R.L., Lowrance, R.R., Hendrickson, O., Asmussen, L.E., Leonard, R., Fail, J. and Herrick, B., 1983. Riparian vegetation as filters of nutrients exported from a coastal plain agricultural watershed In: *Nutrient Cycling in Agricultural Ecosystems,*(Eds.) R.R. Lowrance, R.L. Todd, L.E. Asmussen and R.A. Leomard. The University of Georgia, College of Agriculture Experiment Station Special Publication, 23: 485.

Tree Buffer Protects Groundwater at Contaminated Site. 1994. *Water Env. Technol.,* 6: 15–16.

White, D.A., Labayru, R., Obermoller, H.R. and Deryde, X., 1991. The application of inorganic membranes to electrodialysis process. *Desalination and Water Reuse,* 4: 177–188.

Wilson, W.S., Ball, A.S. and Hinton, R.H., 1999. *Managing Risks of Nitrates to Human and the Environment.* The Royal Society of Chemistry.

Young, G.K., Bungay, H.R, Brown, L.N. and Parsons, W.A., 1964. Chemical reduction of nitrate in water. *JWPCF,* 36: 395–398.

Young, K., 1975. *Geology: The Paradox of Earth and Man.* Houghton Miffin, Boston.

Zawaideh, L.L. and Zhang, T.C., 1998. The effects of pH and addition of an organic buffer (HEPES) on nitrate transformation in Feo water systems. *Wat. Sci. Tech.,* 38: 107–115.

Water and Wastewater Management in 2 Vols (2011) *Pages 368–371*
Editors: D.R. Khanna, R. Bhutiani & Gagan Matta
Published by: DAYA PUBLISHING HOUSE, NEW DELHI

Chapter 56

Physico-chemical and Biological Characterization of the River Kunda at Downstream of Khargone, Madhya Pradesh

S.K. Mahajan[1], Yogesh Garde[2], Rakesh Bhawsar[3] and Sharad Sharma[4]

[1]*Retired Professor of Botany, 31 Jin Mandir Path, Khargone – 451 001, M.P.*
[2]*Research Student, Nutan Nagar Extension, Khargone – 451001, M.P.*
[3]*Research Student, 17, Bhawsar Mohalia, Khargone – 451 001,M.P.*
[3]*Research Student, 48, Kunda Nagar, Khargone – 451 001, M.P.*

ABSTRACT

Present chapter is an attempt to present the results of physico-chemical and biological investigation carried out on the river Kunda at Khargone, M.P., and India. In all 24 taxa were identified.

Keywords: *Physico-chemical, Khargone, Biological, River water, Quality, Characterization.*

Introduction

Khargone is situated between 27° 45' N and 75° 30' E at 250 amsl, about 84 km south of Khandwa central railway station, and at the bank of river Kunda, a tributary of Narmada. Algal growth can be observed round the year on the bank of this river. At present the river is polluted due to the mixing of sewage water flowing through drainage and joins the river at various places.

Earlier Mahajan (1990, 1991,1997 a and b), Mahajan and Sanvalia (1997), Bhawsar *et al.* (2001) have reported algal taxa from river Kunda but no work has been done on the water analysis and algal flora of downstream of the river Kunda. Hence the present investigation has been undertaken.

Materials and Method

Water samples and algal materials were collected during 2000-2001. Physico-chemical characteristics of the water samples were determined following Golterman and Clymo (1969), APHA (1980). Phycological data were carefully identified after consulting the standard literature (Desikachary, 1959; Randhawa, 1959 and Phillipose, 1967).

Results and Discussion

During the present study in all 24 algal taxa were reported, out of which 45.83 per cent belongs to Chlorophyceae, 33.33 per cent to Bacillariophyceae and 20.83 per cent to cyanophyceae. Thus, Chlorophyceae is the dominant group. *Scenedesmus* and *Closterium* are, most common amongst green algae while *Navicula, Gomphonema* and *Fragillaria* are amongst diatoms. Blue-greens like *Oscillatoria* and *Nostoc* are less common. Occurrence of *Johannesbaptistia* sp. is a new record for the West Nimar district of Madhya Pradesh. Besides this, few microscopic invertebrates *viz. Daphni, Vorticella, Rhabditis, Paramecium* have also been identified in the present investigations (Tables 56.1 and 56.2).

Table 56.1: Physico-chemical Data of Water Sample Collected from Downstream of River Kunda at Khargone

Parameters	*Concentration*
Temperature	25.5°C
Turbidity (NTU)	10.72
pH	8.63
Conductivity	1.23 μ mhos/cm
Chlorides	139.9 mg/l
Total Hardness	290 mg/l
Total Alkalinity	90 mg/l
Dissolved O_2	10 mg/l
BOD	2 mg/l
Total Solids	70 mg/l
Calcium	40.08 mg/l
Magnesium	46.02 mg/l

Table 56.2: Algal Species Reported from Downstream of River Kunda at Khargone, Madhya Pradesh

Sl.No.	*Class*	*Name of Algal Taxa Reported*
1.	Chlorophyceae	1. *Chlorelaa* sp.
		2. *Closterium cyclicum*
		3. *Closterium cynthea*
		4. *Cosmarium cyclinum*
		5. *Mougeotia* sp.
		6. *Scenedesmus armatus*
		7. *Scenedesmus brasilienses*
		8. *Scenedesmus denticulatus*
		9. *Spirogyra condensate*
		10. *Spirogyra sinensis*
		11. *Zygnema* sp.

Contd...

Table 56.2–Contd...

Sl.No.	Class	Name of Algal Taxa Reported
2.	Bacillariophyceae	1. *Cymbella aphinis*
		2. *Fragillaria rumpens*
		3. *Fragillaria rumpens* var. *familiaris*
		4. *Gomphonema lanceolatum*
		5. *Gomphonema sperophorum*
		6. *Navicula pupula*
		7. *Navicula raingardii*
		8. *Pinnularia brevicostata*
3.	Cyanophyceae	1. *Aphanocapsa muscicola*
		2. *Johannesbaptistia* sp.
		3. *Nostoc calcicola*
		4. *Oscillatoria princes*
		5. *Spirulina* sp.

The river Kunda is unpolluted at the point when it enters the Khargone city, but gets progressively polluted due to inputs from a number of sewage channels. Consequently the pH, Conductivity, Turbidity, Chlorides, Total hardness, Total alkalinity of the water increases and simultaneously dissolved oxygen of the river water decrease. The average values recorded for Dissolved Oxygen, Chlorides and Total Hardness were 10 mg/l, 139.2 mg/l and 290 mg/l respectively.

Acknowledgements

The authors are thankful to Prof. (Major) A. V. Rao, Principal, Govt. P. G. College, Khargone for encouragement and Prof. Dr. Pushpa Patel for facilities.

References

APHA, AWWA, WPCF, 1980. *Standard Methods for the Examination of Water and Wastewater*, 15th edn. American Public Health Association, Washington D.C., 1133 pp.

Bhawsar, R., Mahajan, S. K. and Dhanse, A., 2001. Phycological studies in the upstream of river Kunda at Khargone, M. P. In: *Nat. Seminar on Freshwater Ecosystem*, Organized by Deptt. of Zoology, Govt P.G. College, Khargone and sponsored by UGC Regional Office, Bhopal, Abst. p. 21.

Desikachary, T.V., 1959. *Cyanophyta.* Pub. I.C.A.R., New Delhi.

Golterman, H.L. and Clymo, R.S., 1969. *Methods for the Chemical Analysis of Freshwater*. B.P. Handbook No. 8, Blackwell. Sci. Pub., Oxford, pp. 166.

Mahajan, S.K., 1990. Cyanobacteria of Khargone Madhya Pradesh. In: *Proc. Nat. Symp. on Cyanobacteria.* Nitrogen Fixation NFBHAC, IARI, New Delhi.

Mahajan, S.K., 1991. Algal flora of Choolgiri Hills, Madhya Pradesh. *Phykos*, 30(1 and 2): 87–89.

Mahajan, S.K., 1997a. A checklist of diatoms from Madhya Pradesh. In: *Proc. 84th Annual Sess. Ind. Sci. Cong. Part–III* (*Bio. Sect.*). Abst. p. 8.

Mahajan, S.K., 1997b. A checklist of blue green algae from Madhya Pradesh. In: *Nat. Symp. on Recent Trends in Algal Research,* Vishakhapatnam. Abst. p. 8.

Mahajan, S.K. and Sanvalia, V., 1997. Taxonomic studies of algae in polluted and non-polluted waters of river Kunda, Madhya Pradesh. In: *Nat. Symp. on Recent Trends in Algal Research* (Sponsored by UGC, CSIR, INSA and Seaweed Research and Utilisation Association). Organized by Deptt. of Bot., Andhra University, Vishakhapatnam, Abst. p. 10.

Philipose, M.T. 1967. *Chlorococcales*. ICAR, New Delhi.

Randhawa, M.S., 1959. *Zygnemaceae*. ICAR, New Delhi.

Water and Wastewater Management in 2 Vols (2011) *Pages 372–375*
Editors: D.R. Khanna, R. Bhutiani & Gagan Matta
Published by: DAYA PUBLISHING HOUSE, NEW DELHI

Chapter 57

Hydrobiological Observations on Virla Reservoir at Khargone, Madhya Pradesh

Pushpa Patel[1], S.K. Mahajan[2] and S.K. Pathak[3]
[1]Botany Department, [3]Zoology Department,
Govt. P.G. College, Khargone – 451 001, M.P.
[2]Retired Professor of Botany, 31, Jain Mandir Path, Khargone – 451 001, M.P.

ABSTRACT

The present chapter is an attempt to present the hydrobiological observation on Virla reservoir of Khargone district. The data revealed that 5 major algal groups *i.e., Cyanophyceae, Bacillariophyceae, Chlorophyceae, Euglenophyceae* and *Chrysophyceae* are present in Virla reservoir.

Keywords: Freshwater biology, Algal taxonomy, Limnology.

Introduction

The pollution of water bodies invariably affects its flora and fauna. A large number of studies were carried out in recent years to emphasize the importance of planktonic algae in the assessment of water quality (Palmer, 1961; Shashikant, 1979; Somashekar and Ramaswamy, 1984). In West Nimar district very little work has been done in this direction (Mahajan, 1990; Mahajan and Sanvalia, 1997 and Garde *et al.*, 2001).

In the present investigation, an attempt has been made to assess the feasibility for application of algal data to evaluate and monitor the water quality and pollution level at Virla reservoir (21°50'30" N

Latitude, 75°23' 30" E Longitude, 309 m amsl) situated over a distance of 25 km from Khargone city. It was constructed during state time before 1947. Its catchment area is 50 sq.km., and 445 hectares area is irrigated with the help of this reservoir.

Materials and Method

Algal material and water samples were collected during the month of May, 2001. Standard methods were followed during the collection and analysis of water samples (Goltennan and Clymo, 1969 and APHA, 1980). The enumeration of algal taxa was made after careful identification with the help of standard literature (Desikachary, 1959, Randhawa, 1959 and Philipose, 1967).

Results and Discussion

It is revealed from Tables 57.1 to 57.3, that out of 35 algal taxa 14 members belong to Cyanophyceae, 9 to Bacillariophyceae, 9 to Chlorophyceae, 2 to Euglenophyceae and 1 to Chrysophyceae. Besides this, 2 microscopic invertebrates *i.e.*, Cyclops and Rhabditis have also been reported. Thus blue-greens dominated over the other alga groups which may be due to low oxygen and high oxidisable organic matter

Table 57.1: Physico-chemical Data of Water Sample Collected from Virla Reservoir during May, 2001

Sl.No.	*Characteristic*	*Results*
1.	Temperature	35°C
2.	pH	8.5
3.	Conductivity	6.58 μmhos
4.	DO	5.2 mg/L
5.	BOD	3.2 mg/L
6.	Chloride	39.99 mg/L
7.	Total hardness	180 mg/L
8	Total alkalinity	20 mg/L
9.	Calcium	32.06 mg/L

Table 57.2: Percentage Composition of Algal Flora at Virla Reservoir, Khargone (MP)

Algal Groups	*Percentage*
Cyanophyceae	40.00
Bacillariophyceae	25.74
Chlorophyceae	25.71
Euglenophyceae	5.71
Chrysophyceae	2.85

Table 57.3: Algal Species Reported from Virla Reservoir

Sl.No.	*Class*		*Name of Algal Taxa Reported*
1.	Cyanophyceae	1.	*Anabaena fertilissima*
		2.	*Aphanocapsa* sp.
		3.	*Aphanothece* sp.
		4.	*Aulosira fertilissima*
		5.	*Chamaesiphon rostaffinskii*
		6.	*Chroococcus minor*
		7.	*Cylindrospermum* sp.
		8.	*Dichothrix fusca*
		9.	*Gomphosphaeria*
		10.	*Merismopedia convoluta*
		11.	*Nostoc commune*

Contd...

Table 57.3–Contd..

Sl.No.	Class		Name of Algal Taxa Reported
		12.	*Osciilatoria princeps*
		13.	*O. vizagapatensis*
		14.	*Phormidium* sp.
2.	Bacillariophyceae	1.	*Cymbella affinis*
		2.	*Fragilaria rumpens* var. *familiaris*
		3.	*Gomphonema sphaerophorum*
		4.	*Navicula mutica*
		5.	*N. rediosa*
		6.	*Nituchia* sp.
		7.	*Pinnularia* sp.
		8.	*Pleuroslgma spenceri*
		9.	*Stauroneis* sp.
3.	Chlorophyceae	1.	*Ankistrodesmus falcatus* var. *acicularis*
		2.	*Chlorella* sp.
		3.	*Mougeotia* sp.
		4.	*Scenedesmus armatus* var. *exaculeatus*
		5.	*S. denticulatus*
		6.	*S. quadrifida*
		7.	*Spirogyra condensata*
		8.	S. *sinensis*
		9.	*Tetraedron trilobulatum*
4.	Euglenophyceae	1.	*Euglena* sp.
		2.	*Phacus* sp.
5.	Chrysophyceae	1.	*Synura* sp.

associated with neutral pH as Prescott (1938) and Rao (1977) have concluded. Temperature as a factor does not seem to play significant role in promoting the growth of blue-greens. The presence of less quantity of desmids in the Virla reservoir shows the polluted nature of the water. Blue-greens and euglenoid flagellates are mostly associated with organically rich water and low dissolved oxygen (5.2 mg/l). Chloride concentration is 39.99 mg/l which is less in comparison with the stipulated standards. Dissolved oxygen is an important indicator of water quality which also determines the distribution and abundance of algal population. Water is chemically distinct for desmids and other green algae than water with diatoms and Cyanophycean members. The average value of total hardness of water (180 mg/l) of Virla reservoir is within the prescribed limit of ISI and ICMR.

Acknowledgement

The authors are thankful to Prof. (Major) A.V. Rao, Principal, Govt. P.G. College for providing laboratory facilities.

References

APHA, AWWA, WPCF, 1980. *Standard Methods for the Examination of Water and Wastewater*, 15th edn. American Public Health Assocication, Washington DC, pp. 1133.

Desikachary, T.V., 1959. *Cyanophyta*. ICAR, New Delhi.

Golterman, H.L. and Clymo, R.S., 1969. *Methods for the Chemical Analysis of Freshwater*. B.P. Handbook No. 8, Blackwell Sci. Pub. Oxford, pp. 166.

Mahajan, S.K., 1990. Biomonitoring indicators of water quality at Khargone, Madhya Pradesh. In: *Natl. Symposium on Biomonitoring Indicators in an Aquatic Ecosystem*, Organized by Vellavar College for Women, Erode, Tamil Nadu.

Mahajan, S.K. and Sanvalia, V., 1997. Taxonomic studies of algae in polluted and non-polluted waters of river Kunda, Madhya Pradesh. In: *Natl. Symposium on Recent Trends in Algal Research*, Sponsored by UGC, CSIR, INSA and Seaveal Research Associations, Organized by Deptt. of Botany, Andhra University, Vishakhapatnam.

Palmer, C.M., 1969. A composite rating of algae tolerating organic pollution. *J. Phycol.*, 5: 78–82.

Philipose, M., 1967. *Chlorococcales*. ICAR, New Delhi.

Prescott, G., 1938. Objectionable algae and their control in lakes and reservoirs. *Lousiana Municipal Rev.*, 1: 2–3.

Randhawa, M.S., 1959. *Zygnemaceae*. ICAR, New Delhi.

Rao, V.S., 1977. An ecological study of three freshwater ponds of Hyderabad, India. IVth Phytoplankton. *Hydrobiologia*, 53: 13–22.

Shashikant, 1979. Diatoms as indicators of water quality. *Proc. Symp. Environ. Biol.*, p. 303–308.

Somashekar, R.K. and Ramaswamy, S.N., 1984. Biological assessment of water pollution: A Study of the river Kapila. *Intern. J. Environ. Studies*, 23: 261–267.

Yogesh, Garde, Mahajan, S.K., Sohani, V.S., Gangrade, K. and Atre, Manish, 2001. Hydrobiological studies of Dejla–Devada dam, Khargone, M.P. In: *Natl. Seminar on Freshwater Ecosystem in and Around Urban Area: Its Management and Aquaculture in New Millennium*, Organized by Deptt. of Zoology, Govt. P.G. College, Khargone.

Water and Wastewater Management in 2 Vols (2011) *Pages 376–380*
Editors: D.R. Khanna, R. Bhutiani & Gagan Matta
Published by: DAYA PUBLISHING HOUSE, NEW DELHI

Chapter 58

Potential of Remote Sensing and Geographical Information System in Water Resource Management

Samir Kumar Banger
DP Engineer, Hope Technologies Ltd.,
1 Navjeevan Vihar, New Delhi–110017

ABSTRACT

Water the source of life, is passionate. Too passionate to manage. Excess of it leads to flood and lack of it results in drought and famine. Overgrowing demand of water and increasing shortage of it may lead to major conflicts. It is important to make people aware of values of water more so that they will waste less and pollute less amount of water. Remote sensing technology and Geographical Information system have been extensively used in exploitation of ground and surface water in many areas. Now, time has come to use these technologies to realize the concepts such as artificial recharge, water harvesting and watershed management. These technologies have the potential to help in reaching the ultimate goal of water resource management so that every one can get sufficient amount of water. This article explores the potential of remote sensing and Geographical Information System in water resources management, development and its capability to handle the issues related to water management and development.

Keywords: Remote sensing, Geographical Information System, Water resources.

Introduction

Water, the elixir of life, has highly uneven availability in space, time and quantity. Water shortage or excess may have disastrous consequences. On the one hand there are droughts, on the other, flood devastates vast tract of land year after year. The myriad needs of the thronging population, burgeoning economic ambitions, demanding agriculture, expanding urbanization, increasing industrialization have caused great challenges in the best use of the available water resources. Consequently, the need for planning, conservation, optimum utilization and management of this natural resource is paramount for the economic upheaval of a country.

Water resources world over are under pressure due to land use change, urbanization, overuse and environmental polluting activities. The rapid advances made in spatial data acquisition technology over the last two decades, has given a unique advantage in synoptic, temporal coverage with vastly improved scales of observation.

The most important areas of application of Remote Sensing and GIS in water resource management are:

- ☆ Rainfall monitoring, including severe storm and flash flood forecasting and evaluation. (Andrawis *et al.*, 1978; Kumar, 1993)
- ☆ Surface water inventory and monitoring. (Chaurasia *et al.*, 1993)
- ☆ Snow mapping, measurement and monitoring.
- ☆ Soil moisture measurement through mainly for un-vegetated or sparsely vegetated localities. (Brucker and Stenage, 1988).
- ☆ Evapo-transpiration estimation and monitoring.
- ☆ River morphology (Kumar *et al.*, 1993).
- ☆ Sedimentation patterns rate in reservoirs and deposition in rivers (Rao and Mahabaleswara, 1990; Jain and Goel, 1993).
- ☆ Groundwater mapping and assessment (Aggarwal and Mishra, 1992; Ravindran and Manchanda, 1993) Water pollution monitoring (Baman, 1993).
- ☆ Hydrological forecasting of the above parameters and phenomena and numerous dependant variables.
- ☆ Watershed conservation, planning and management (Ja and Walsh, 1992).
- ☆ Drought monitoring.
- ☆ Estuary studies and coastal zone management.

Remote Sensing Sensors Used for Water Resource Management Studies

Currently aircraft and spacecraft employ a whole range of sensors for collection of water resources information. Polar orbiting satellites, such as IRS, land sat, SoPT and other satellites have scanners to provide panchromatic and multi spectral information of varying spectral, spatial, and temporal and radiometric resolution. The NoAA series of satellites in addition to weather related data provide terrain related information of use in snow studies, sea surface temperature, flood mapping and vegetation monitoring. Satellites such as ERS and Radarsat are carrying microwave instruments, like altimeter; scatter meter and synthetic aperture radar to study ocean and related phenomena.

Remote Sensing for Monitoring Water Availability

Remote sensing represents a very supportive tool in the following areas

- ☆ Detection and monitoring of water bodies

- Water volume estimation.
- Detection of potential groundwater sites.
- Population estimation for water supply planning.
- Monitoring agricultural uses in the boundaries of the city for water supply planning.
- Estimating hydrological parameters like runoff, soil moisture and drainage network for estimating water availability.

Remote Sensing for Updating Water Infrastructure

Remote sensing techniques may represent an economic tool for supporting water infrastructure design and maintenance. Particularly, remote sensing is successfully used in different applications

- Runoff estimation for supporting sewage system design.
- Geological studies for Dam construction.
- Monitoring urban expansion for updating the water supply and sewage systems.

Remote Sensing for Monitoring Wastewater

Remote sensing provides an alternative means for obtaining relatively low cost spatial and temporal data about surface water condition from a large geographical area. Landsat TM and MSS, Spot multi-spectral images, Air borne Imaging spectrometer AVIRIS can be used for detecting spatial and temporal variation in surface suspended sediments, organic yellow matter, chlorophyll and green algae and bacteria.

Remote Sensing for Monitoring Solid Waste Disposal

Solid waste is not efficiently collected and disposed off. Runoff and leaches can pollute and contaminate surface water and groundwater resources. Therefore, efficient and economic techniques are needed that allow water authorities to identify and monitor illegal dumping sites. Here remote sensing automatic classification techniques find a potential application to detect and monitor solid waste landfills (Karthiketeyan *et al.*, 1993). It is to be noted that large landfills could not be resolved due to limited spatial resolution of current sensors.

Integration of SAR data and passive data represent an efficient alternative for increasing accuracy in the detection of landfills by combining high spatial resolution of SAR data with the spatial information contained in the optical remote sensing imagery.

Remote Sensing for Uncontrolled Extraction of Natural Resources

Remote sensing is an efficient tool to monitor environmental degradation. The use of change detection maps, generated by using multi temporal remotely sensed images, allows changes in landscapes to be identified and monitored. This information of land degradation, deforestation and desertification is used for identifying lakes, rivers or wetlands particularly exposed to mud-slides, sediment discharge or floods.

In particular, the main characteristics, which make remote sensing techniques suitable to support water resource management, are:

1. Remote sensing data allows a wide variety of information concerning land and water to be extracted

2. Data can be obtained and analyzed with the necessary frequency.
3. A single remote sensing image covers all regions and provides information of the whole area.
4. Remote sensing technique avoids loss of data or gaps in information.
5. Automatic remote sensing techniques reduce the possible human errors that can occur when manipulating large quantities of data.
6. The integration of remote sensing and GIS provides a very supportive tool for monitoring water resources in large and complex geographical areas.

Main Advantages of Integrated GIS and RS

1. An efficient GIS needs a continuous updating of the various data elements held in its system and RS is an important information source.
2. Remotely sensed data are the only data sources that make a real time information system possible.
3. RS benefits from access to highly accurate ancillary data, available in a GIS, for increasing classification accuracy.
4. RS techniques allow a wide variety of geological, hydrological and urban information to be extracted from imagery data.
5. The capabilities of GIS for displaying geographic information prove very supportive tools for decisions makers.
6. The integration of RS and GIS provide a solid basis for developing a shared database among all the different entities involved in the water management activities.

In this context, an efficient integration of both Geographic Information System and remote sensing data represent a powerful tool for management of water resources and support decision making of water resources experts.

References

Aggarwal, A.K. and Mishra, D., 1992. Evaluation of groundwater potential in the environs of Jhansi city, U.P. using hydrogeomorphological assessment by satellite remote sensing technique. *Photonirvachak,* 20 (2 and 3): 121–128.

Andrawis, A.S., Moore, D.G. and Doka, A., 1978. Evaluation of landsat data for disaster assessment and planning, sudan flood. In: *Proceeding of the 14th Symposium on Remote Sensing of Environment,* Ann Arbor, Michigan, 3: 1443–1447.

Baman, D.K., 1993. Remote sensing contribution for augmentation of urban water supply in Greater Guwahati area. In: *Proc. Nat. Symp. on Remote Sensing Application for Resource Management with Special Emphasis on N E Region,* Guwahati, Nov. 25–27, pp. 490–496.

Bhadra, S.K. and Bhavanarayana, M., 1993. A study on the effect of soil moisture on spectral reflectance characteristics of Soils. In: *Proc. Nat. Symp. on Remote Sensing Application for Resource Management with Special Emphasis on N E Region,* Guwahati, Nov. 25–27, pp. 370–375.

Brucker, Wintnop and Stenge, P., 1988. Near surface soil moisture estimation from microwave measurement. *Remote Sensing of Environment,* 26: 101–121.

Chaurasia, R., Joshai, D.C. and Sharma, P.K., 1993. Shifting of water channel of river Satluj in Ludhiana District. In: *Proc. Nat. Symp. on Remote Sensing Application for Resource Management with Special Emphasis on N E Region*, Guwahati, Nov. 25–27, pp. 465–472.

Ja, E.M. and Walsh, S.J., 1992. GIS implications for hydrological modeling: Simulation of non-point pollution generated as a consequences of watershed development scenarios. *Computer, Environment and Urban Systems*, 16: 43–63.

Jain, S.K. and Goel, M.K., 1993. Reservoir sedimentation using digital image processing of IRS–1, IISS–1 data. In: *Proc. Nat. Symp. on Remote Sensing Application for Resource Management with Special Emphasis on N E Region*, Guwahati, Nov. 25–27, pp. 504–510.

Karthiketeyan, K.G., Elliott, H.A. and Brandt, R.C., 1993. Using a geographic information system for citing water treatment sludge minefills. In: *Proc. Int. Symp. on Integrated Resource Management and Landscape Modification for Environmental Protection*, pp. 315–325.

Kumar, A., 1993. Flood damage assessment in the area of the Ganga confluence with Ghanghra, Gandak and Slone. In: *Proc. Nat. Symp. on Remote Sensing Application for Resource Management with Special Emphasis on N E Region*, Guwahati, Nov. 25–27, pp. 46–49.

Rao, H.S.S. and Mahabaleswara, H., 1990. Prediction of rate of sedimentation of Tungabhadra reservoir. In: *International Symposium on Water Erosion, Sedimentation and Resource Conservation*, Oct. 9–13, Dehradun.

Ravindran, K.V. and Manchanda, M.I., 1993. Groundwater investigation in JNU campus, New Delhi using remote sensing technique. In: *Proc. Nat. Symp. on Remote Sensing Application for Resource Management with Special Emphasis on N E Region*, Guwahati, Nov. 25–27, pp. 161–166.

Vinod, K., Arun, P., Chakraborti, A.K. and Bhan, S.K., 1993. Geomorphological mapping in lower Ganga plain using Remote Sensing. In: *Proc. Nat. Symp. on Remote Sensing Application for Resource Management with Special Emphasis on N E Region*, Guwahati, Nov. 25–27, pp. 115–119.

Water and Wastewater Management in 2 Vols (2011) *Pages 381–387*
Editors: **D.R. Khanna, R. Bhutiani & Gagan Matta**
Published by: **DAYA PUBLISHING HOUSE, NEW DELHI**

Chapter 59

Zooplankton Diversity in Relation to Certain Physico-chemical Parameters of Swamp of Kishanganj District, Bihar

B.N. Pandey[1], O.P. Ambasta[2], A.K. Thakur[3], Sanjay Kumar[4] and Ritu Kumari[3]

[1]*P.G. Department of Zoology, Purnia College, Purnea*
[2]*Department of Botony, R.L. College, Madhav Nagar*
[3]*Department of Zoology, M.A.M. College, Naugachia*
[4]*P.S. Rahua School, Madhepura*

ABSTRACT

The zooplankton fauna of Kishanganj swamps in relation to certain physico-chemical parameter was studied. Out of 28 species recorded, 18 belong to *Rotifera*, 6 to *Cladocera* and 4 to *Copepoda*. The total zooplankton density was more in summer and least in rainy season. The quantitative relationship amongst the group of zooplankton was *Rotifera* > *Cladocera* > *Copepoda*. Zooplankton also comprised of some pollution tolerant species like *Brachionus*, *Keratella* and *Mesocyclops etc*. The Rotifers showed negative correlation with pH, sulphate and dissolved oxygen Cladocerans showed negative correlation with pH, sulphate and phosphate while Copepods revealed negative correlation with pH, sulphate and water temperature.

Keywords: *Zooplankton, Rotifera, Cladocera, Copepoda, Physico-chemical.*

Introduction

Zooplankton are the microscopic free swimming animals. They are found almost universally in all the aquatic environment comprising the first trophic level of heterotrophic food chains and form a link between phytoplankton and aquatic animals. They provide main food to fishes and can be used as indicators of trophic level of water body zooplankton play an integrated role in transferring energy to consumer, hence they make a higher trophic level in energy flow after phytoplankton. They respond to change in water quality (Holland *et al.*, 1983). Zooplankton play an important role in the trophic dynamic of aquatic ecosytems (Venkataraman, 1981). The occurrence and diversity of planktonic organisms are almost universal in all aquatic habitats and the greatest concentration of zooplankton occurs in upper layers of water (Cable, 1966). The extent of degradation of water bodies can be reliably evaluated with plankton (Vareethiah and Haniffa, 1998).

Investigation of biological diversity and primary productivity rates are of great importance to determine the potential for organic production and subsequent exploitation. The need for knowledge of zooplankton diversity is often stressed as a reliable indicator for the integrity of aquatic ecosystem (Barbosa *et al.*, 1995). The occurrence and physiological condition of zooplankton can be an indicator of environmental conditions. Hence in the present chapter an attempt has been made to study the population density of zooplankton of Kishanganj swamps in relation to certain physico-chemical factors.

Materials and Method

The monthly observation of physico-chemical factors and zooplankton population was made. Surface water and zooplankton were collected on every 15th day of the month at a fixed time. Water temperature was recorded with ordinary thermometer and free CO_2 was measured on the spot. The pH, transparency, dissolved oxygen, carbonate, bicarbonate, chloride, sulphate and phosphate were analyzed according to methods of APHA (1975) and Trivedy and Goel (1984). Zooplankton were collected by filtering 50.00 liters of water through a plankton net made up of bolting silk (no. 21) and preserved in 5 per cent formalin. Quantitative and qualitative analysis of zooplankton were made by Lackey (1938), as modified by Edmondson (1974). Correlation between zooplankton and certain physico-chemical parameters were computed.

Results and Discussion

The physico-chemical characteristics of the swamp water and zooplankton populations are shown in the Tables 59.1 and 59.2 respectively. The zooplankton taxa collected from the swamps water belong to three dominant groups *viz. Rotifera, Cladocera* and *Copepoda*. There were 28 zooplankton species identified from which 18 belong to *Rotifera*, 6 belong to *Cladocera* and 4 belong to *Copepoda*. Rotifers were the most dominant and abundant group showing highest percentage (56.85 per cent) composition and diversity followed by *Cladocera* (27.47 per cent) and *Copepoda* (15.68 per cent).

Following zooplankton were found in the swamps:

Rotifera

Brachionus angularis, B. rubens, B. caudatum, B. calyciflorus, B. forticula, B. falcatus, B. bicenata, Keratella tropica, K. vulga, Filinia longiseta, F. terminalis, Monostyla lunaris, M. bulla, Hordella sp., *Lepadella sp., B. plicatilis, Rotaria* sp., *Polyarthra* sp.

Cladocera

Moina micrura, M. dubia, Alonella sp., *Daphnia* sp., *Ceriodaphnia*, sp., *Chdorus gibba.*

Copepoda

Cyclops sp., *Nauplius* sp., *Diaptomus, Cyclops* sp.

Table 59.1: Physico-chemical Characteristics of Swamps water

Month	*Temp.*		*pH*	*DO*	*Free CO_2*	*CO_3^{2-}*	*HCO_3^-*	*NO_3^-*	*PO_4^{2-}*	*Cl^-*	*SO_4^{2-}*
	Atm.	*Water*									
January	20.90	19.10	7.50	8.20	8.50	0.00	125.00	0.36	0.50	14.90	49.60
February	23.80	21.10	7.40	7.80	6.70	0.00	122.00	0.37	0.52	16.50	60.40
March	25.50	22.40	7.20	7.30	9.80	0.00	126.00	0.38	0.60	20.40	72.20
April	30.10	26.80	7.10	7.10	9.60	0.00	130.00	0.42	0.84	22.20	75.60
May	33.90	31.70	7.90	6.80	8.80	0.00	135.00	0.49	0.94	28.70	71.80
June	32.80	31.90	6.80	6.10	7.20	0.00	118.00	0.52	0.72	26.20	70.90
July	33.50	30.80	6.60	5.80	9.90	0.00	105.00	0.58	0.74	19.90	78.50
August	30.80	29.40	6.70	5.10	12.80	0.00	98.60	0.62	0.72	16.40	73.20
September	29.20	28.60	6.40	4.80	11.40	0.00	106.70	0.56	0.70	12.40	80.50
October	29.30	26.20	6.90	5.20	10.50	0.00	110.20	0.52	0.73	15.10	63.80
November	26.80	25.40	7.10	5.90	9.60	0.00	112.40	0.49	0.71	16.20	57.40
December	21.40	22.10	7.30	7.80	8.10	0.00	114.20	0.41	0.68	17.10	60.5

Table 59.2: Monthly Variation of Zooplankton Groups in Swamps Water

Month	*Rotifera*	*Cladocera*	*Copepoda*
January	390.00	95.00	90.00
February	450.00	125.00	95.00
March	475.00	230.00	130.00
April	460.00	240.00	120.00
May	470.00	695.00	105.00
June	395.00	205.00	75.00
July	78.00	125.00	135.00
August	210.00	90.00	70.00
September	295.00	55.00	88.00
October	395.00	80.00	72.00
November	398.00	130.00	120.00
December	442.00	85.00	130.0

The quantitative zooplankton analysis showed that total plankton density is more in summer and least in rainy (Table 59.2; Figures 59.1 and 59.2). The quantitative relationship amongst the groups of zooplankton is *Rotifera* > *Cladocera* > *Copepoda*. *Rotifers* showed superiority over other groups both in terms of number of species, genera and population density. Dominance of *Rotifera* in seasonal data of zooplankton in the present study is in accordance with the earlier finding of Ferneska and Lewkowiez, 1966; Schindler and Noven, 1971; Pandey *et al.*, 1994, 2004. Dominance of Rotifers over other groups is an indication of congenial conditions of the system (Arora, 1966). Daggett and Davis (1974) suggested that a rich Rotifer community requires a stable medium as they depend upon

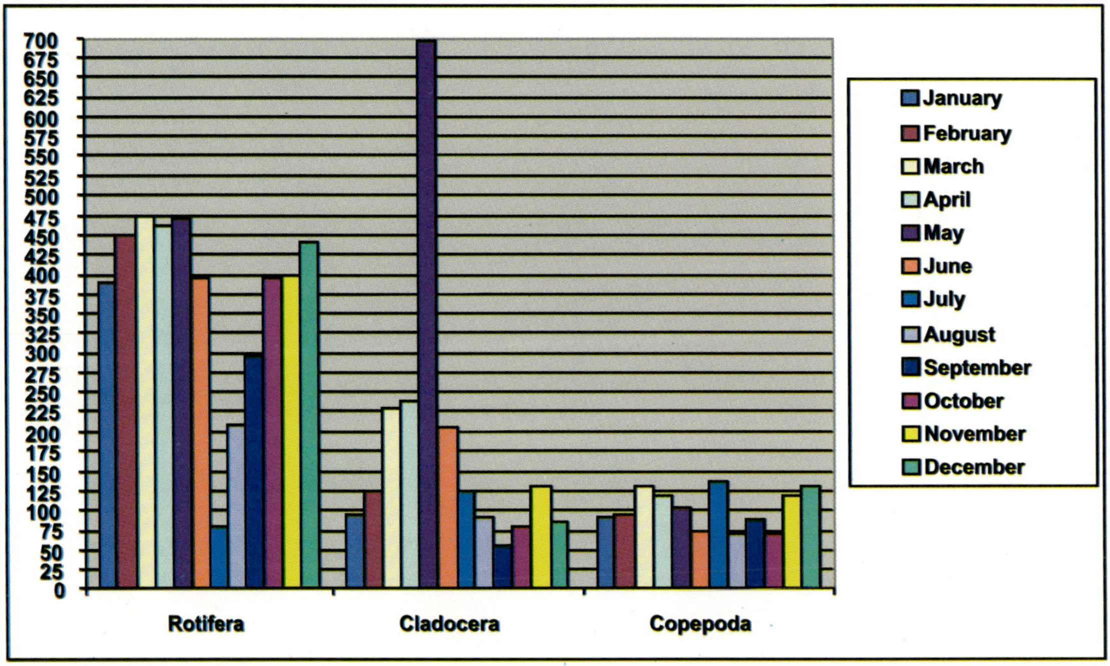

Figure 59.1: Showing Month-wise Distribution of Zooplankton Groups in Swamp Water

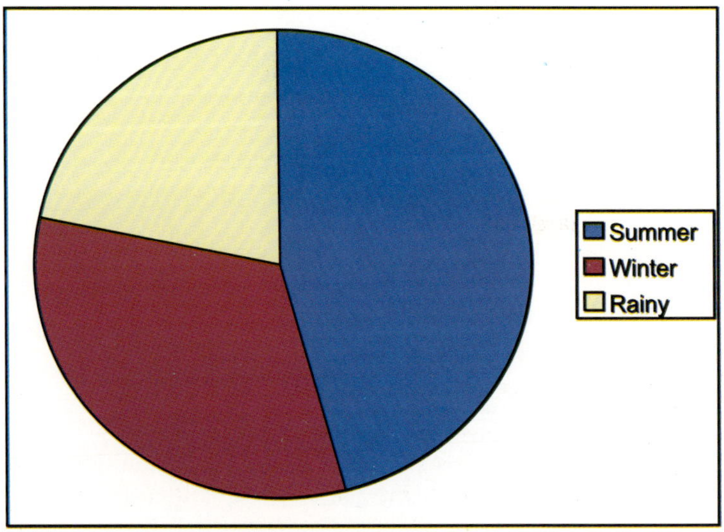

Figure 59.2: Seasonal Variation of Zooplankton

certain species of phytoplankton, whereas cladocerans and copepods due to their vast adaptability can withstand a wide range of environmental stress and utilize generally the phytoplanktonic cells as their natural food items. The rotifers communities have some association with water quality and any variation in the dissolved oxygen, organic matter, suspended solids would immediately affect their distribution (Holland *et al.*, 1983). The observed rotifers- *Branchionus calyciflorus, B. rubens, B. forticula, B. falcatus* can be considered to indicate the eutrophicated nature of the water body under the present investigation as they are most abundant in the present study. George (1966) and Bansei (1976) have reported summer peak of rotifers while Nasar (1977), Baker (1979) and Edmondson (1996) have shown winter peak of rotifers. But in the present investigation peak of rotifers was observed in summer season. The abundance of rotifers in any ecosystem as compared to other groups is an indication of eutrophication (George, 1966). Laal and Karthikeyan (1993) have reported the maximum rotifers at polluted zone of different rivers. Biswas and Konar (2000) have reported more rotifers at station of mixing zone of waters in Damodar river water. However, in the present investigation no such correlation was observed.

Table 59.3: Karl Pearsons's Correlation Coefficient (r)

Variables	*Rotifera*		*Cladocera*		*Copepoda*	
	r	*r²*	*r*	*r²*	*r*	*r²*
Water Temp. (°C)	0.22	0.494	0.053	0.0028	–0.094	0.0089
pH	–0.075	0.0057	–0.323	0.1045	–0.079	0.00062
DO (mg/l)	–0.349	0.0057	0.149	0.0223	0.527	0.2774
Free CO_2 (mg/l)	0.327	0.1070	0.620	0.3846	0.495	0.2455
Bicarbonate (mg/l)	0.124	0.248	0.434	0.1887	0.223	0.0499
NO_3^- (mg/l)	0.183	0.336	0.519	0.2693	0.432	0.1866
PO_4^{-3} (mg/l)	0.198	0.0393	–0.009	0.001	0.079	0.0063
Cl^- (mg/l)	0.078	0.006	0.454	0.206	0.530	0.2806
SO_4^{2-} (mg/l)	–0.233	0.0543	0.278	0.00775	–0.287	0.082

The seasonal occurrence and abundance of different species of rotifers showed that *B. angularis* was dominant in number over other species. Such result has been reported by Pandey *et al.* (1992). *B. calyciflorus* and *Keratella tropica* were observed throughout the year while other species showed fluctuations, *Keratella* sp. was regarded as an indicator of eutrophication. *Polyarthra* was abundant during winter indicating its preference for clear water. Seasonality was not shown by any other species.

The cladoceran population was less scanty in comparison to the rotifers. The cladoceran peak was observed during winter followed by summer and rainy seasons. Prasadam (1977) has reported the abundance of cladocerans in the month of May. Copepods were abundant during rainy season. Rainy peak of copepods has been also reported by Maruthanyagam *et al.* (2003) and Pandey *et al.* (2004). *Mesocyclops* were mainly represented by *Mesocyclops* sp. They indicate presence of particular matter in water and form primary food of planktivorous fish (Ivlev, 1961). Presence of copepods indicate rich trophic status of the water body (Pejler, 1983).

Presence of maximum zooplankton population in summer (Figure 59.2) might be due to the presence of higher population of bacteria. The lowest number recorded during rainy season may be related to the flood and fast currents. This finding support to Bilgrami and Datta Munshi (1985), Yousuf (1989). Seasonal fluctuation was observed in zooplanktonic community during the course of present investigation. Patra and Dutta (2004) have noted the seasonal fluctuation of zooplankton which is governed by abiotic and biotic factors. To assess the importance of abiotic interaction an attempt was made to analyze the data statistically (Table 59.3). Rotifers showed positive correlation with all the studied parameters except pH sulphate and dissolved oxygen, cladocerans showed negative correlation with pH and phosphate while copepods revealed negative correlation with water temperature, pH and sulphate.

References

APHA, 1975. *Standard Methods for Examination of Water and Wastewater*, 14th edn. American Public Health Association, Washington, D.C.

Arora, H.C., 1966. Rotifers as indicators of trophic nature of environment. *Hydrobiol.*, 27: 146–159.

Baker, R.L., 1979. Specific status of *Keratella cochlearis* (Goose) and *K. Eurlinare* Ahlaofrom (*Rotifera* and *Brachionidae*): Morphological and ecological consideration. *Can. J. Zool.*, 57: 1719–1722.

Bansei, I., 1976. Zooplankton investigation on the dammed river. *Tisza Reaches, Tiscia (Szeged)*, 11: 119–124.

Barbosa, F.A.R., Breudo, C.E.M. and Huszer, V.L.M., 1995. Phytoplankton studies in Brazil community structure, variation and diversity. In: *Limnology in Brazil*, (Eds.) J.S. Tundisi, C.E.M. Bicudo and T. Matstmura Tundisi. Brazilian Limnological Society, Sao Paulo, p. 18–36.

Bilgrami, K.S. and Datta Munshi, J.S., 1985. *Ecology of River Ganges: Impact of Human Activities and Conservation of Aquatic Biota (Patna to Farakka)*. Allied Press, Bhagalpur, India.

Biswas, B.K. and Konar, S.K., 2000. Influence of Nunia Nallah (canal) discharge on plankton abundance and diversity in the river Damodar at Narankuri (Raniganj) in west Bengal. *Indian J. Environ. and Ecoplan.*, 3: 209–217.

Cable, I.E., 1966. Plankton United States Department of Interior Fish and Wildlife Service. Bureau of Commercial Fishes, Fishery Leaflet, pp. 583.

Dagget, R.F. and Davis, C.C., 1974. A seasonal quantitative study of littoral cladocera and copepoda in a big pond an acid marsh in Newfoundland. *Int. Rev. Res. Hydrobiol.*, 59(5): 667–683.

Edmondson, W.T., 1974. A manual on methods for measuring primary production in aquatic environment. Acad. Press, N.Y., U.S.A.

Edmondson, W.T., 1996. *Freshwater Biology*. John Willey and Sons Inc., New York.

Ferneska, M. and Lewkowiez, S., 1966. Zooplankton in pond in relation to certain chemical factors. *Acta. Hydrobiol.*, 8: 127–153.

George, M.C., 1996. Comparative plankton ecology of five fish tank in Delhi, India. *Hydrobiol.*, 32: 47–60.

Holland, C.E., Brayan, C.F. and Neman, Jr. J.P., 1983. Water quality and the rotifer populations in the Alchafalaya river basin, Louisiana. *Hydrobiol.*, 98: 55–69.

Ivlev, V.S., 1961. *Experimental Ecology of the Feeding of Fishes*. Translated from Russian by D. Scott, Val. University Press, New Haven.

Laal, A.K. and Karthikeyan, M., 1993. Rotifer population or productivity indicators. *Curr. Sci.*, 5: 874–875.

Lackey, J.B., 1938. *The Manipulation and Counting of River Plankton and Changes in some Organism Due to Formation Preservation*. Publ. Health Deptt., 53: 20–80.

Maruthanayagam, M., Sasikumar, M. and Senthikumar, C., 2003. Studies on zooplankton population in Thirukkulum pond during summer and rainy seasons. *Nat. Env. and Poll. Tech.*, 2(1): 13–19.

Nasar, S.A.K., 1977. Investigations the seasonal periodicity of zooplankton freshwater pond in Bhagalpur India. *Acta Hydrochem. Hydrobiol.*, 5(6): 577–584.

Pandey, B.N., Mishra, A.K., Jha, A.K., and Lal, R.N., 1992. Studies on qualitative composition and seasonal fluctuation in plankton composition of river Mahananda, Katihar (Bihar). *J. Ecotoxicol. Environ. Mont.*, 2(2): 93–97.

Pandey, B.N., Jha, A.K., Das, P.K.L and Pandey, K., 1994. Zooplanktonic community in relation to certain physico–chemical features of Kosi swamps, Purnea, Bihar. *Env. and Ecol.*, 12(3): 563–567.

Pandey, B.N., Hussain, S., Jha, A.K. and Shyamanand, 2004. Seasonal fluctuation of zooplanktonic community in relation to certain physico-chemical parameters of river Ramjan of Kishanganj, Bihar. *Nat. Env. and Poll. Tech.*, 3(3): 325–330.

Patra, S.B. and Dutta, N.C., 2004. Seasonal fluctuations of different zooplanktonic group of a rainfed wetland in relation to some abiotic factors. *Ind. J. Env. and Ecoplan.*, 8: 7–12.

Pejler, B., 1983. Zooplanktonic indicators of trophy and their food. *Hydrobiol.*, 101: 111–114.

Prasadam, R.D., 1977. Observation of zooplankton population of some freshwater impoundments in Karnataka. In: *Proc. Symp. Warm. Wat. Zooplankton*, NIO, 609, pp. 214–225.

Reid, G.K. and Wood, R.D., 1976. *Ecology of Inland Waters and Estuaries*, 2nd edn. Van Nostrand Co., N.Y. pp. 485.

Schindler, D.W. and Noven, B., 1971. Vertical distribution and seasonal abundance of zooplankton in two shallow lakes at Ontario. *J. Fish Res. Bd.Can.*, 28: 245–256.

Trivedy, R.K. and Goel, P.K., 1984. *Chemical and Biological Methods for Water Pollution Studies*. Env. Pub. Karad, India.

Vareethiah, K. and Haniffa, M.A., 1998. Phytoplankton pollution indicators of Coir retting. *J. Env. Poll.*, 5(2): 117–122.

Venkataraman, K., 1981. Field and laboratory studies on *Daphnia carinta* (Cladocera : Daphinidae) from a seasonal tropical pond. *Hydrobiol.*, 30(1): 129–136.

Yousuf, A.R., 1989. Zooplankton study in India with special reference to North India. A critical review. In: *Management of Aquatic Ecosystem*. Society of Biosciences, Muzaffarnagar, India, pp. 309–324.

Water and Wastewater Management in 2 Vols (2011) *Pages 388–399*
Editors: D.R. Khanna, R. Bhutiani & Gagan Matta
Published by: DAYA PUBLISHING HOUSE, NEW DELHI

Chapter 60

Waste Management Strategy of a Fertilizer Plant

R. Bhutiani, D.R. Khanna and P. Sarkar
Department of Zoology and Environmental Sciences,
Gurukul Kangri Vishwavidalaya, Haridwar, Uttarakhand

ABSTRACT

This chapter presents the study to understand evaluate and examine the different pollution abatement strategies of Indian Farmer's Fertilizer Cooperative Ltd. It also includes the identification, types and means of various pollution hazards and to study the abatement strategy of different pollutants.

Keywords: Fertilizer, Management, Effluent, Pollution.

Introduction

Indian Farmer's Fertilizer Cooperative Ltd. (IFFCO) is one of the fastest growing fertilizer industry of India. At present its four plants are operating in the country. IFFCO was established on 3rd November 1967. The plant is situated at a distance of 28 km South-West of Bareilly on Bareilly-Aonla road in Uttar Pradesh.

The total capacity of Aonla Unit including both phases is 8,91,000 MTPA for Ammonia and 14,52,000 MTPA for Urea (EMIAU, 2000). India uses about 16 kg per hectare of fertilizer while the world average is 55 kg per hectare (Sharma and Kaur, 1997). A typical fertilizer unit in India produces 300 tons per day of ammonia, 300 tons per day of Urea, 1100 tons per day complex fertilizer and 130 tons per day of Methanol (Mahajan, 1985). According to eminent soil chemist, Dr. H.H. Koepf, modern agriculture can honestly claim two notable crops "disease and pests" but now a third factor" Poison" (as NO_2, NO_3 *etc.*) can be frequently added. Therefore due to high level of toxicity proper management and disposal of effluent is necessary. The Knitting of production with environmental protection is the unique feature of Aonla unit. This new technique has eased out various problems of environmental management. The technology of the plant is based on recycle and reuse of the effluent to achieve Zero effluent discharge from plant.

Considering these problems study has been made as Environment Management Strategy and' results are presented in this chapter.

Materials and Method

Management of environment confers utilization of resources prudently so that waste generation is minimized. The environment management at IFFCO aims for abatement of pollution at the source itself.

The various plants and sectors which are running at IFFCO Aonla unit with their capacity are as follows:

Sl.No.	*Plants*	*Capacity*
1.	Ammonia plant (2 no.)	2 × 1350 MTPD
2.	Urea plant (4 no.)	4 × 1100 MTPD
3.	Utilities	
	(*a*) Steam Generation service boiler	1 × 1500 MT/hr.
	(*b*) Gas turbine Generator (2 no.)	2 × 25 MW (ISO)
	(*c*) Water treatment plant	6 × 140 M^3/hr.
	(*d*) Inert gas Generation	600 Nm^3/hr.
	(*e*) Ammonia storage tank	4 × 10,000MT
4.	Centralized effluent treatment plant	
	(*a*) Cooling tower blow down treatment	480 m^3/hr.
	(*b*) Ammonia bearing occasional waste treatment	120 m^3/hr.
5.	Urea bagging plant	8 slats of 250 MT/hr./slat

The study has been made to examine the various abatement strategies of pollution at IFFCO Aonla unit with the following objectives:

1. To identify various pollution hazards.
2. To identify the types and means of pollution *i.e.* sources of pollution
3. To study the abatement strategy of different pollutants.
4. To study the environment strategy at IFFCO Aonla.
5. To develop an alternative strategy in existing system, and
6. To ensure awareness among factory employees about the environmental pollution.

Liquid Waste Management at IFFCO Aonla

To achieve zero effluent from Ammonia plant based on Halder Tapsone Technology has process condensate stripper, Turbine condensate, Boiler blow down, Disc oil separator.

The urea plant is based on Snam Progetti technology having waste water treatment section.

Process Condensate Stripping Section

This section treats process condensate from two separators and excess condensate from CO_2 removal section. The condensate stripping removes a substantial part of NH_3, CO_2 and Methanol from condensate before the treated condensate is passed to demineralised plant (Figure 60.1).

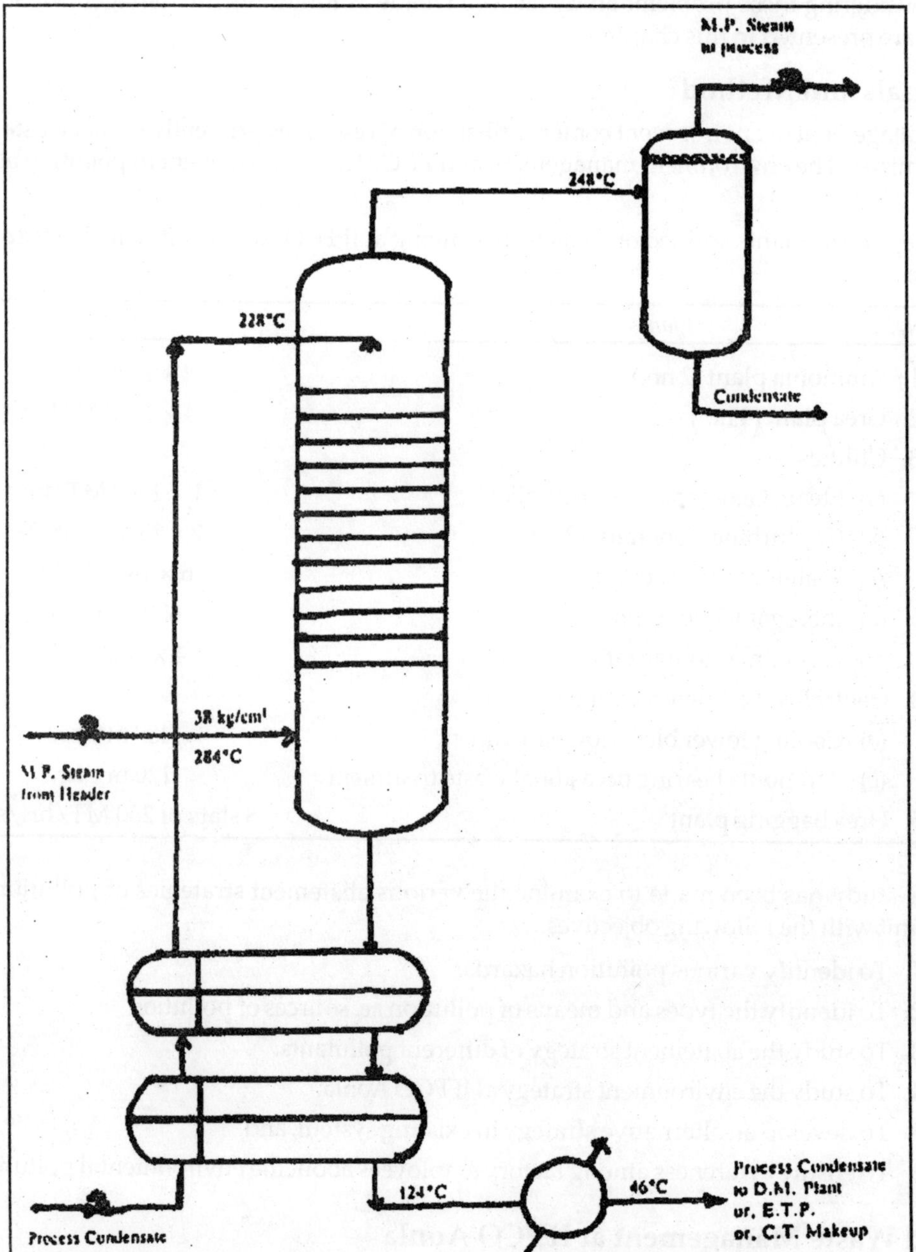

Figure 60.1: Flow Diagram of Process Condensate Stripping Section

Both NH_3 and Methanol together with CO_2 enters the process condensate according to following reactions:

(*i*) $NH_3 + H_2O = NH_4^+ + OH^-$

(*ii*) $CO_2 + H_2O = HCO_3 + H^+$

(*iii*) $2HCO_3^- = 2CO_3 + H_2$

(*iv*) $NH_3 + HCO_3^- = NH_2COO + H_2O$

In the reforming section methanol and ammonia will undergo chemical reaction and end up as N_2, H_2 and CO_2. The stripped condensate is finally cooled to 45°C before sent to demineralised plant.

Treatment of Turbine Condensate

Turbine condensate which comes from condenser is contaminated with small quantity of dissolved and undissolved solids. It is treated in condensate polishing section which consists of:

Activated Carbon Filter Section

Two no. of activated carbon filter are provided for removal of methanol, oil, grease, diethylamine and other undissolved and organic matter. Regeneration of carbon filter is done once in every 48 hrs.

Mixed Bed Unit

Two no. of mixed bed unit are provided for treating return condensate. Cation and anion (mainly Na^+, K^+, Ca^{++} *etc.*) are removed in this unit.

Treatment of Boiler Blow Down

The boiler blow down in flashed in a blow down vessel, where flashed saturated low pressure steam is separated from steam condensate.

Treatment of Oily Water

Here the spilled oil is collected through underground PVC pipes to oily water basin where oil and water forms two phases with oily layer at top. Here oil is skimmed off with the help of Disc oil separator and is transferred to slope oil tank. Waste water is pumped to neutralization pit.

Treatment of Ammonia and Urea Bearing Water in Urea Plant

The exhaust solution from stripping column containing mainly Urea of fed to deep hydrolyser operating at 36 atm pressure

On hydrolysis urea is broken down into NH_3 and CO_2.

$$NH_2.CO.NH_2 \xrightarrow{H_2P} 2NH_3 + CO_2$$

The hydrolyzed solution is fed to distillation tower where ammonia and carbon dioxide is distilled out by low pressure steam. The treated water is used as cooling tower make up.

Treatment of Acidic and Alkaline Waste

Ammonia and water treatment plants are having ion exchangers. The water which comes from cation exchangers contains lot of free mineral acids, salts of corresponding acids *viz.* HCl, H_2CO_4, H_2CO_3, NaCl, $CaCl_2$ *etc.* Similarly, the water which comes through anion exchanger contains sodium salts of weak and strong acid *viz.* NaCl, Na_2SO_4, and Na_2SiO_3. To neutralize acid waste 5 per cent HCl solution is used. Neutralization pits are connected to centralized effluent treatment plant.

$$2\,ReH + CaSO_4 \longrightarrow Re_2Ca + H_2SO_4$$

$$ReH + NaCl \longrightarrow ReNa + HCl$$

$$Re_2Ca + 2HCl \longrightarrow 2ReH + CaCl_2$$

$$ReNa + HCl \longrightarrow ReH + NaCl$$

Formation of sodium salts of weak and strong acid in anion exchanger:

$$ReOH + HCl \longrightarrow ReCl + H_2O$$

$$2ReOH + HZSO_4 \longrightarrow Re_2SO_4 + 2H_2O$$

$$2ReOH + H_2CO_3 \longrightarrow Re_2CO_3 + 2H_2O$$

$$ReCl + NaOH \longrightarrow ReOH + NaCl$$

$$Re_2SO_4 + 2\,NaOH \longrightarrow 2\,ReOH + Na_2SO_4$$

$$Re_2SiO_3 + 2NaOH \longrightarrow 2\,ReOH + Na_2SiO_3$$

Treatment of Cooling Tower Blow Down and Occasional Waste by Centralized Effluent Treatment Plant

Ammonia and urea plant are having separate cooling towers which uses chromate based corrosion inhibitors. The chromated cooling tower blow down is treated in effluent treatment plant which is designed to:

Remove Hexavalent Chromium from Cooling Tower Blow Down

Cooling tower blow down contains following contaminants:

Ca as Ca CO_3	144 ppm
Mg as Mg CO_3	96 ppm
Na as Na_2CO_3	1448 ppm
SiO_2 as SiO_2	120 ppm
Cr as Cr^{+6}	14–16 ppm
Free chlorine as Cl_2	1.0 ppm

Impurities are treated in chromate reduction plant measured amount of conc. H_2SO_4 and ferrous sulphate at suitable conc. is dosed in to blow down water (H_2SO_4 is added to control the blow down pH in the range of 2–3.5. Because this process progresses rapidly at low pH and reaction end point is more distinct at low pH.

Finally in chromate reduction pond the hexavalent chromium reduces to trivalent chromium.

The effluent is then collected in precipitation pond. In this pond, alkali (lime) is added to elevate the pH to 9.0, where trivalent chromium and iron is precipitated out as hydroxide of corresponding metal.

Reduction Pond

$$Na_2Cr_2O_7 + 6FeSO_4 + 7H_2SO_4 \longrightarrow Cr_2(SO_4)_3 + 3Fe_2(SO_4)_3 + Na_2SO_4 + 7H_2O$$

Precipitation Pond

$$Cr_2(SO_4)_3 + 3Ca(OH)_2 \longrightarrow 2Cr(OH)_3 + 3CaSO_4$$

$$Fe(SO_4)_3 + 3Ca(OH)_2 \longrightarrow 3Fe(OH)_3 + 3CaSO_4$$

Chemical Consumption

Ferrous Sulphate	1614 kg/day
Lime	1625 kg/day
98 per cent conc. H_2SO_4	1222 kg/day

Used for flow of 480 m^3/hr. of cooling tower blow down to chromate treatment plant. Now the precipitated liquid is transferred to clarlflocculator where the solids can be agglomerated in to fast settling particles by adding coagulating agent *i.e.* Alum. The clear water is taken to guard pond and sludge separated in centrifuge is dumped in sludge pits.

Treatment of Occasional Waste from Ammonia and Urea Plant

During normal operating conditions, there is zero effluent generated from urea plant except floor washing. But during abnormal condition ammonia bearing waste from urea or ammonia plant is collected in to contaminated effluent from demineralising plant and lime solution is added at controlled rate to maintain pH about 11. The effluent after filteration is sent for stripping of free ammonia. The air stripped effluent is further heated and residual ammonia is further removed by steam stripping with low pressure steam stripper. The treated effluent is sent to "*Treated effluent lagoon*" These ponds are quite safe from environmental angles as they are double line PCC with polyethylene sheets.

Treatment of Sewage Waste

Two no. of oxidation ponds are installed to treat sewage from township. In this aeration is carried out to ensure the dissolved oxygen content of waste water. The water thus obtained is used for irrigation purpose.

Air Pollution Management at IFFCO Aonla

Analysis of Gaseous Pollutants

There are 5 primary pollutants which together contribute more than 90 per cent of global air pollution these are (*i*) Carbon monoxide (CO), (*ii*) Nitrogenoxides $(NO)_x$, (*iii*) Hydrocarbons $(HC)_x$, (*iv*) Sulphuroxides $(SO)_x$, (*v*) Particulates.

At IFFCO, the concentration of sulphur being negligible in natural gas obtained from Bombay high basin the SO_2 concentration in flue gases remains in traces. .

To assess concentration of pollutants IFFCO Aonla, separately incorporated SO_2 and $(NO)_x$ analyzer, apart from this CO, O_2 analyser and High volume air sampler are also incorporated. The stack height of steam generation service boiler flue gas stack is kept at 120m for efficient dispersion of plume.

Dedusting System at Urea Prilling Tower

The air loaded with urea sublimate from prilling tower enters the dedusting system at point A, and escapes the plant at point B with clear air quality (Figure 60.2).

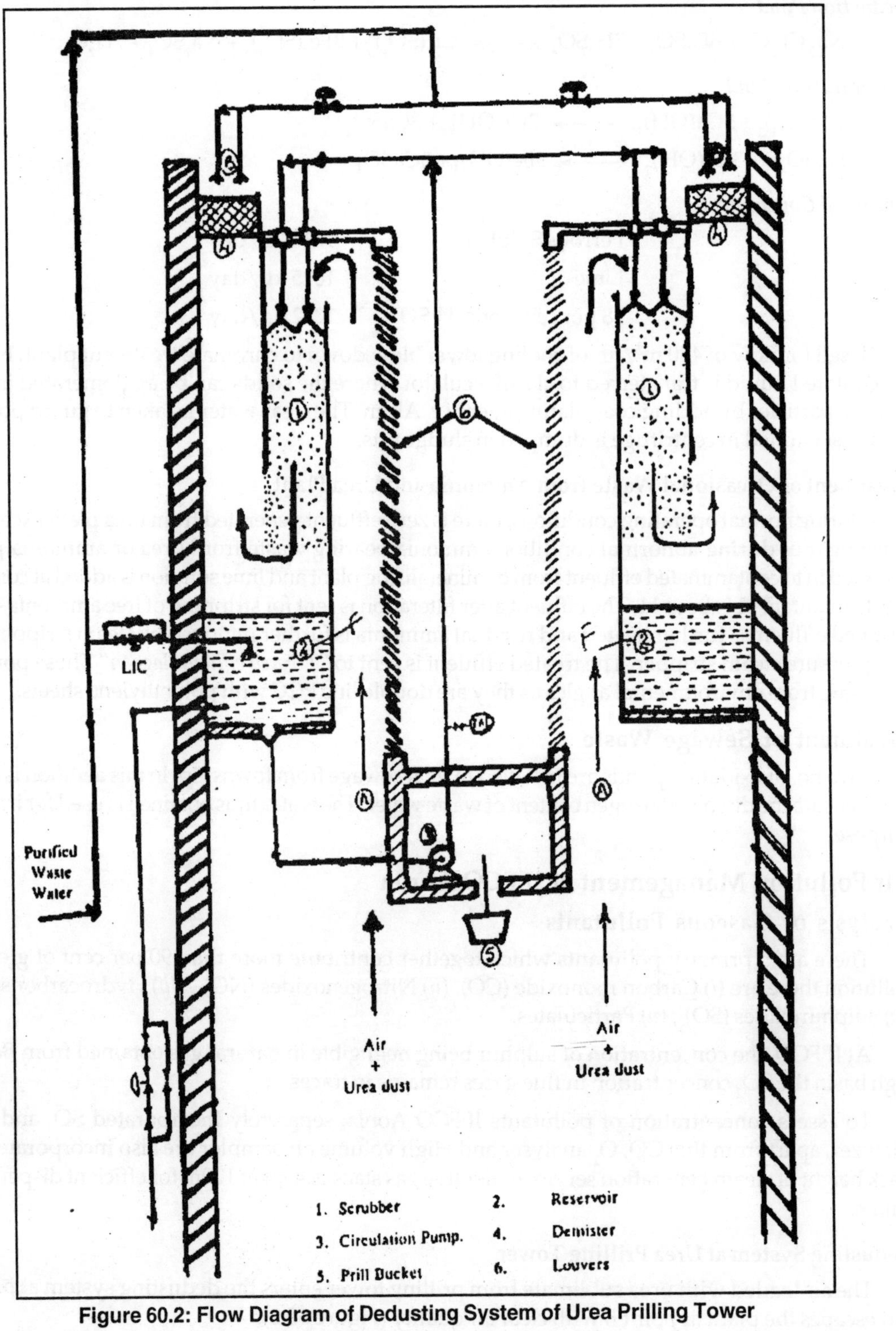

Figure 60.2: Flow Diagram of Dedusting System of Urea Prilling Tower

The urea dedusting containing gas is scrubbed by lean urea solution which is circulated by pumps. The scrubbing liquid enters the scrubber and is distributed by means of nozzles to become drops with a large absorption surface. This at the same time produces a pressure gain causing at suction connection (A) of the scrubber a suction pressure which sucks off exactly whereby the desired absorption surface is obtained. After escapes at point B into atmosphere containing 5ppm of urea dust. The liquid is collected in a reservoir thus, urea solution can be recovered and extracted at point F. Process condensate water is fed in at point E.

Land Management at IFFCO Aonla

Land management includes both the soil development and solid waste management.

Soil Development and Afforestation

Initially the soil of the land of IFFCO was highly alkaline in nature with higher content of clay and barren for agricultural purpose but after treating the soil with pyrite along with Gypsum and finally made for plantation. They used 336 acres of total land out of 1273 acres for plantation. IFFCO Aonla developed their social forestry farm. As a result at present there is 80 m wide green belt around the factory and 250 m towards township and additional afforestation is in progress with about 1.5 lac tree plantation.

Solid Waste Disposal and Management

Solid Waste from Plant

One of the solid waste from plant is hexavalent chromium which after treatment is dumped in to sludge pit which is safe from environmental point of view.

Solid Waste from Township

Solid waste collected from township is dumped into low line area to maintain the level of the land.

Noise Pollution Management at IFFCO Aonla

In IFFCO to control the pollution through noise there is a regular checking of heavy machines such as pumps, motors, compressor and structure support of different machines. There is also a regular checking of vibrations and knocking through machines. To avoid noise pollution there is regular maintenance of producing sources such as greasing, oiling of machines *etc*.

Control Methods as described by Gupta and Singh (1988) were used for the above study.

Standard Parameters like pH, free ammonia, urea, Total Kjeldahl nitrogen, BOD, Oil and grease, SO_2, NO_x chromium were selected for the study. Methods as described by APHA (1985) and Trivedy and Goel (1986) were used.

Results and Discussion

Flow sheet showing interconnection among plants and treatment facilities is shown in Figure 60.3.

The values of various important parameters which were studied are shown in Tables 60.1 and 60.2.

The Figure 60.5 shows the amount of pollution decrease during the year 1995–96 to 1998–99.

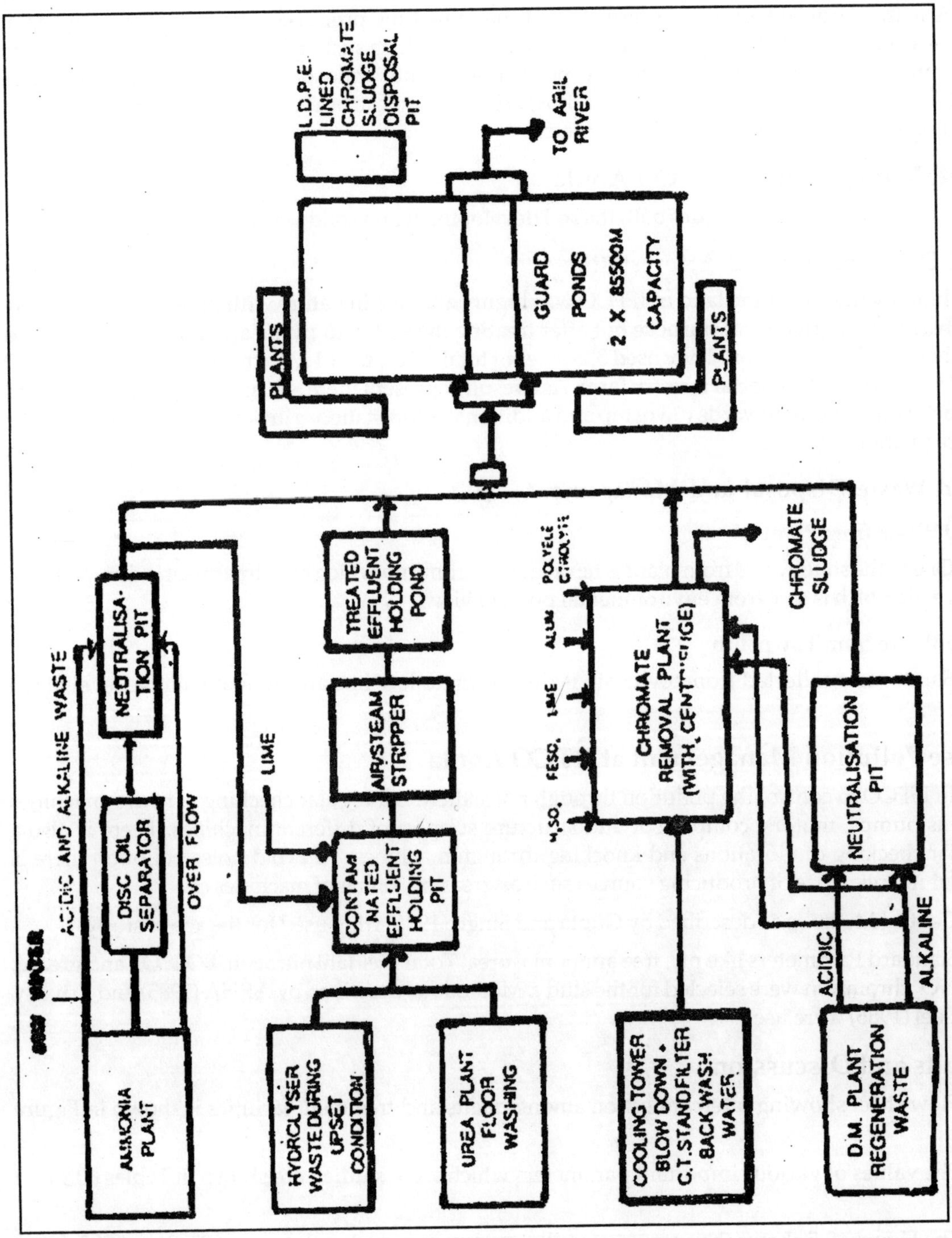

Figure 60.3: Flow Diagram of Pollution Control Measure at IFFCO Aonla

Table 60.1: Average Value of Treated Effluent Analysis (in μgm/Nm³)

Parameter	*Result of Analysis*
pH	7.6
T.KN.	23.69
Free ammonia	0.99
B.O.D.	12.6
Hexavalent chromium	NT
Oil and grease	NT
Nitrate	1.67

Table 60.2: Average Value of Ambient Air Quality (in gm/Nm³)

Parameter	*Result of Analysis*
Sulphur dioxide	NT
Nitrogen oxides	17.10
Ammonia	50.50
Carbon monoxide	NT
Suspended Particulate	135.05
Urea dust	180.0

The strategies of the present study have already been described above. The value of free ammonia in treated effluent was found to be 0.99 mg/l. The value of pH was recorded as 7.6.

The quantity of raw water demand and effluent discharged was found to be 4056400m^3 and 393085.8m^3. While according to Gupta (1997) the quantity of raw water discharged was 437111m^3 and 393085.8m^3 for quantity of discharged effluent. Here the loss in the quantity of discharged effluent showed the better utilization of effluent from plant (Figures 60.4 and 60.5).

The value of BOD was found to be 12.6 ppm while the MINAS (1985) was given as 30.0 ppm. Saxena and Mehra (1989) found Kjeldahl Nitrogen in fertilizer plant effluent as 150 mg/l while compairing to IFFCO value it is only 23.69 mg/l which is within the prescribed limit.

Due to effective environmental management system, the treated effluent and ambient water quality is maintained.

From the above data, we can conclude that with the effective pollution and environmental management system at IFFCO Aonla unit, the quality of final effluent after treatment and ambient air quality is well with in the limits.

Conclusion

After evaluating and analyzing the various pollutants and pollution control measures at IFFCO Aonla, the following conclusions can be drawn

1. The utilization of treated waste water in irrigation practices is unique of its kind. The philosophy of utilization can be extended throughout the country, thereof utilizing the precious natural water resources.

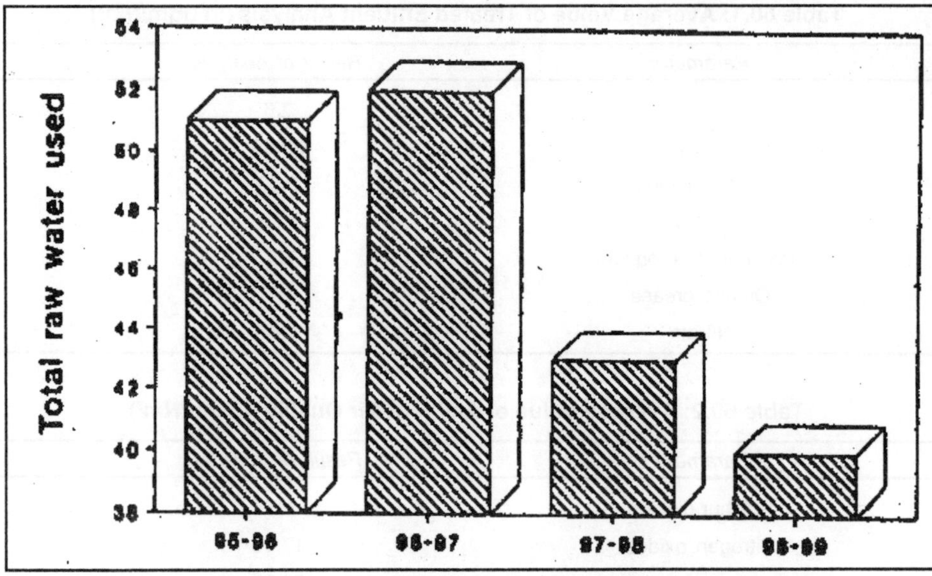

Figure 60.4: Quantity of Raw Water Demand

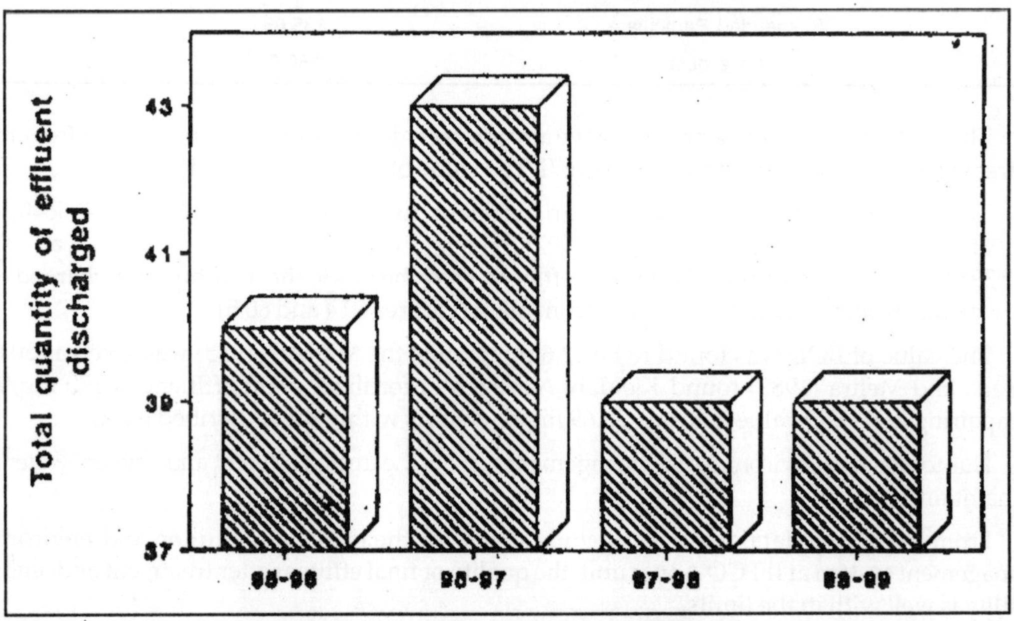

Figure 60.5: Quantity of Discharged Effluents

2. A study is needed for gainful utilization of solid waste (*i.e.* chromium hydroxide, $CaCO_3$, NaOH) as raw material for dye and pigment industries.
3. Acid and alkali, utilized for pH balancing can be replaced with CO_2 gas.
4. Development of green belt around the factory to reduce the percentage of carbon dioxide in the ambient air.

References

APHA, 1985. *Standard Methods for Examination of Water and Wastewater*, 16th edition. American Public Health Association, Washington, D.C.

EMIAU, 2000. *Environment Management at IFFCO Aonla Unit*, A book by IFFCO Aonla Unit.

Gupta, A., 1997. *A Case Study of IFFCO Aonla, Bareilly*. Indian Institute of Ecology and Environment, New Delhi.

Gupta, S.P. and Singh, A.P., 1988. Environment management system by Indian farmers Fertilizer Cooperative Limited. *Process and Plant Engineering Journal*, pp. 71–73.

Mahajan, S.P., 1985. *Pollution Control in Process Industries*. Tata McGraw Hill Publication Company Limited, New Delhi, p. 204–218.

MINAS, 1985. *Straight Nitrogenous Fertilizer Industry: Minimal National Standards*. Central Pollution Control Board, New Delhi, p. 29.

Saxena, S.K. and Mehra, K., 1989. Nitrogenous fertilizer industries in India: Need for an effective pollution control strategy. *IJEP*, 9(3): 175–177.

Sharma, B.K. and Kaur, H., 1997. *Environmental Chemistry*. Goel Publishing House, Meerut.

Water and Wastewater Management in 2 Vols (2011) Pages 400–426
Editors: D.R. Khanna, R. Bhutiani & Gagan Matta
Published by: DAYA PUBLISHING HOUSE, NEW DELHI

Chapter 61

Ecology of Plankton in Some Paddy Fields Near Nagri Bus Stand, Kathua, J&K

S.P.S. Dutta, V. Sharma and S. Kour
Department of Environmental Sciences, University of Jammu, Jammu

ABSTRACT

Physico-chemical characteristics of water and plankton were analysed from four paddy fields, irrigated by a tributary of the river Ravi, near Nagri bus stand, Kathua, J&K, during July, 2005 to September, 2005 have been described. Phytoplankton has shown the presence of Chlorophyceae (14 genera), Bacillariophyceae (9 genera) and Cyanophyceae (6 genera). The order of qualitative dominance of zooplankton is observed as Protozoa (10 genera and 31 species) > Rotifera (8 genera and 9 species) > Arthropoda (8 genera, 8 species and 3 larvae) > Annelida (1 genus and 1 species) and > Nematoda. Planktonic diversity and density remained low during first two observations (preparation of fields and seedlings transplantation). Coefficient of correlation (r) of phytoplankton and zooplankton, with the various parameters of water, is insignificant.

Keywords: *Planktonic ecology, Paddy fields, Nagri bus stand, Kathua.*

Introduction

Paddy fields, which support a diversified aquatic biota, have received the scientific attention of Goyal *et al.* (1984), Rather and Mir (1987), Pandoh (1998) and Dutta *et al.* (2002 a,b; 2004; 2005 and 2006 a,b,c) from J&K state. However, there is no record of any work for seasonal wetlands from Kathua district. The present planktonic study from paddy fields, irrigated by a tributary of the river Ravi, near

Nagri bus stand, was undertaken to add to the existing knowledge of biotic characteristics of seasonal wetlands from Jammu region of J&K state.

Topography of the Area

Kathua district, of Jammu region, situated at 32°17′ to 32°55′ North Latitude and 75°70′ to 76°16′ East Longitude has an area of 2651 km^2. The District is surrounded by Punjab in the S-E, Himachal Pradesh in N-E, Doda and Udhampur in North and N-W, Jammu in the West and Pakistan in S-W (Figure 61.1). The main drainage of the area is by the river Ravi. The area being plain is suitable for agriculture. The texture of the soil is soft and loamy. Annually two crops *viz.* wheat during winter and paddy during summer-monsoon season are cultivated. Three varieties of paddy *viz.* China, Ratna and local basmati are cultivated in the area. For the present study, four paddy fields (about 1 acre), near Nagri bus stand, were selected. In these fields, local basmati is cultivated. Irrigation of fields in the area is done through a tributary of the river Ravi called as Kathua canal or Kashmir canal.

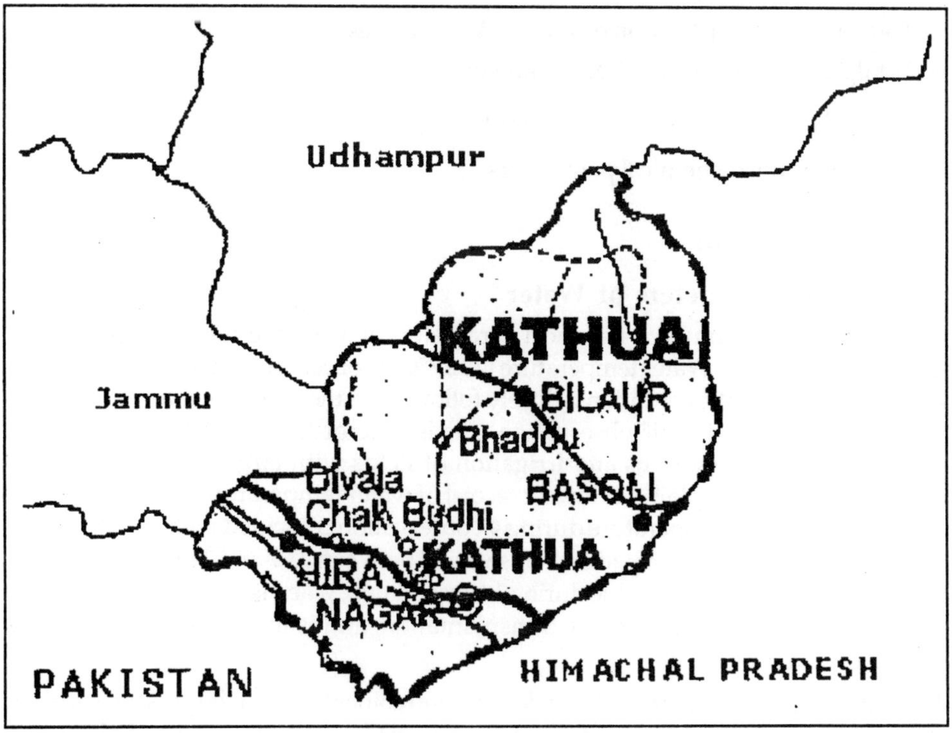

Figure 61.1: Study Area

Materials and Method

Sampling of Water

Water samples, from four selected paddy fields, were collected during preparation of fields for seedlings transplantation, during seedlings transplantation and weekly thereafter, during paddy growing season *viz.* July to September (2005) in plastic containers and analysed for various abiotic characteristics by standard methods (Indian Standard Method, 1973; Trivedy *et al.* 1987 and APHA, 1998). Depth was recorded by using a meter rod and temperature by mercury bulb (°C) thermometer.

Sampling and Analysis of Plankton

Planktonic samples from each paddy field were collected by filtering two litres of water through a planktonic net (No. 25) and analysed in the laboratory (Smith, 1950; Nair *et al.*, 1971; Kant, 1977; Kant and Anand, 1978; Dutta, 1983; Adoni, 1985; Kudo, 1986; Battish, 1992; Edmondson, 1992 and Biswas and Raut, 2000) and counted by drop count method. The results are expressed as number per litre (nil).

Coefficient of correlation (r) of phytoplankton and zooplankton, with various abiotic parameters of water, was calculated by the formula:

$$\text{Coefficient of correlation (r)} = \frac{\Sigma xy - \overline{X}\Sigma y}{\sqrt{\left(\Sigma x^2 - \overline{X}\Sigma x\right)\left(\Sigma y^2 - \overline{Y}\Sigma y\right)}}$$

where,

x: Total no. of phytoplankton or zooplankton species.

y: Total no. of physico-chemical parameters.

$\overline{X}$: Mean of phytoplankton or zooplankton.

$\overline{Y}$: Mean of physico-chemical parameters.

Results and Discussion

Physico-chemical Parameters of Water

The results of various physico-chemical parameters of water are summarized in Table 61.1 and depicted in Figures 61.2a–h. Water temperature (22.17°C, 1st observation and 29.22°C, 4th observation), due to shallowness and lentic conditions, closely followed by the air temperature and varied between (25°C, 1st observation and 34°C, 4th observation) and is in accordance with the findings of Dutta *et al.* (2004 and 2006 a,b,c). Heavy rains and irrigation of fields with large quantities of water, during preparation of fields, may explain the lowest record of water temperature noticed on 1st observation. Maximum record of water temperature during 4th observation coincided with the highest observation of air temperature. Depth in these paddy fields varied between 2.87 and 11.92 cm (Table 61.1, Figure 61.2b) and observed its maximum value during 1st observation, when large amount of water is available for field preparation, before seedlings + transplantation, and is in agreement with the findings of Dutta *et al.* (2004 and 2006 a,b,c).

pH varied between 6.38 and 8.18. Its wide variation coincided with the presence or absence of free CO_2 and CO_3. Direct relationship of pH with CO_3 may explain highest record of pH (8.18) during 4th observation, when carbonate was present in two fields. Similarly, an inverse relationship of pH with free CO_2 may explain the low record of pH during 2nd observation, with free CO_2 was comparatively high. An inverse relationship of pH with free CO_2 and direct with CO_2 is already on record (Welch, 1952; Reid and Wood, 1976; Goldman and Home, 1983; Jhingran, 1991 and Wetzel, 2000). Dissolved oxygen fluctuated between 2.49 mg/l to 7.15 mg/l. Irrigation of fields with large quantities of water and agitation of water during preparation of fields, before seedlings transplantation, may account for rise in DO during 1st observation. Free CO_2 recorded its highest (17.44 mg/l) and lowest (6.93 mg/l) value during 9th and 4th observation, respectively. Presence of water in pools and records of decaying dead organic matter may explain highest record of free CO_2 during 9th observation. Carbonate is seen

Table 61.1: Mean Observations of Various Physico-chemical Characteristics of Water in Four Paddy Fields near Nagri Bus Stand, Kathua (6th July 2005 to 26th September 2005)

Water Parameters	*1*	*2*	*3*	*4*	*5*	*6*	*7*	*8*	*9*	*10*	*11*	*12*
	06.07.05	*08.07.05*	*18.07.05*	*25.07.05*	*02.08.05*	*09.08.05*	*17.08.05*	*22.08.05*	*29.08.05*	*12.09.05*	*19.09.05*	*26.09.05*
Air temperature (°C)	25	28	30	34	31	29.5	27	28	29	30	30.5	29.5
Water temperature (°C)	22.17	26.55	28.2	29.22	28.75	27.42	26	23.5	25.97	25.95	25.72	25.47
Depth (cm)	11.92	7.9	6	7.72	4.97	4.85	4.86	5.9	4.67	2.87	3.42	3.95
pH	6.69	6.38	7.71	8.18	7.43	7.39	7.69	7.99	7.67	7.66	8.06	7.52
DO (mg/l)	7.15	4.37	4.95	3.74	6.16	3.26	5.63	6.34	6.3	2.73	2.49	6.67
Free CO_2 (mg/l)	11.98	7.27	12.47	4.56	16.2	6.93	12.65	9.08	17.44	14.84	11.01	8.46
CO_3^- (mg/l)	–	–	–	21.3	–	–	–	–	–	–	–	–
HCO_3^- (mg/l)	133.3	106.09	84.16	117.08	104.78	107.9	93.4	131.93	99.5	142.56	151.85	67.05
Cl^- (mg/l)	16.63	12.45	8.85	9.33	9.19	12.15	11.74	11.34	11.87	17.43	15.63	13.5
Ca^{++} (mg/l)	22.43	17.65	25.66	23.76	27.5	31.23	23.36	24.37	33.22	30.5	28.1	16.8
Mg^{++} (mg/l)	5.57	5.35	6.97	4.83	1.57	5.79	5.87	9.93	11.36	8.69	10.96	6.8
Total hardness (mg/l)	74.77	65.6	92.67	79.12	75.01	101.57	69.27	103.14	130.83	107.88	115.12	69.53
COD (mg/l)	73.73	64.24	74.52	39.8	105.4	30.94	101.66	47.52	50.83	80.71	28.92	23
SO_4^- (mg/l)	80.62	27.12	14.92	23.5	14.87	9.25	8.44	14.4	39.3	14.62	10.85	14.82
SiO_3^- (mg/l)	15.4	9.82	3.82	1.21	6.12	0.94	0.65	7.43	1.56	2.5	4.02	1.62
NiO_3^- (mg/l)	0.79	1.5	0.125	1.125	2.04	0.57	1.57	0.405	0.725	2.5	0.85	1.89
Electrical conductivity (mMho/cm)	0.104	0.203	0.131	0.152	0.129	0.122	0.09	0.158	0.189	0.292	0.256	0.146

Figures 61.2a–h: Showing Mean Graphical Variations in Various Physico-Chemical Parameters of Water in Four Paddy Fields Near Nagri Bus Stand, Kathua

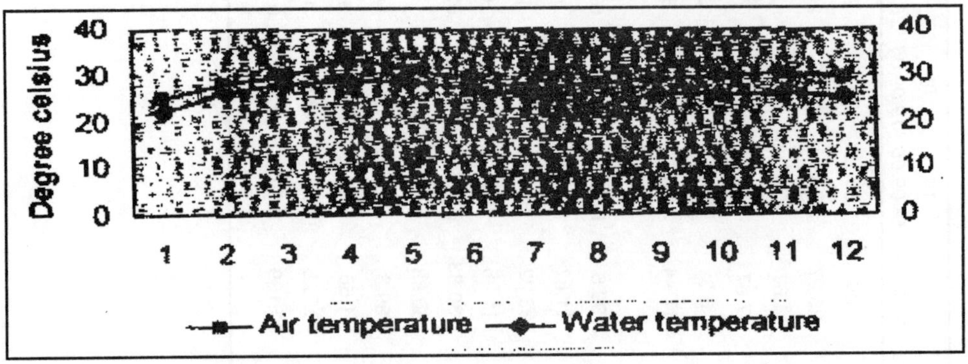

Figure 61.2a

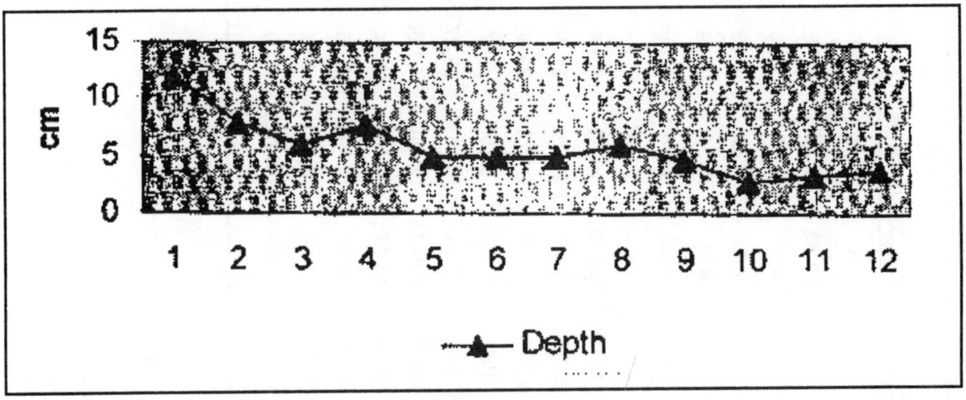

Figure 61.2b

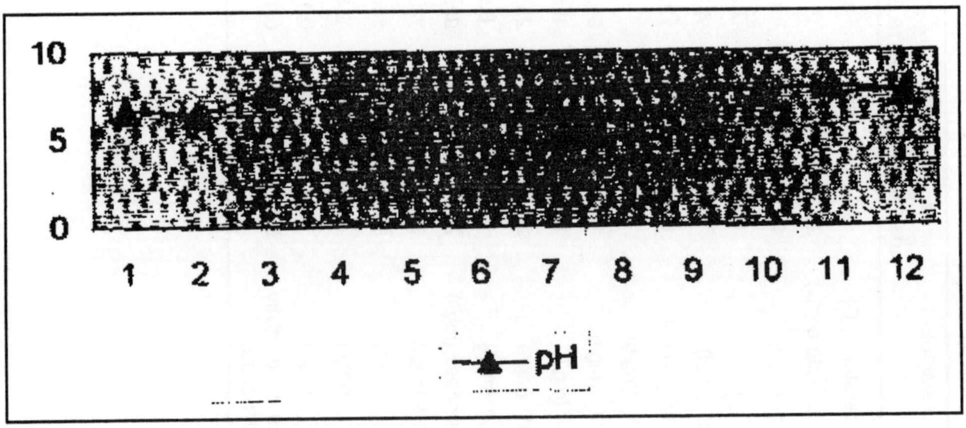

Figure 61.2c

Contd...

Figures 61.2a–h–Contd...

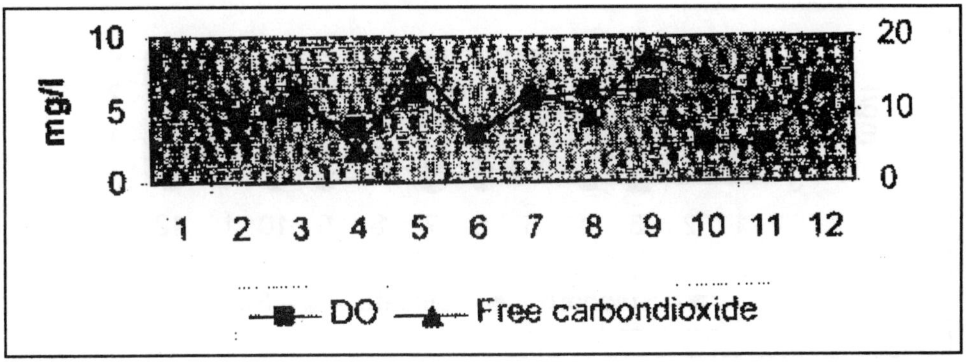

Figure 61.2d

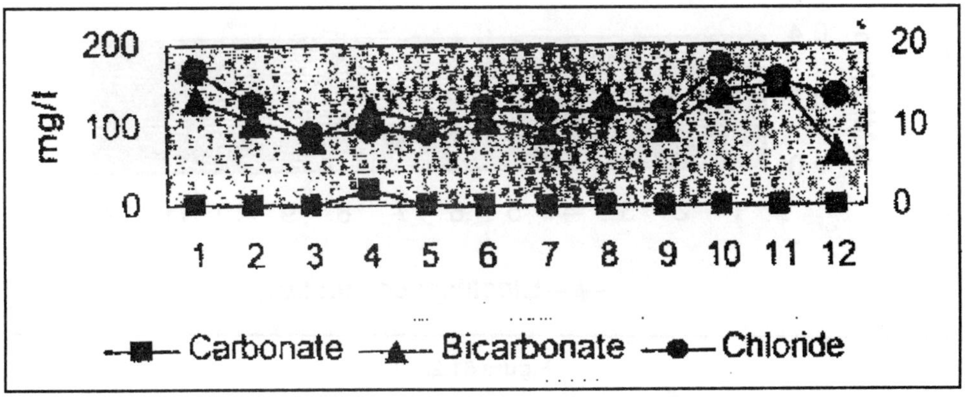

Figure 61.2e

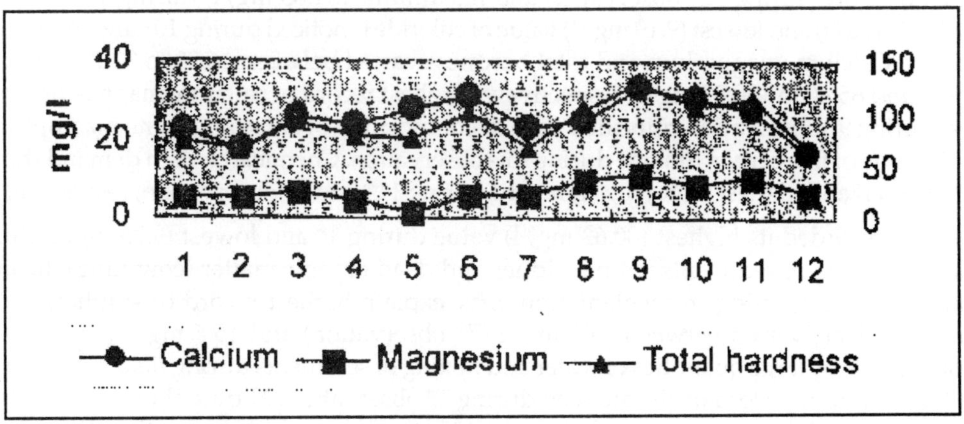

Figure 61.2f

Contd...

Figures 61.2a–h–Contd...

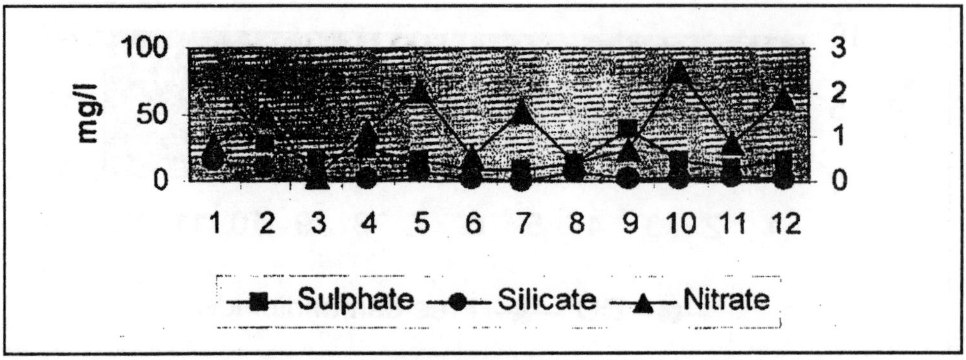

Figure 61.2g

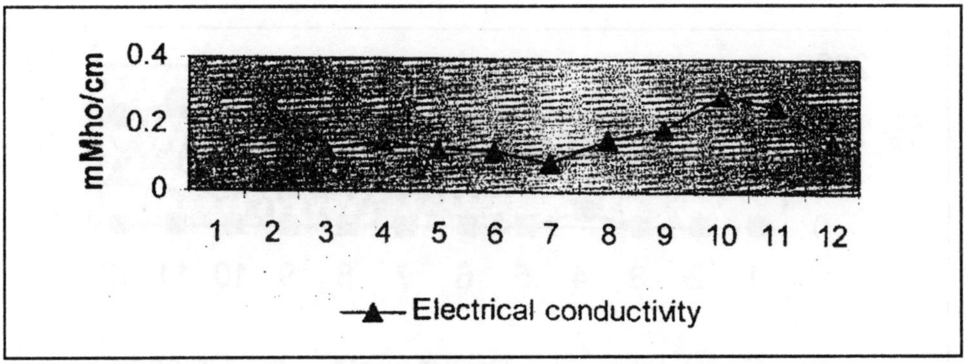

Figure 61.2h

only once during 4th observation (21.3 mg/l), in two paddy fields. Bicarbonate showed its minimum (67.05 mg/l) record during 12th observation and maximum (151.85 mg/l) during 11th observation. Highest (17.43 mg/l) and lowest (9.19 mg/l) value of chloride is noticed during 10th and 5th observation, respectively. Calcium, magnesium and total hardness varied between 16.80 to 33.22 mg/l, 1.57 to 11.36 mg/l and 65.60 to 130.83 mg/l, respectively. Highest record of calcium, magnesium and total hardness during 9th observation coincided with decomposition of algae and some macrophytes and maximum value of free carbon dioxide during this observation. Chemical oxygen demand showed its lowest (23 mg/l) and highest (105.4 mg/l) record during 11th and 5th observation, respectively.

Sulphate recorded its highest (80.62 mg/l) value during 1st and lowest (9.25 mg/l) during 6th observation. Mixing of sediments, crop residues and dead organic matter (cowdung) during field preparation, before seedlings transplantation, may explain highest record of sulphate during 1st observation. Silicate varied between 0.65 mg/l (7th observation) and 15.4 mg/l (1st observation). Irrigation of fields by large quantities of water and mixing of sediments during field preparation may explain the maximum record of silicate seen during 1st observation. Nitrate showed its highest (2.5 mg/l) value during 10th observation and lowest (0.125 mg/l) during 6th observation. Decomposition of algae and macrophytes in the pools of water in these paddy fields may account for nitrate enrichment during 10th observation.

Maximum (0.292 µMhos/cm) and minimum value (0.090 µMhos/cm) of electrical conductivity is noticed on 10th and 7th observation, respectively. Low record of various salts like bicarbonate, chloride, calcium, magnesium and total hardness may explain lowest record of conductivity during 7th observation.

Planktonic Analysis

The results of phytoplanktonic analysis are shown in Table 61.2 and depicted in Figures 61.3a–61.3i. Phytoplankton, qualitatively, comprising of 29 genera has shown the dominance of Chlorophyceae (14 genera) followed by Bacillariophyceae (9 genera) and Cyanophyceae (6 genera) and is in accordance to the findings of Dutta *et al.* (2006b) for paddy fields in Maralia Morh, Jammu. Qualitatively, various phytoplanktonic genera showed their irregular presence. Among the various genera of *Chlorophyceae, Spirogyra* made its appearance 10 times; *Chlorococcum* and *Volvox* nine times; *Cosmarium, Eudorina* and *Closterium* seven times, each; *Pediastrum* five times; *Lepocynclis* and *Pandorina* four times; *Pleodorina, Euastrum, Spirotenia* and *Dictyospherium* thrice and *Jonium* twice (Table 61.2).

Figures 61.3a–i: Showing Mean Graphical Variations in Various Phytoplanktonic Genera in Four Paddy Fields Near Nagri Bus Stand, Kathua

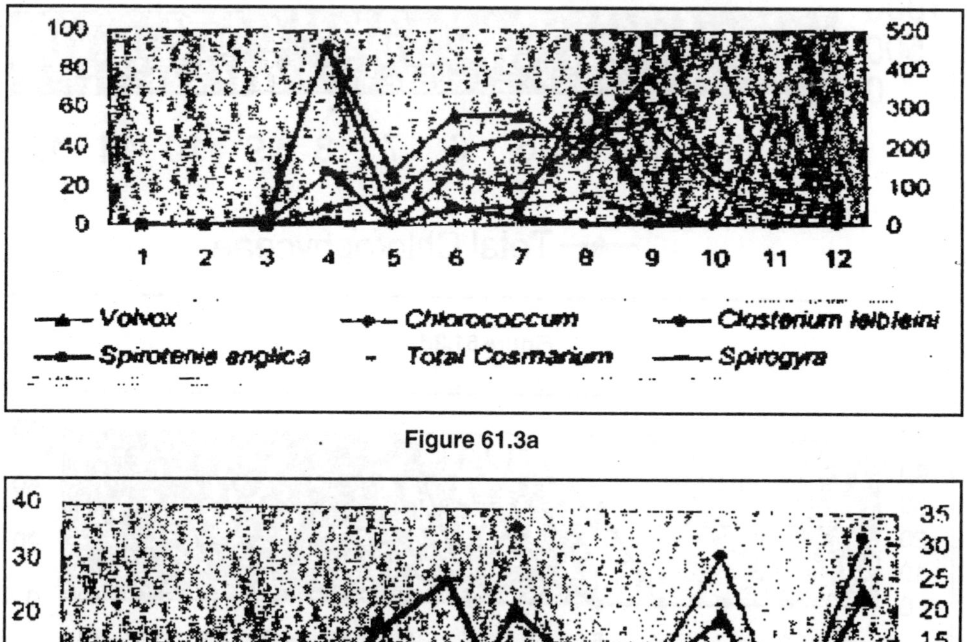

Figure 61.3a

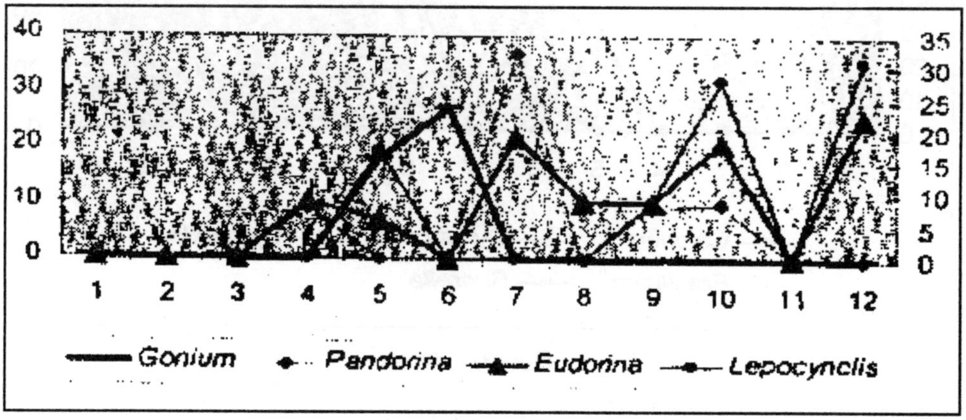

Figure 61.3b

Contd...

Figures 61.3a–i–Contd...

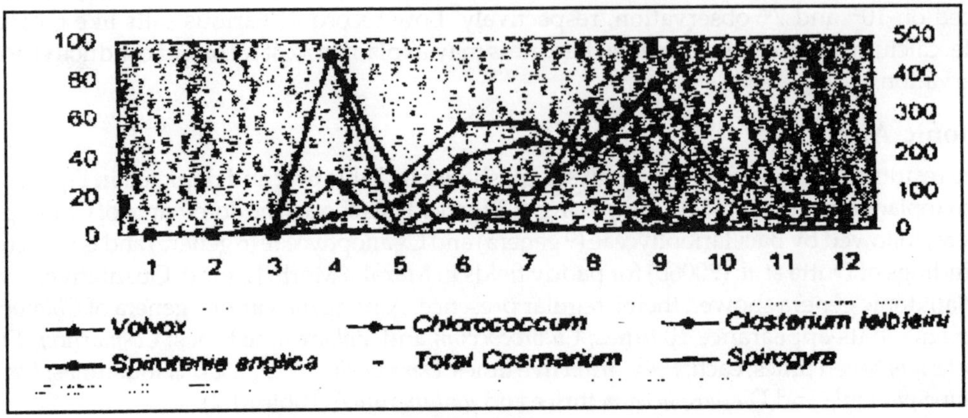

Figure 61.3c

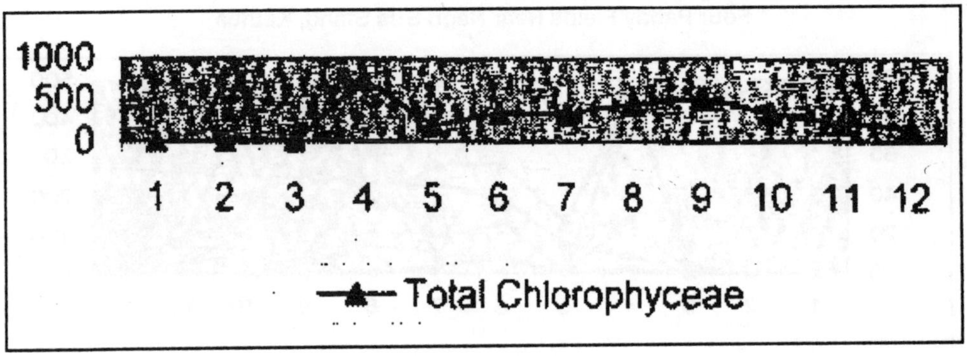

Figure 61.3d

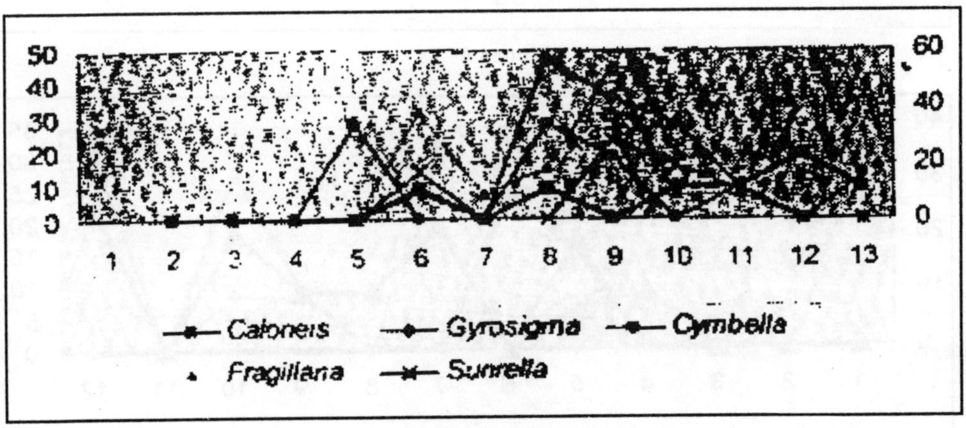

Figure 61.3e

Contd...

Figures 61.3a–i–Contd...

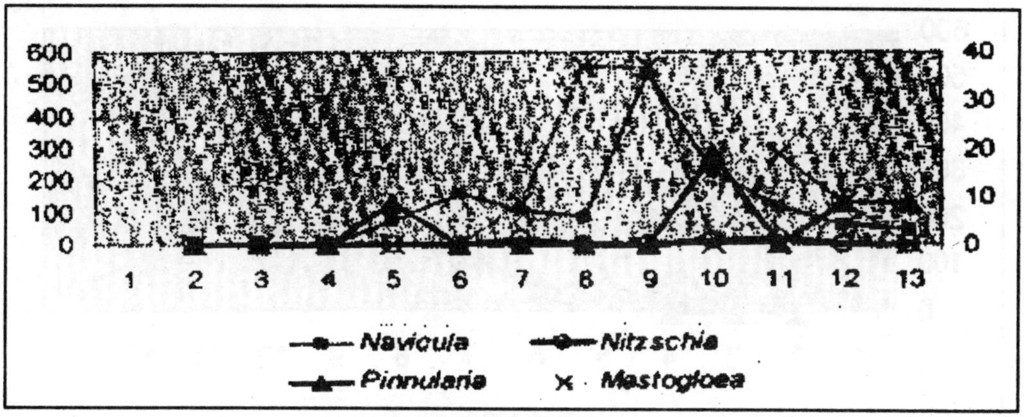

Figure 61.3f

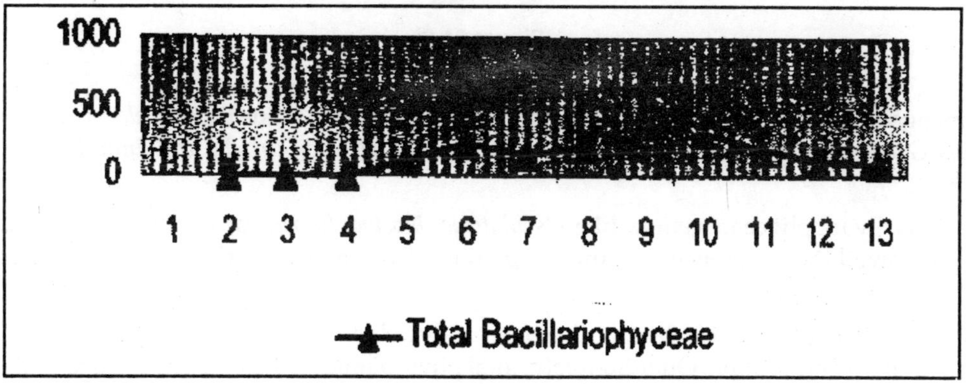

Figure 61.3g

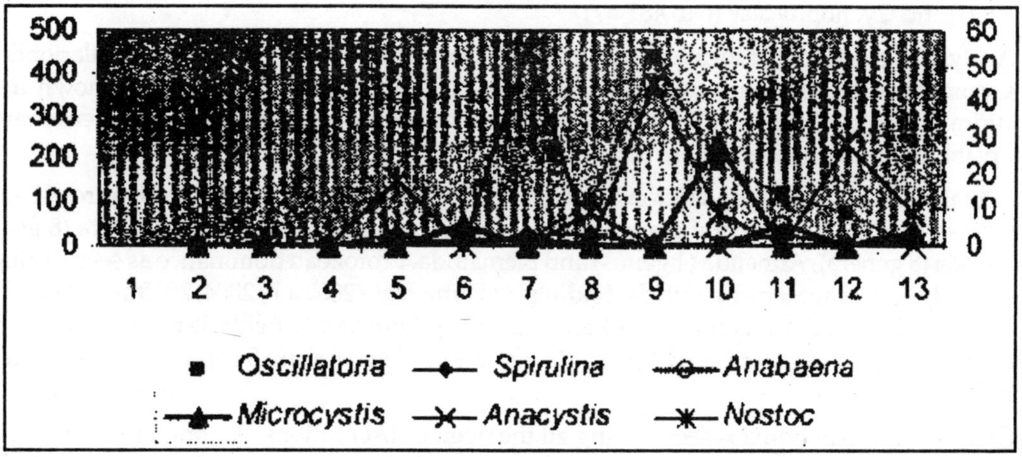

Figure 61.3h

Contd...

Figures 61.3a–i–Contd...

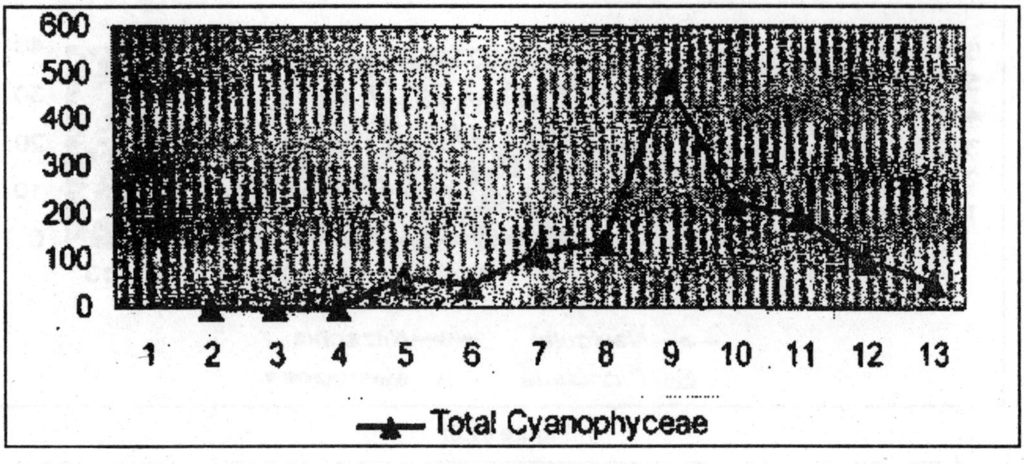

Figure 61.3i

Among various genera of Bacillariophyceae, *Navicula* is noticed nine times; *Caloneis* and *Fragillaria* six times; *Gyrosigma, Cymbella* and *Mastogloea* five times, each and *Surirella* and *Pinnularia* four times (Table 61.2).

Various Cyanophycean genera like *Oscillatoria, Nostoc, Spirulina, Anabaena, Microcystis* and *Anacystis* showed their presence nine times, six times, five times, four times, thrice and only once, respectively (Table 61.2).

Maximum phytoplanktonic diversity in these paddy fields is seen during 8th observation minimum during 1st and 2nd observation. Quantitatively, total phytoplankton varied between 0 to 1650 n/l and recorded a bimodal increase *viz.* during 4th and 8th observation. The order of quantitative dominance of various phytoplanktonic groups is recorded as Chlorophyceae (0 to 751 n/l), Bacillariophyceae (0 to 704 n/l) and Cyanophyceae (0 to 481 n/l).

Analysis of coefficient of correlation (r) of total phytoplankton, Chlorophyceae, Bacillariophyceae and Cyanophyceae, with various physico-chemical characteristics of water, has shown mostly insignificant results (Table 61.4). This indicates that no single factor is a strong determinant for phytoplanktonic abundance in these paddy fields.

A total of 28 genera of zooplankton, seen in these paddy fields (Table 61.3 and Figures 61.4a–m), have shown the quantitative dominance of Protozoa (10 genera), followed by Rotifera (8 genera), Arthropoda (8 genera), Annelida (1 genus) and Nematoda. Protozoan dominance as seen during the present analysis is in agreement with the findings of Dutta *et al.* (2002 a,b; 2004; 2005 and 2006 a,b and c). Protozoa, the most dominant zooplanktonic group in these paddy fields, is represented by three classes *viz.* Sarcodina (6 genera and 24 spp.) Mastigophora (2 genera and 5 spp.) and Ciliata (2 genera and 2 spp.).

Qualitatively, Sarcodina is seen during all the twelve observations. Among its various genera, *Centropyxis* is observed during all the twelve observations; *Difflugia* and *Arcella* eleven times; *Lesqueresia* seven times; *Nebela* five times and *Cucurbitella* four times (Table 61.3). Along ciliates, *Chilodonella* and *Paramecium* are seen five and four times, respectively. *Euglena* and *Phacus*, among mastigophores,

Table 61.2: Mean Phytoplanktonic Variations (n/l) in Four Paddy Fields Near Nagri Bus Stand, Kathua (6th July 2005 to 26th September 2005)

Phytoplankton	*Observations*											
	1	*2*	*3*	*4*	*5*	*6*	*7*	*8*	*9*	*10*	*11*	**12**
	06.07.05	*08.07.05*	*18.07.05*	*25.07.05*	*02.08.05*	*09.08.05*	*17.08.05*	*22.08.05*	*29.08.05*	*12.09.05*	*19.09.05*	*26.09.05*
CHLOROPHYCEAE												
Gonium Muller	–	–	–	7	18	27	–	–	–	–	–	–
Volvox Linnaeus	–	–	–	93	27	56	56	37	75	93	18	9
Pleodorina Shaw	–	–	–	–	–	–	–	–	19	18	–	18
Pandorina Bory	–	–	–	9	–	–	37	–	9	10	–	–
Eudorina Ehren	–	–	–	9	6	–	19	9	9	18	–	22
Lepocynclis Perty	–	–	–	–	17	–	–	–	9	28	–	31
Clorococcum Fries	–	–	–	9	15	38	46	44	76	29	18	21
Dictyospherium pulchellum Wood	–	–	–	–	–	18	–	–	–	10	9	–
Closterium leibleini Kutzing	–	–	–	27	–	9	9	65	15	–	55	9
Spirotenia anglica Brebisson	–	–	–	9	–	46	19	–	–	–	–	–
Euastrum gemmatum Ehren	–	–	–	–	–	18	–	–	9	19	–	–
Pediastrum Meyer	–	–	–	–	–	–	–	9	19	19	19	4
Cosmarium Corda	–	–	–	119	67	–	36	37	28	–	53	38
C. pachydermum West	–	–	–	–	–	–	19	9	–	–	–	–
C. pseudobroomei Wolle	–	–	–	–	15	–	–	–	–	–	–	–
C. monomazum Lundell	–	–	–	18	9	–	–	18	–	–	18	–
C. granatum Brebisson	–	–	–	–	18	–	–	9	–	–	–	–
Total *Cosmarium*	–	–	–	137	109	–	55	73	28	–	71	38
Spirogyra Link	–	–	18	458	9	129	94	219	268	104	27	19
Zygote	–	–	–	–	–	9	9	9	–	–		
Total Chlorophyceae	–	–	18	751	201	350	344	465	530	348	217	171

Contd...

Table 61.2–Contd...

Phytoplankton	*Observations*											
	1	*2*	*3*	*4*	*5*	*6*	*7*	*8*	*9*	*10*	*11*	*12*
	06.07.05	*08.07.05*	*18.07.05*	*25.07.05*	*02.08.05*	*09.08.05*	*17.08.05*	*22.08.05*	*29.08.05*	*12.09.05*	*19.09.05*	*26.09.05*
BACILLARIOPHYCEAE												
Caloneis Bory	–	–	–	–	9	–	9	–	9	10	19	9
Gyrosigma (Grun.)	–	–	–	–	8	–	47	36	28	9	–	–
Cymbella (Hemp. and Ehren)	–	–	–	28	–	–	28	18	–	10	–	13
Fragillaria Desmazieres	–	–	–	–	38	9	18	54	9	9	–	–
Surirella Turpin	–	–	–	–	17	–	–	28	19	10	–	–
Navicula Kutz	–	–	–	74	159	111	84	531	233	114	63	49
Nitzschia Nitzsch	–	–	–	–	–	18	–	–	9	18	–	–
Pinnularia Ehren	–	–	–	9	–	–	–	–	19	–	9	9
Mastogloea Thwaites	–	–	–	–	–	9	37	37	–	19	9	–
Total Bacillariophyceae	–	–	–	111	231	147	223	704	326	199	100	80
CYANOPHYCEAE												
Oscillatoria Vaucher	–	–	–	18	34	18	102	436	187	116	73	4
Spirulina Turpin	–	–	–	28	–	27	28	–	–	29	–	13
Anabaena Bory	–	–	–	–	9	18	–	–	–	47	–	22
Microcystis Kutzing	–	–	–	–	6	–	–	–	28	–	–	4
Anacystis Meneghini	–	–	–	–	–	–	9	–	–	–	–	–
Nostoc Vaucher	–	–	–	18	–	54	–	45	9	–	28	9
Total Cyanophyceae	–	–	–	64	49	117	139	481	224	192	101	52
TOTAL PHYTOPLANKTON	–	–	18	926	481	614	706	1650	1080	739	418	303

Table 61.3: Mean Zooplanktonic Variations (n/l) in Four Paddy Fields Near Nagri Bus Stand, Kathua (6th July 2005 to 26th September 2005)

Zooplankton	*Observations*											
	1	*2*	*3*	*4*	*5*	*6*	*7*	*8*	*9*	*10*	*11*	*12*
	06.07.05	*08.07.05*	*18.07.05*	*25.07.05*	*02.08.05*	*09.08.05*	*17.08.05*	*22.08.05*	*29.08.05*	*12.09.05*	*19.09.05*	*26.09.05*
PHYLUM–PROTOZOA												
CLASS–SARCODINA												
SUBCLASS–RHIZOPODA												
Order–Testacida												
Difflugia lebes Penard	–	38	9	55	94	37	65	129	66	66	63	63
D. urceolata Carter	–	–	–	–	–	–	–	–	–	–	10	–
D. pyriformis Perty	–	–	9	18	–	28	–	–	28	29	–	–
D. tuberculata Wallich	–	–	–	–	9	9	–	–	–	19	–	–
D. rubescens Penard	–	–	–	–	9	27	–	–	9	9	–	–
D. acuminata Ehren.	–	–	–	–	18	–	9	–	9	–	–	9
D. corona Wallich	–	–	–	9	–	–	9	–	18	–	–	–
D. oblonga Ehren.	–	–	9	–	–	–	46	27	28	10	9	8
D. lobostoma Leidy	–	–	–	–	–	–	–	–	–	–	9	9
Total *Difflugia*	–	38	27	82	130	101	129	156	158	133	91	89
Centropyxis aculeatea (Ehren.)	–	–	–	–	27	–	–	–	19	18	–	–
C. arcelloides Penard	–	–	–	18	–	–	18	9	–	–	–	–
C. constricta (Ehren.)	–	9	–	–	9	18	19	19	9	–	9	8
C. ecomis (Ehren.)	9	–	9	9	32	–	–	18	18	–	–	8
C. aerophila Deflandu	–	–	–	–	–	27	–	–	19	9	9	8
C. stellata Wailes	–	–	–	–	–	–	–	–	–	9	–	9
Total *Centropyxis*	–	9	9	27	68	45	37	46	65	36	18	33

Contd...

Table 61.3–Contd...

Zooplankton	*Observations*											
	1	*2*	*3*	*4*	*5*	*6*	*7*	*8*	*9*	*10*	*11*	*12*
	06.07.05	*08.07.05*	*18.07.05*	*25.07.05*	*02.08.05*	*09.08.05*	*17.08.05*	*22.08.05*	*29.08.05*	*12.09.05*	*19.09.05*	*26.09.05*
Arcella discoides Ehren.	–	–	–	–	–	9	18	37	19	9	9	21
A. vulgaris Ehren.	28	9	9	–	37	9	18	56	38	37	18	–
A. polypora Penard	–	9	–	–	–	19	–	–	–	–	10	9
A. dentata Ehren.	–	–	–	–	–	–	9	–	–	–	9	4
A. megastoma Penard	–	–	–	–	–	9	–	–	9	–	9	–
Total *Arcella*	28	18	9	–	37	46	45	93	66	46	55	34
Cucurbitella Penard	–	–	9	–	–	9	–	–	9	10	–	–
Lesqueresia modesta Rhumbler	–	–	–	–	35	18	19	9	19	–	9	8
L. spiralis Ehren.	–	–	–	–	–	9	9	–	28	–	–	4
Total *Lesqueresia*	–	–	–	–	35	27	28	9	47	–	9	12
Nebela Leidy	–	–	–	–	15	–	19	32	–	–	19	37
TOTAL–SARCODINA	37	65	54	109	285	228	258	336	345	225	192	205
CLASS–CILIATA												
Chilodonella (Ehren.)	–	–	–	–	–	–	19	19	–	9	9	9
Paramecium Ehren.	–	–	–	66	–	–	–	54	–	–	19	9
TOTAL CILIATA	–	–	–	66	–	–	19	73	–	9	28	18
CLASS–MASTIGOPHORA												
Order–Euglenoidida												
Euglena acus (Ehren.)	–	–	–	–	9	19	129	–	9	37	9	–
E. gracilis Klebs	–	–	–	9	–	–	–	18	9	–	–	8
E. viridis Ehren.	–	–	–	46	30	36	28	38	47	9	63	–
E. spirogyra (Ehren.)	–	–	–	–	–	9	–	9	–	–	–	–
Total *Euglena*	–	–	–	55	39	64	157	65	65	46	72	8

Contd...

Table 61.3–Contd...

Zooplankton	*Observations*											
	1	*2*	*3*	*4*	*5*	*6*	*7*	*8*	*9*	*10*	*11*	*12*
	06.07.05	*08.07.05*	*18.07.05*	*25.07.05*	*02.08.05*	*09.08.05*	*17.08.05*	*22.08.05*	*29.08.05*	*12.09.05*	*19.09.05*	*26.09.05*
Phacus Dujardin	–	–	–	9	–	–	9	18	–	29	–	8
TOTAL MASTIGOPHORA	–	–	–	64	39	64	166	83	65	75	72	16
TOTAL PROTOZOA	37	65	54	239	324	292	443	432	410	309	292	239
PHYLUM–ROTIFERA												
CLASS–MONOGONONTA												
Order–Ploima												
Lecane luna Muller	–	–	–	–	–	–	28	18	–	–	–	31
Lepadella ovalis Muller	–	–	–	–	18	–	–	–	9	–	–	–
Monostyla bulla Ehren	–	–	–	9	–	9	–	18	–	–	9	–
Platyias platulus Harring	–	–	–	66	–	9	9	–	–	–	9	–
Colurella Bory	–	–	–	–	9	9	–	–	19	–	–	9
Brachionus quadridentata Pallas	–	–	–	9	9	–	–	–	–	10	–	–
B. bidentata Pallas	–	–	–	9	–	–	28	–	–	–	–	–
Total *Brachionus*	–	–	–	18	9	–	28	–	–	10	–	–
Asplanchna priodonta Gosse	–	–	–	–	–	–	–	19	19	9	27	18
Total Ploima	–	–	–	93	36	27	65	55	47	19	45	58
Order–Bdelloida												
Philodina roseola Ehren.	–	–	–	9	–	18	–	–	–	–	9	–
TOTAL ROTIFERA	–	–	–	102	36	45	65	55	47	19	54	58

Contd...

Table 61.3–Contd...

Zooplankton	*Observations*											
	1	*2*	*3*	*4*	*5*	*6*	*7*	*8*	*9*	*10*	*11*	*12*
	06.07.05	*08.07.05*	*18.07.05*	*25.07.05*	*02.08.05*	*09.08.05*	*17.08.05*	*22.08.05*	*29.08.05*	*12.09.05*	*19.09.05*	*26.09.05*
PHYLUM–ARTHROPODA												
CLASS–CRUSTACEA												
Order–Copepoda												
Cyclops Muller	–	–	–	–	–	–	19	–	–	9	–	–
Mascocylops hyalinus Rehberg	–	–	–	–	–	–	–	–	37	19	–	–
Nauplius larva	–	–	–	9	16	–	18	27	27	133	45	8
Metanauplius larva	–	–	–	9	15	–	–	–	9	29	–	9
Total Copepoda	–	–	–	18	31	–	37	27	73	190	45	17
Order–Cladocera												
Alona Sars	–	–	–	–	–	–	–	–	–	–	9	–
Alonella Sars	–	–	–	–	–	29	–	–	–	–	18	–
Moina Baird	–	–	–	9	26	–	–	–	–	–	–	–
Ceriodaphnia Dana	–	–	–	9	7	9	9	–	–	–	9	–
Daphnia Claus	–	–	–	9	–	–	–	–	–	9	–	–
Total Cladocera	–	–	–	27	33	38	9	–	9	36	–	
Order–Ostracoda												
Cypris Muller	–	–	–	–	–	–	9	9	–	19	18	–
Total Crustacea	–	–	–	45	64	38	55	36	73	218	99	17

Contd...

Table 61.3–Contd...

Zooplankton	*Observations*											
	1	*2*	*3*	*4*	*5*	*6*	*7*	*8*	*9*	*10*	*11*	*12*
	06.07.05	*08.07.05*	*18.07.05*	*25.07.05*	*02.08.05*	*09.08.05*	*17.08.05*	*22.08.05*	*29.08.05*	*12.09.05*	*19.09.05*	*26.09.05*
CLASS–INSECTA												
Order–Diptera												
Chironomous larva Meigen	–	–	–	–	–	–	9	–	19	–	–	18
TOTAL ARTHROPODA	–	–	–	45	64	38	64	36	92	218	99	35
PHYLUM–ANNELIDA												
CLASS–OLIGOCHAETA												
Nais Muller	–	–	–	–	–	–	–	–	–	–	–	9
PHYLUM–NEMATODA	–	–	–	–	–	9	–	92	47	–	19	–
Eggs	–	–	–	–	–	–	28	9	–	19	18	–
TOTAL ZOOPLANKTON	37	65	54	386	424	384	600	684	596	565	482	341

Figures 61.4a–m: Showing Mean Graphical Variations in Various Zooplanktonic Genera in Four Paddy Fields Near Nagri Bus Stand, Kathua

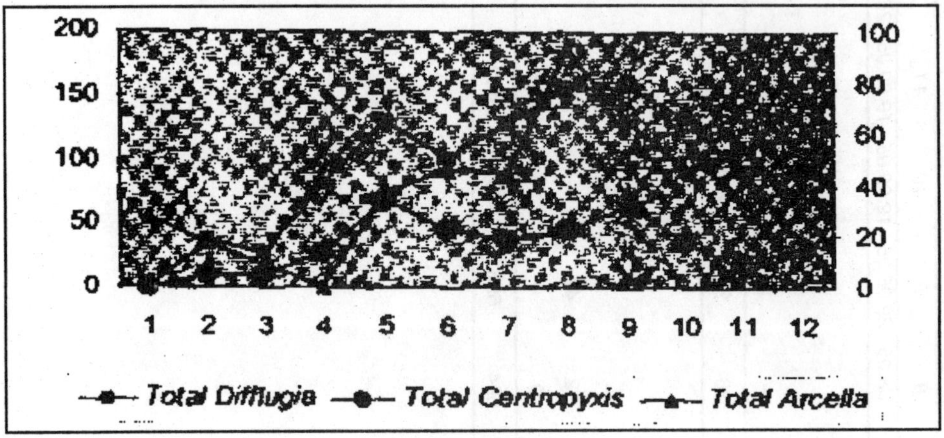

Figure 61.4a

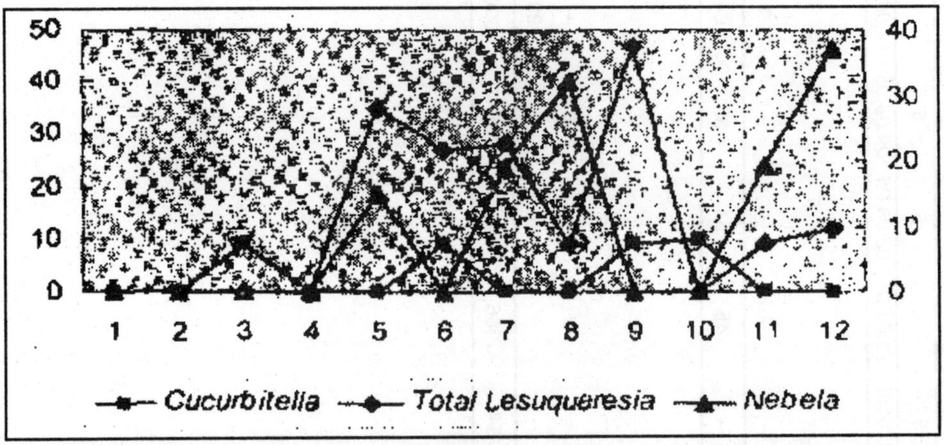

Figure 61.4b

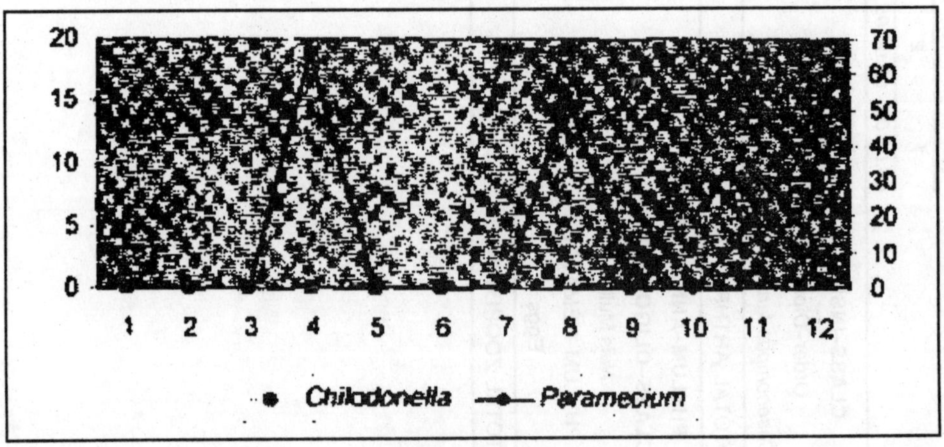

Figure 61.4c

Contd...

Figures 61.4a–m–Contd...

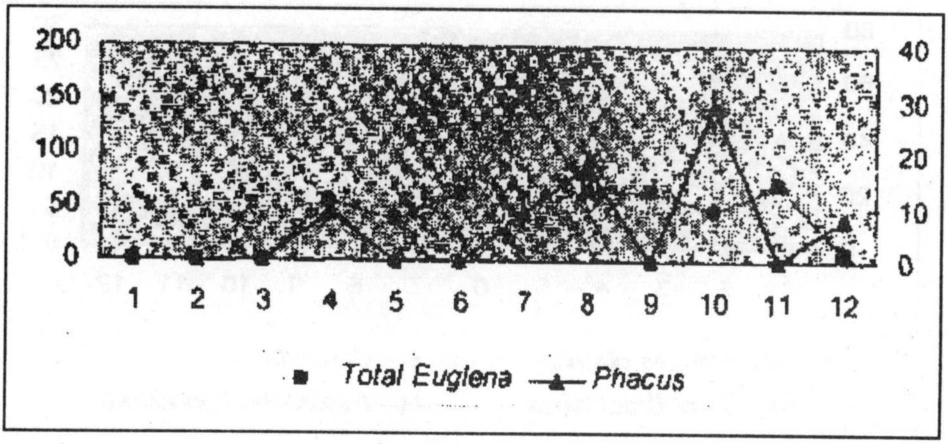

Figure 61.4d

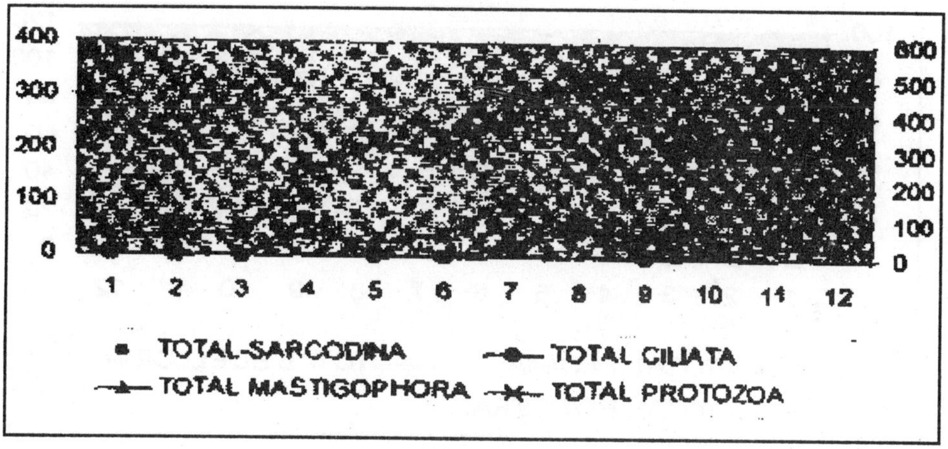

Figure 61.4e

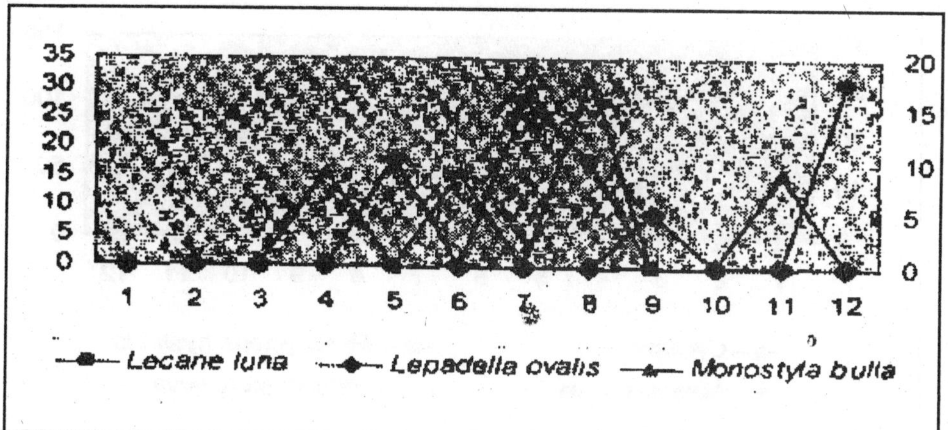

Figure 61.4f

Contd...

Figures 61.4a–m–Contd...

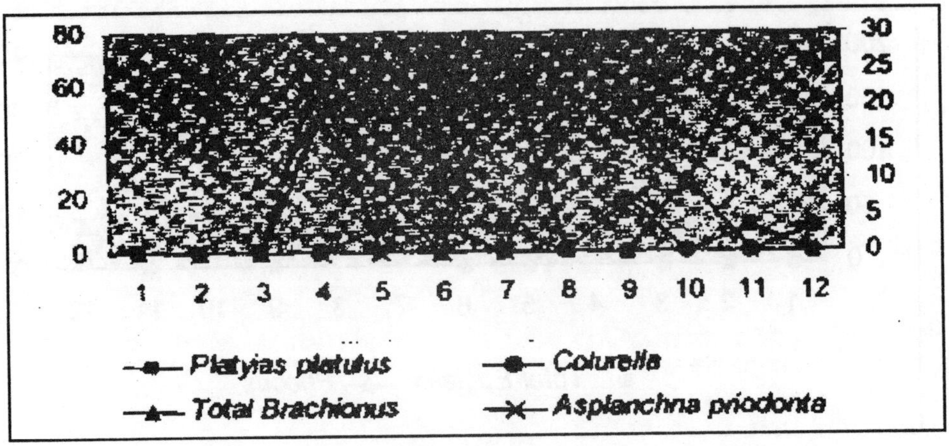

Figure 61.4g

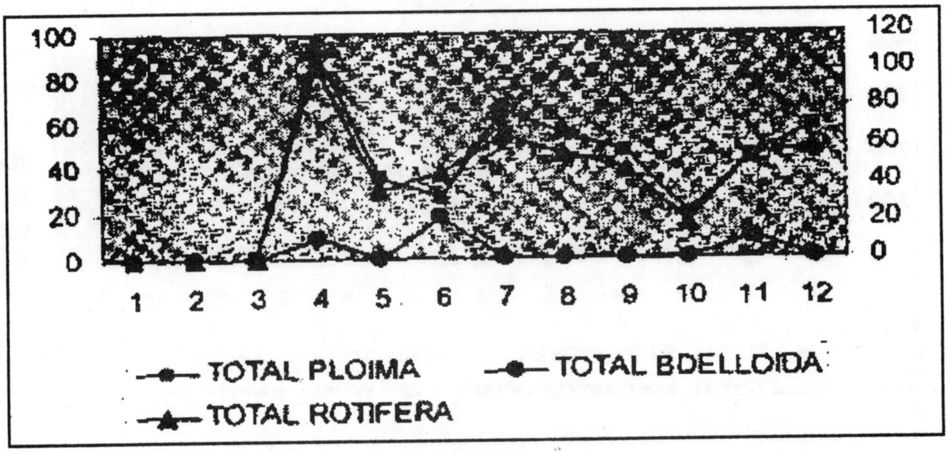

Figure 61.4h

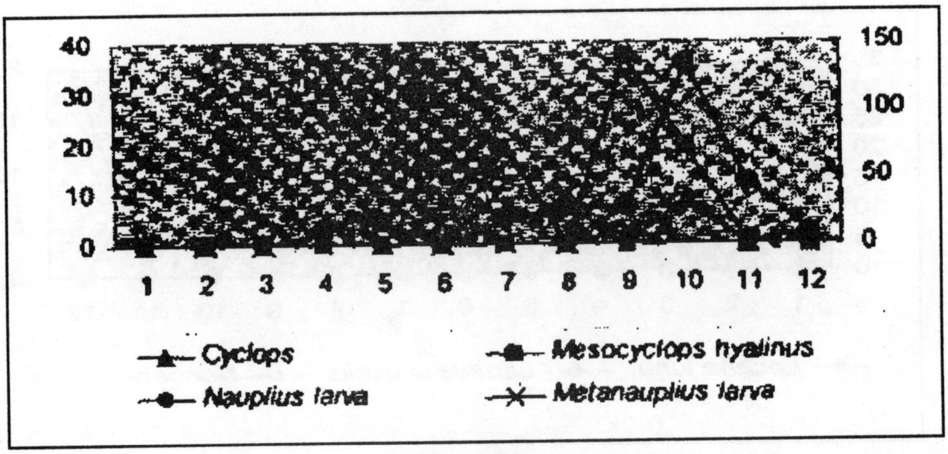

Figure 61.4i

Contd...

Figures 61.4a–m–Contd...

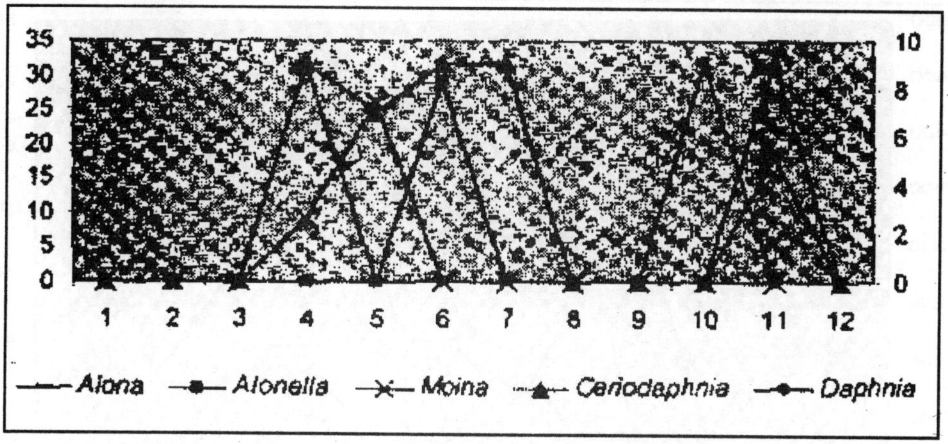

Figure 61.4j

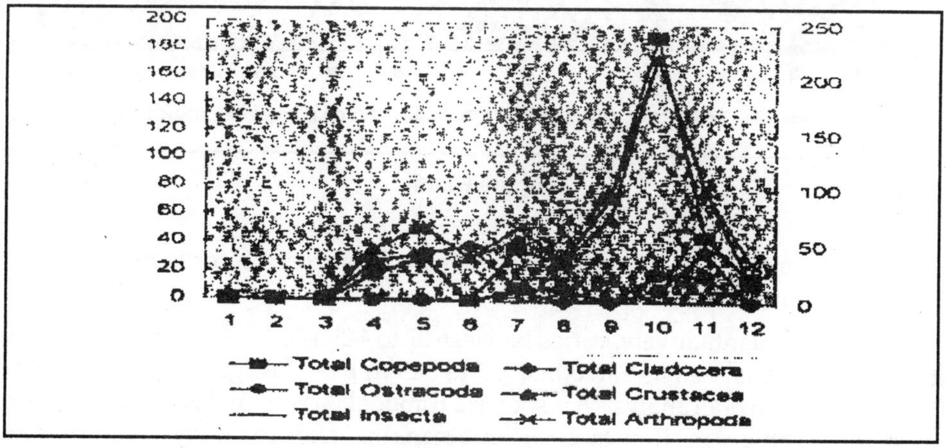

Figure 61.4k

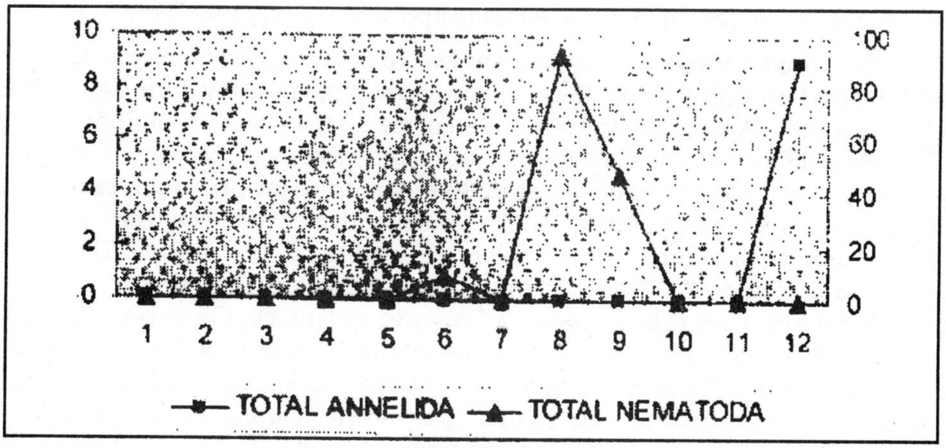

Figure 61.4l

Contd...

Figures 61.4a–m–Contd...

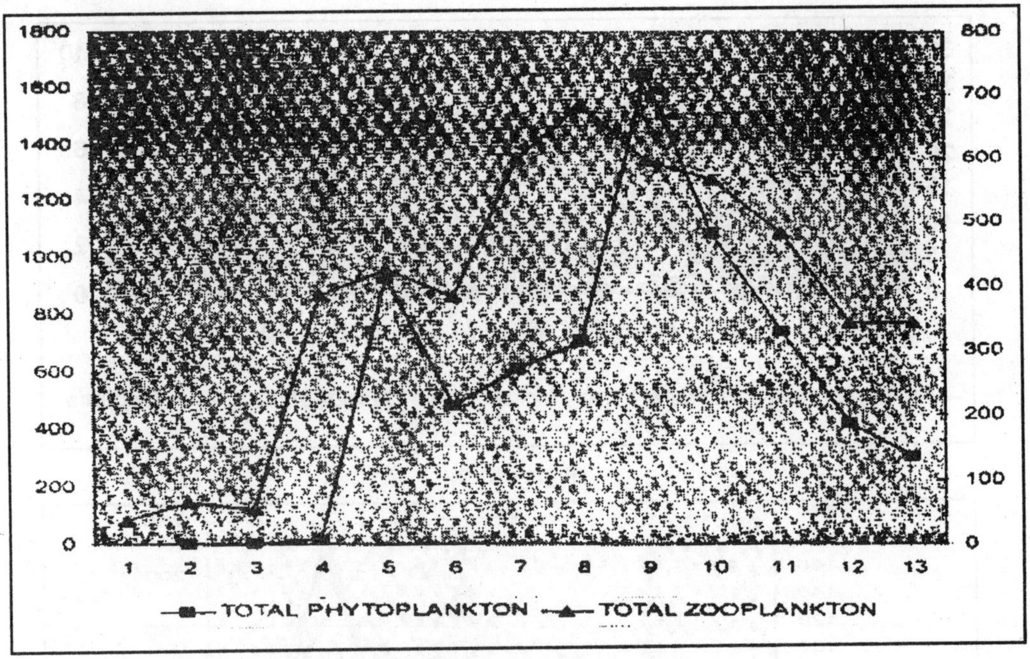

Figure 61.4m

showed their presence nine and five times, respectively. Protozoans recorded maximum diversity during 8th and minimum during 1st observation.

Total protozoans, quantitatively, varied between 37 to 443 n/l. Among protozoans, the order of quantitative dominance is seen as Sarcodina (37 to 345 n/l) > Mastigophora (0 to 166 n/l) and> Ciliata (0 to 73 n/l). An overall analysis has shown a trimodal increase of Protozoa *viz.* during 2nd, 5th and 8th observation (Table 61.3). Resistance of Testacean rhizopods to the environmental conditions *viz.* wide fluctuations in water level, drying of fields after paddy harvest and other abiotic characteristics, due to the presence of test/shell may account for their presence during 1st observation.

Rotifers are noticed during nine times, out of twelve observations, in the planktonic samples collected from these four paddy fields. These showed maximum diversity during 4th and minimum during 10th observation. Among the various genera of class Monogononta, of Rotifera, *Monostyla, Platyias, Colurella* and *Brachionus* showed their presence four times, each; *Lecane* and *Philodina*, thrice, each; *Lepadella* twice and *Asplanchna* only once (5th observation) in the planktonic samples collected from four paddy fields in Kathua District. Rotifers, quantitatively, varied between 0 to 102 n/l and showed maximum diversity during 4th and minimum during 10th observation. These recorded a trimodal increase *viz.* during 4th, 7th and 12th observation, respectively. Arthropods in these paddy fields have shown the presence of two classes *viz.* Crustacea and Insecta. Crustaceans are represented by three orders *viz.* Copepoda, Cladocera and Ostracoda. Among copepods, *Mesocyclops hyalinus* and *Cyclops* are seen only twice. *Nauplius* and *Metanauplius* larva of Copepoda are noticed eight and five times, respectively. An overall Copepod analysis has shown their eight times presence. Among *Cladocerans, Ceriodaphnia* recorded its five times presence; *Alonella, Daphnia* and *Moina* twice, each,

Table 61.4: Coefficient of Correlation (r) Between Phytoplankton and Physico-chemical Parameters of Water in Four Paddy Fields Near Nagri Bus Stand, Kathua

Phytoplanktonic Groups	Water Parameters														
	Water Temperature	Depth	pH	DO	Free CO_2	HCO_3^-	Cl^-	Ca^{++}	Mg^{++}	Total Hardness	COD	SO_4^-	SiO_3^-	NO_3^-	Electrical Conductivity
Chlorophyceae	0.017	0.46	0.492	–0.08	–0.3	0.304	–0.075	0.181	0.11	0.172	–0.22	0.506	–0.25	–0.04	0.056
Bacillariophyceae	–0.59	024	0.191	0.456	0.16	0.215	–0.164	0.129	0.33	0.283	0.142	0.14	0.66	–0.41	–0.086
Cyanophyceae	–0.73	0.1	0.281	0.245	0.08	0.35	0.109	0.191	0.33	0.453	–0.02	0.119	0.52	–0.46	0.125
Total Phytoplankton	–0.5	0.26	0.387	0.205	–0.1	0.406	0.048	0.205	0.35	0.327	–0.1	0.337	0.26	–0.15	0.081

Table 61.5: Coefficient of Correlation (r) Between Zooplankton and Physico-chemical Parameters of Water in Four Paddy Fields Near Nagri Bus Stand, Kathua

Zooplanktonic Groups	Water Parameters														
	Water Temperature	Depth	pH	DO	Free CO_2	HCO_3^-	Cl^-	Ca^{++}	Mg^{++}	Total Hardness	COD	SO_4^-	SiO_3^-	NO_3^-	Electrical Conductivity
Sarcodina	–0.051	–0.641	0.461	0.151	0.41	0.007	–0.071	0.5041	0.342	0.4851	–0.002	–0.41	–0.452	0.147	0.101
Ciliata	0.069	0.85	0.766	0.2	–0.7	0.204	–0.7	–0.059	–0.03	0.072	–0.342	0.5611	0.529	–0.8	–0.3
Mastigophora	–0.22	0.07	0.218	–0.09	0.11	0.163	0.025	0.0901	0.114	–0.05	0.4715	–0.25	–0.199	–0.11	–0.02
Total Protozoa	–0.5	–0.6	0.607	0.04	0.25	0.11	–0.08	0.4398	0.353	0.419	–0.007	–0.47	–0.515	0.118	0.076
Total Rotifera	0.287	0.79	0.622	0.06	–0.7	–0.2	–0.538	–0.499	–0.16	–0.38	–0.333	0.138	–0.279	–0.37	–0.43
Crustacea	–0.06	–0.58	0.025	–0.561	0.49	–0.2	0.708	0.4929	0.278	0.412	0.3396	–0.01	0.002	0.516	0.824
Insecta	–0.46	–0.59	–0.51	0.9	0.13	–0.25	0.4787	0.205	0.699	0.58	–0.902	0.725	0.989	–0.34	0.937
Total Arthropoda	–0.1	–0.63	–0.02	–0.5	0.54	0.525	0.728	0.4737	0.324	0.426	0.3291	0.066	–0.058	0.539	0.822
Nematoda	–0.92	0.74	0.475	0.84	0.14	0.062	–0.4681	–0.578	0.421	–0.05	0.7861	0.242	0.79	–0.64	–0.14
Total Zooplankton	–0.05	–0.65	0.668	–0.1	0.26	0.228	0.044	0.4882	0.439	0.499	–0.052	–0.47	–0.549	0.193	0.254

and *Alona* only once. An overall Cladoceran analysis has shown their six times presence. Ostracoda, another order of class Crustacea, is represented by genus *Cypris* and is seen during four observations only. An overall analysis of Crustaceans (Table 61.3), belonging to Phylum Arthropoda, has shown their nine times presence. Class Insecta of Arthropoda, is qualitatively represented by *Chironomous* larva and is noticed presence thrice.

An observation of the Table 61.3 reveals nine times presence of arthropods in the planktonic samples collected from four paddy fields of Kathua district. Maximum and minimum diversity of arthropods is noticed on 10th and 12th observation, respectively.

Quantitatively, arthropods varied between 0 to 218 n/l (Table 61.3).These recorded a trimodal increase *viz.*, during 5th, 7th and 10th observation. Annelida, represented by *Nais* spp. is seen only once (12th observation) during the present study. Water nematodes and eggs showed their four times presence in the present study area.

Total zooplankton recorded maximum, qualitative diversity during 8th and minimum during 1st observation. Total zooplankton quantitatively, varied between 37 to 684 nil and observed a trimodal increase. Zooplanktonic first rise is seen during 2nd, 2nd during 5th and 3rd during 8th observation. Maximum and minimum quantitative count of zooplankton is seen during 8th and 1st observation, respectively. An overall study has indicated mostly insignificant correlation(r) of protozoans, rotifers, arthropods, nematodes and total zooplankton with various physico-chemical parameters of water (Table 61.5).

Insignificant results of coefficient of correlation of phytoplankton and zooplankton (Tables 61.4 and 61.5), with various physicochemical parameters of water, indicate that any of the physico-chemical water parameter alone does not appear to be a strong determinant factor for the planktonic abundance in these four paddy fields of Kathua District, J&K. A similar type of conclusion has been drawn by Dutta *et al.* (2002 a,b; 2004; 2005 and 2006 a,b and c). Present analysis has indicated that the range of various physico-chemical parameters of water *viz.* temperature, depth, pH, free CO_2, bicarbonate, chloride, calcium, magnesium, total hardness, COD, silicate, sulphate, nitrate and electrical conductivity, in four paddy fields, near Nagri bus stand, Kathua, J&K, are within the optimum limits for fish culture. Biotic analysis has shown that there is sufficient planktonic food in these paddy fields that can be utilized by the fish. However, for the exploitation of these paddy-fields for aquaculture practices, depth above 15 cm has to, be maintained during the paddy growing period through regular irrigation. Fish larvae of some fishes, however, can be reared in these shallow paddy fields through regular irrigation. These may be released in these fields during seedlings transplantation and allowed to grow before drying of fields for harvesting. Young fishes, after collection, may be transferred into stocking ponds for their further growth. Some fishes like common carps, through depth management, are known to reach marketable size during this paddy growing period of three months (Nath and Dey, 1990).

Acknowledgements

Thanks are due to authorities of University of Jammu for providing necessary laboratory facilities in the Department of Environmental Sciences, University of Jammu. Dr. Meenakshi Khajuria and Mrs. Deepika Salathia are acknowledged for their help.

References

Adoni, A.D., 1985. *Workbook on Limnology*. Pratibha publishers.

APHA, 1998. *Standard Methods for the Examination of Water and Wastewater*, 20th edn. American Public Health Association Inc., New York.

Battish, S.K., 1992. *Freshwater Zooplankton of India.* Oxford and IBH Publishing Co. Pvt. Ltd., New Delhi, Bombay, Calcutta, p. 323.

Biswas, A. and Raut, S.K., 2000. Bioindicators for freshwater mussel, *Lamellidens marginalis* (Lamarck) farming. In: *Proceedings of National Seminar on Environmental Biology,* April 03–05, 1998, (Eds.) A.K. Aditya and P. Haldar. Daya Publishing House, Delhi, p. 38–53.

Dutta, S.P.S., 1983. On some freshwater Rhizopoda from Jammu Part–I. *Jammu Univ. Review (Sciences),* 1: 91–94.

Dutta, S.P.S., Dogra, K. and Tak, A., 2002a. Qualitative and quantitative study of plankton in some paddy fields, Maralia Morh, Jammu (J&K). In: *National Seminar on Environmental Pollution and Fishery Management*, August 8–10, Organised by School of Studies on Zoology, Vikram University, Ujjain (M.P.), India. (Abstract No. 28).

Dutta, S.P.S., Khajuria, M., Dogra, P. and Singh, A., 2005. Ecology of zooplankton in some ground water irrigated paddy fields, Indira Nagar, Miran Sahib, Jammu. In: *1st Jammu and Kashmir State Science Congress*, February 7–9, University of Jammu, Jammu (Abstract No. 165).

Dutta, S.P.S., Khajuria, M., Dogra, P. and Singh, A., 2006c. Plankton ecology in two groundwater irrigated paddy fields, Indira Nagar, Miran Sahib, Jammu. *Env. Con. J.*, 7(1–2): 1–9.

Dutta, S.P.S., Kiran, K., Sharma A., Khatoon, N. and Mir, A.M., 2002b. Ecology of plankton in some paddy fields, Miran Sahib, Jammu; (J&K). In: *National Seminar on Recent Trends in Life Sciences,* August 26–28. Organized by Centre of Research for Development. The University of Kashmir, Srinagar, p. 32 (Abstract no. 27).

Dutta, S.P.S., Kour, S. and Khajuria, V., 2006a. Plankton ecology of a paddy field at Maralia Morh, Miran Sahib, Jammu, (*Communicated*).

Dutta, S.P.S., Kour, S. and Sharma, V., 2006b. Qualitative and quantitative analysis of plankton in some paddy fields at Maralia Morh, Miran Sahib, Jammu. In: 2nd *J&K Science Congress*, Organized by Univ. of Kashmir, in collaboration with J&K State DST and DST GOI, p. 262 (Abstract no. 45).

Dutta, S.P.S., Mir, A.M., Kiran, K., Sharma, A., Khatoon, N. and Salathia, D., 2004. Ecology of plankton in some paddy fields of Miran Sahib, Jammu, (J&K). *Aquacult.* 5(1): 1–10.

Edmondson, W.T., 1992. *Freshwater Biology,* 2nd edn. International Books and Periodicals Supply Services, New Delhi, p. 1248.

Goldman, C.R. and Horne, A.J., 1983. *Limnology*. McGraw Hill International Book Company, New Delhi.

Goyal, S.K., Sharma, B.M. and Gupta, R.S., 1984. Algal flora of rice field soils of J&K state. *Phykos,* 23(1 and 2): 57–64.

Indian Standard Methods, 1973. *Indian Standard Methods for Sampling and Test (Physical and Chemical) for Water Used in Industry*. Manak Bhawan, 9, New Delhi.

Jhingran, V.G., 1991. *Fish and Fisheries of India.* Hindustan Publishing Co., New Delhi.

Kant, S., 1977. Notes on the diatoms of Jammu. *The University Review*, 6: 6–13.

Kant, S. and Anand, V.K., 1978. New records for *Scendesmus* species from Jammu (J&K). *Phykos*, 17(12): 81–85.

Kudo, R.R., 1986. *Protozoology*. Books and Periodicals Corporation (India), p. 1174.

Nair, K.N., Das, A.K. and Mukherjee, B.N., 1971. On some freshwater Rhizopoda and Helizoa (protozoa) from Calcutta and its environs, Part I. *Res. Zool. Surv.*, India, 65(1–4): 1–16.

Nath, P. and Dey, S.C., 1990. *Fish and Fisheries of North-East India: Volume on Arunachal Pradesh*. A.K. Choudhary, Barrackpore, West Bengal, India.

Nickolsky, G.V., 1962. *The Ecology of Fishes*. Academic Press, London and New York.

Rather, M.D. and Mir, A.M. 1987. Non-symbiotic nitrogen fixation by some BGA from paddy soil of Kashmir. *Phykos*, 26: 63–66.

Reid, G.K. and Wood, R.D., 1976. *Ecology of Inland Waters and Estuaries*. D.Van Northland Company, New Northland Company. New York, p. 485

Smith, G.M., 1950. *The Freshwater Algae of United States*, 2nd Edn. McGraw Hill Book Co., New York, p. 799.

Trivedy, R.K., Goel, P.K. and Trisal, C.L., 1987. *Practical Methods in Ecology and Environmental Sciences*. Environmental Publication, Karad, India.

Welch, P.S., 1952. *Limnology*. McGraw Hill Book Co. Inc., New York, p. 539.

Wetzel, R.G., 2000. *Limnology: Lake and River Ecosystem*, 3rd Edn. Academic Press, p. 1–841.

Water and Wastewater Management in 2 Vols (2011) *Pages* 427–432
Editors: **D.R. Khanna, R. Bhutiani & Gagan Matta**
Published by: **DAYA PUBLISHING HOUSE, NEW DELHI**

Chapter 62

Impact of Stored Effluents of Slaughter Houses on Groundwater Quality of Hand Pumps Situated Nearby Slaughter House at Khurja Town, District Bulandshahar (U.P.), India

J.P. Singh[1], S.K. Singh[2] and Shakun Singh[3]
[1]U.P. Pollution Control Board, Aligarh
[2]U.P. Pollution Control Board, Lucknow
[3]Department of Zoology and Environmental Science, Gurukula Kangri University, Haridwar

ABSTRACT

Khurja is a tehsil of district Bulandshahar (U.P.) India which is famous for manufacturing of crockery products. There is a large scale slaughter house at Mundakhera road which discharges its blood red effluents in nearby ponds having area more than 5 acre. The city Khurja is situated on G.T. road between Aligarh and Bulandshahar cities and approximately 75 km away from Delhi towards south. This chapter presents a case study of water quality of hand pumps situated around the ponds which receives more than 80 per cent blood red effluents of slaughter house. The study reveals that the groundwater quality of hand pumps selected within range of 2 km. from slaughter house is not suitable for drinking purpose with respect to colour, odour, pH, TDS, conductivity, hardness, DO, BOD and chloride at sampling points 1–3 while at sampling point 4 and 5 the parameter colour, odour, pH, hardness, BOD and chloride were found within limit of drinking purpose where as TDS, conductivity, and DO were observed beyond the limit.

Keywords: Conductivity, TDS, Hardness, DO and BOD.

Introduction

Khurja is a tehsil of district Bulandshahar (U.P.) India which is famous for crockery manufacturing industries. This town is situated on G.T. road between Bulandshahar–Aligarh and approximately 75 km away from Delhi. It has more than 2 lakh population of different communities. There is a big slaughter house between Mundakhera road and Bulandshahar road from where a large quantity of blood red effluents are discharge into near by ponds. The pond covers an area of 5-6 acres. More than 2 thousand animals are slaughtered in open area everyday and red effluents as well as their wastes are discharged into this pond. This pond is named as, Khooni Talab by the people. This slaughter house comes under the jurisdiction of Nagar Palika but it seems that Government has never looked into the matter.

Inhabitants of the vicinity of the slaughter house are facing many problems due to highly polluted pond causing pungent, untolerable, bad smell and also groundwater of hand pumps situated around this slaughter house. Therefore, it was proposed to assess the impact of stored effluents of slaughter house on groundwater quality of hand pumps situated in the vicinity of slaughter house in respect of physico-chemical characteristics.

Several researchers/workers have carried out studies on water of different sources in respect of physicochemical parameters (Vaishya and Agarawal, 1993; Abbasi *et al.*, 1999; Soren and Julian, 1977; Pande and Hasan, 1979; Singh *et al.*, 1988, 1989, 1991, 1993 and 1994 and Khanna *et al.*, 2003). So far it is reviewed that no such type of investigation have been conducted to assess the impact of stored effluents of slaughter house on groundwater quality of hand pumps situated near slaughter house. Hence, a study has been carried out to assess the water quality of hand pumps selected in the vicinity of slaughter house in terms of physico-chemical characteristics and to forecast impact of such polluted water on human beings. Study has been made in the month of August, 2005.

Materials and Method

The groundwater samples were collected in a neat clean two liter capacity white plastic Jericanes for general parameters and samples for DO were taken in 300 ml capacity borosil glass bottles and DO was fixed by using $MnSO_4$ and alkaline azide reagents. Methods of analysis, sampling and preservation of samples were adopted as per standard methods of APHA-AWWA-WPCF (1992), Trivedi and Goel (1984), Kotiah and Kumaraswamy (1994).

Parameters studies were colour, odour, pH, conductivity, dissolved solids, DO, BOD, total hardness, calcium hardness, magnesium hardness and chloride.

Slaughter house situated on Mundakhera road and its pond of stored effluents also touches the main GT road Khurja–Bulandshahar. Hand pumps were selected to collect the samples of groundwater situated within the range of 2 km at different locations and distances from the slaughter house. Sampling sites were selected as per following points and map (Figure 62.1).

1. Hand pump situated adjacent to Slaughter house at plot of Mr. Padmi, Mundakhera road Khurja, District Bulandshahar (U.P.). This sampling point is represented as–A.
2. Hand pump situated at 100 metre away from slaughter house at Khurja-Bulandshahar (U.P.). This sampling point is represented as–B.
3. Hand pump situated 1.0 km away near house of Mr. Mahendra Singh, Murari Nagar from Khooni Talab (redish pond of slaughter house), Mundakhera road, Khurja, District Bulandshahar (U.P.). This sampling point is represented as–D.

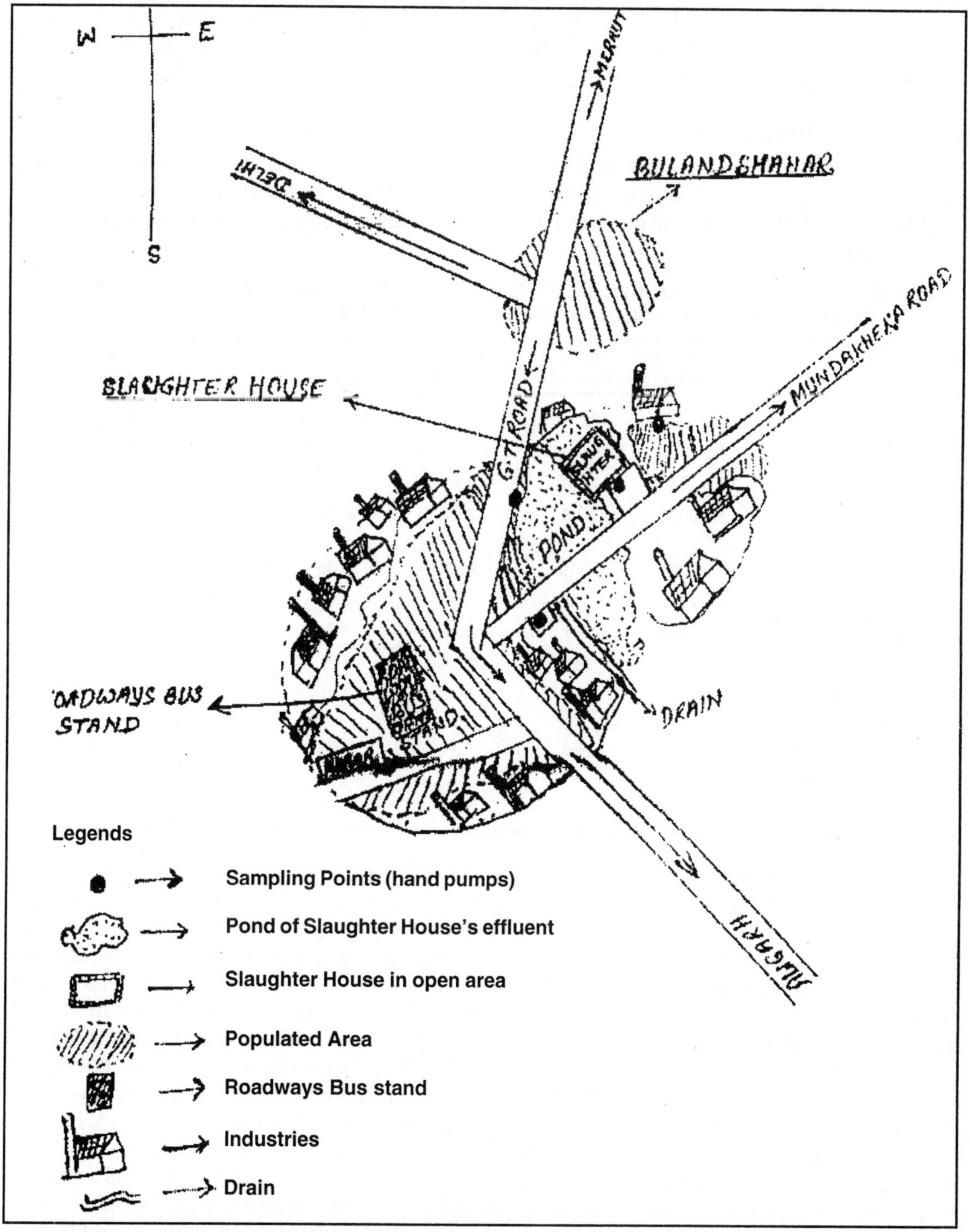

Figure 62.1: Map Showing Situation of Sampling Points (Handpumps) in the Vicinity of Slaughter House at Khurja, District Bulandshahar (U.P.)

4. Hand pump situated 2.0 km. away at Khurja- Bulandshahar road from Khooni Talab Mundakhera road, Khurja, District Bulandshahar (U.P.). This sampling point is represented as–E. All above hand pumps have been installed at depth of 30–45 feet approximately.

Results and Discussion

During the study period, five samples were collected from selected sites of hand pumps and parameters were analysed. The data obtained are shown in Table 62.1 and prescribed limit of BIS (1991) and CPCB (1997) have been given in Table 62.2 to compare the results of studies characteristics with respect to drinking purpose.

Table 62.1: Results of Physico-chemical Characteristics of Groundwater Quality of Hand Pumps Situated Near Slaughter House in Khurja Town, District Bulandshahar (U.P.) India

Parameters	*Date of Sample Collection*	*Sampling Point (A)*	*Sampling Point (B)*	*Sampling Point (C)*	*Sampling Point (D)*	*Sampling Point (E)*
Colour	17.8.2005	Straw	Straw	Light straw	Colourless	Colourless
Odour	17.8.2005	Untolerable	Untolerable	Unpleasant	No specific	Odourless
pH	17.8.2005	9.04	8.91	8.80	7.4	7.2
Conductivity (μmhos/cm)	17.8.2005	2.95	2.62	2.01	1.68	1.08
TDS (mg/l)	17.8.2005	2913	2604	2438	910	689
DO (mg/l)	17.8.2005	1.9	2.4	2.3	5.4	5.8
BOD (mg/l)	17.8.20Q5	3.9	3.5	3.5	0.8	00
Total Hardness (mg/l)	17.8.2005	710	681	744	286	263
Calcium Hardness (mg/l)	17.8.2005	420	390	388	198	170
Magnesium (mg/l) Hardness (mg/l)	17.8.2005	290	291	256	88	93
Chloride (mg/l)	17.8.2005	679	312	366	143	134

As it is clear from the results that the values of studied parameters ranged as pH 7.2–9.04, conductivity 1.08–2.95 μmhos/cm, DO 1.9–5.8 mg/l, BOD 0.0–3.9 mg/l, Total hardness 263–710 mg/l, Calcium hardness 170–420 mg/l, Magnesium hardness 88–356 mg/l, Chloride 134–679 mg/l, and Total Dissolved Solids 689–2913 mg/l, within the study area. Minimum values of these parameters obtained at sampling point–E except DO and Magnesium hardness where as maximum values of the characteristics were observed at sampling point–A except DO, and Total hardness. Minimum value of DO and Magnesium hardness were found at sampling point–A and while maximum value of DO and Total hardness observed at sampling point–E and (Table 62.1). The parameters of colour, odour, pH conductivity, DO, BOD, Total Hardness, Calcium Hardness, Magnesium Hardness and Total Dissolved Solids were not found as per prescribed standard of drinking purpose at sampling points–A, B and C as compared to other points but values of conductivity, DO and Total Dissolved Solids were also observed beyond the prescribed limit of drinking purpose BIS (1991) and CPCB (1997) at all sampling points which may be due to contamination of groundwater quality through percolation of stored effluents of slaughter house. However, the groundwater quality of sampling points–D and E is far better than other sampling points with respect to colour, odour, pH, BOD, Total hardness and chloride which may be due to less contamination of groundwater by percolation of stored effluents of slaughter house. Second reason may be distance factor of situated hand pumps location. It is evident from the Table 62.1 that as distance of the sampling points increases from the slaughter house, the values of the studied parameters also decreases besides DO. While an enhancement in DO is obtained. However it is not found as per standards limit. Therefore, groundwater quality of selected hand pumps is alkaline in respect of pH and most contaminated groundwater quality was found at sampling points–A, B and C which may be due to seepaging of stored effluents of slaughter house. Groundwater rich in carbonic acid and dissolved oxygen usually possesses a high solubilizing potential towards soil or rocks that contain appreciable amount of minerals litre, like calcite, gypsum and dolomite and consequently hardness level may increased (WHO, 1984). That's why the values of conductivity, TDS and DO were observed beyond the limit of drinking purpose (Table 62.2) at all sampling points. Hence, the groundwater quality of selected hand pumps situated within 500 metre range from slaughter house are not at all suitable for drinking purpose. Groundwater quality of selected hand pumps situated beyond is 500 metre–2 km range *i.e.* Sampling points–D and E from slaughter house is better than other points because BOD values obtained at these points are found negligible. It is necessary to present such contaminations which are polluting groundwater quality.

Table 62.2: Standards of Water Quality for Drinking Purpose in Terms of Physico-chemical Characteristics (BIS, 1991 and CPCB, 1997)

Parameters	*Standard Limits*
Colour	Colourless
Odour	Odourless
pH	6.5–8.5
TDS (mg/l)	500.0
Conductivity (mmhos/cm)	1.0
DO (mg/l)	>6.0
BOD (mg/l)	2.0
Total hardness (mg/l)	300.0
Calcium hardness (mg/l)	200.0
Magnesium hardness (mg/l)	100.0
Chloride (mg/l)	250.0

References

APHA, AWWA, WPCF, 1992. *Standard Methods for the Examination of Water and Wastewater*, 2nd edn. American Public Health Association, Washington DC.

Abbasi, S.A., Khan, E.I., Sentilvelan, K. and Shabudden, A., 1999. Modelling of Bunkingham canal water quality. *Indian J. Environ. Hlth.*, 41(3): 176–183.

BIS, 1991 and CPCB, 1997. *Drinking Water Quality Standards*. Bureau of Indian Standard, IS: 10500.

Khanna, D.R., Singh, Shakun, Gautam, Ashutosh and Singh, J.P., 2003. Assessment of water quality of River Ganga in district Bulandshahar (U.P.) India. *J. Nat. Con.*, 15(1): 167–175.

Kotaiah, B. and Kumaraswami, N., 1994. *Environmental Engineering: Laboratary Manual*. Charotar Publishing House, Anand, India, p. 115.

Pande, S.P., Narayanaswamy, V.S. and Hasan, M.Z., 1979. Quality of well waters of Nagpur with regard to Nitrates, nitrites (Phase-I). *Ind. J. Environ Hlth.*, 21(1): 35.

Singh, J.P., Yadava, P.K. and Singh, L., 1988. Pollution Status of Sangam and its adjoining rivers before Kumbh Mela at Allahabad. *Indian J. Environmental Protection*, 8(11): 839–842.

Singh, J.P., Yadava, P.K. and Singh, L., 1989. The assessment of water quality of Sangam and its adjoining rivers Ganga and Yamuna after Maha Kumbh Mela at a Allahabad. *Indian J. Environmental Protection*, 1(5): 372–375.

Singh, J.P., Yadava, P.K. and Singh, L., 1989. Mass bathing effect on water quality of Sangam during Maha Kumbh Mela at Allahabad. *Indian J. Environmental Protection*, 9(3): 189–193.

Singh, J.P., Yadava, P.K., Singh, Shakun and Prasad, S.C., 1991. BOD contamination in Kali river at Sadhu Ashram in Aligharh. *Indian J. Environmental Protection*, 11(5): 325–326.

Singh, K.P., Upadhyay, R. and Singh, Shakun, 1994. A study of water quality of river Yamuna at Agra with special reference to coliform bacteria. *Res. J.P.I. En. Vi.*, 10: 53–57.

Singh, J.P., Yadava, P.K., Singh S.K., Singh, Shakun and Madhu, Joshi, 1993. Assessment of BOD load in River Yamuna in respect of potable water at Agra. *Indian J. Environmental Protection*, 13(2): 119–123.

Soren and Julian, 1977. Groundwater quality near the water table in Suffolk Country, Long Island, New York, Suffolk Country. *Department of Environmental Control, Long Island Water Resour. Bull.*, 8: 33.

Trivedi, R.K. and Goel, P.K., 1984. *Chemical and Biological Methods for Water Pollution Studies.* Karad (India).

Vaishya, R.C. and Agarwal, I.C., 1993. Removal of arsenic (III) from contaminated groundwaters by sand. *Indian Water Works Assoc.*, 21: 249–253.

WHO, 1984. *Guidelines for Drinking Water Quality*. World Health Organization, Geneva, 2: 264.

Water and Wastewater Management in 2 Vols (2011) *Pages 433–439*
Editors: **D.R. Khanna, R. Bhutiani & Gagan Matta**
Published by: **DAYA PUBLISHING HOUSE, NEW DELHI**

Chapter 63

Impact of Fabric Dyeing Effluents on the Physical Parameters of Wunna River in Wardha District of Maharashtra

S.S. Gandhewar[1], K. Vasudevan[2] and S.B. Zade[3]
[1]*Department of Fisheries, Jankidevi Bajaj College of Science, Wardha*
[2]*Department of Zoology, DDE, Annamalai University, Annamalai Nagar, Tamil Nadu*
[3]*P.G. Department of Zoology, M.B. Patel College, Sakoli*

ABSTRACT

The physical parameters like temperature, odor, electrical conductivity, turbidity and pH of Wunna River in Maharashtra (India) at three different stations (upstream region-S_1, effluent mixed region-S_2 and 700 m away from effluent mixed region-S_3) were studied during the pre, mid and post-monsoon season. Marginal variations were recorded in temperature and turbidity while rigorous effect of effluent discharge on odor, electrical conductivity and pH was observed at the effluent mixed region as compared to the other stations. The flow of water in the river in different seasons was found to be an important factor responsible for variable impact of pollution. The data analysis confirms greater extent of pollution of the riverine eco-system at the site of release of textile industry effluents.

Keywords: Analysis, Ecosystem, Pollution, Textile industry effluent, Wunna River.

Introduction

Rivers are the lifeline of living beings as a source of drinking water and fish culture. There are 14 major rivers in India that share 83 per cent of the total drainage basin and contribute 85 per cent of the total surface flow (Chaudhari, 1982). Water used by the consumers must be free from microbial contamination, toxic substance and excessive amount of minerals and organic matter (Godi *et al.*, 2003). The pollution of water may be attributed to anthropogenic activities like rapid industrialization, urbanization and improper waste management techniques (Rao and Rao, 1995; Todd, 1995, Ranga Raj *et al.*, 1996).

The effluents of one of the textile industries involved in fabric dyeing are being released in Wunna River. At the moment, very scanty information on study of physical parameters of Wunna River is available. Hence to know the changes in physical parameters of the river due to release of industrial effluents, the present investigation was undertaken.

Materials and Method

Wunna River

It originates from Pilkapar rows of Mahargarh valley in Tahsil Katol, District Nagpur of Maharashtra State. The latitude of the river is 20°32′58", the longitude is 78°49′00" and altitude is 214.2 metre. It is the main tributary of river Wardha that joins the Pranhita River which ultimately

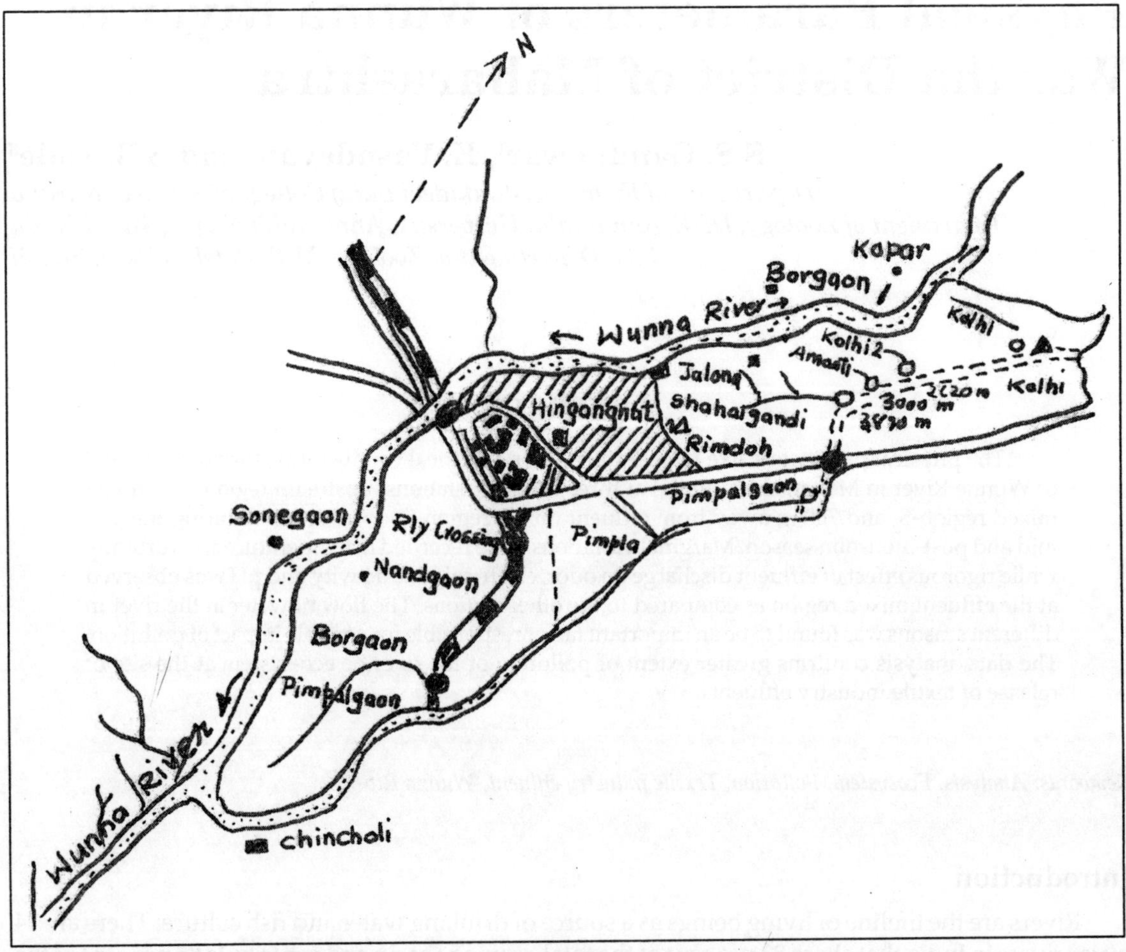

Figure 63.1: Location Map of River Wunna at Hinganghat

flows into the Godavari River. People residing in the vicinity of Wunna River heavily depend on it for drinking and other domestic purposes.

Study Stations

Water samples from three stations of the Wunna River were collected monthly from Station-1 (upstream region), Station-2 (effluent mixed region) and Station-3 (700 m away from effluent mixed region) during June 2007 to March 2008 between 10 a.m. to 11 a.m. The samples were collected 2 meter away from the bank and 0.3 meter below surface to prevent the surface micro layer.

Sample Collection and Analysis

The samples were collected in pre-cleaned plastic bottles. The spot tests for temperature and pH were done instantly after collection of samples at the sampling sites. The samples were brought to the laboratory and stored at 4°C till the analytical work was carried out. The physical analysis of samples were undertaken as per American Public Health Association (APHA, 1989) and Bureau of Indian Standards (IS-3025).

Results and Discussion

The physical analysis of samples of three different locations. Distribution of these parameters for the three stations of the river was analyzed using descriptive statistics and the results are given in the Table 63.1.

The present study was undertaken to evaluate changes in the river water temperature along the course of Wunna River, the changes in the recorded levels of temperature are presented in Table 63.1.

Temperature

The maximum temperature recorded at station 1 was 30 °C in the month of June 2007, while minimum temperature 20 °C recorded in the month of December 2007. At station 2, the maximum temperature recorded was 31.4 °C in June and July 2007 while minimum temperature was 21.1 °C in the month of December 2007. For station 3, the maximum temperature recorded was 30.4 °C in the month of June 2007 while the minimum temperature recorded was 20.2 °C in the month of December 2007 (Figure 63.3).

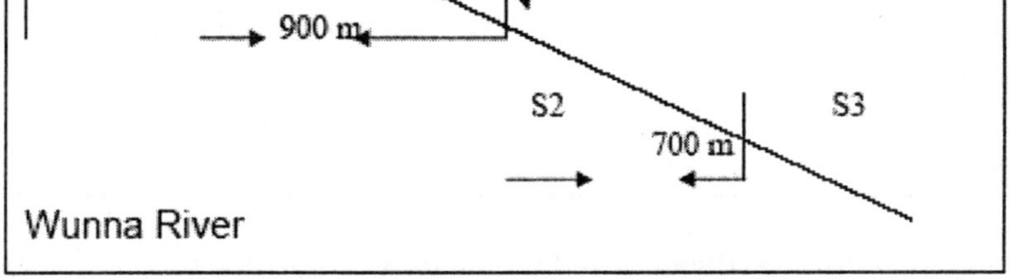

Figure 63.2: Sampling Locations in the Study Area

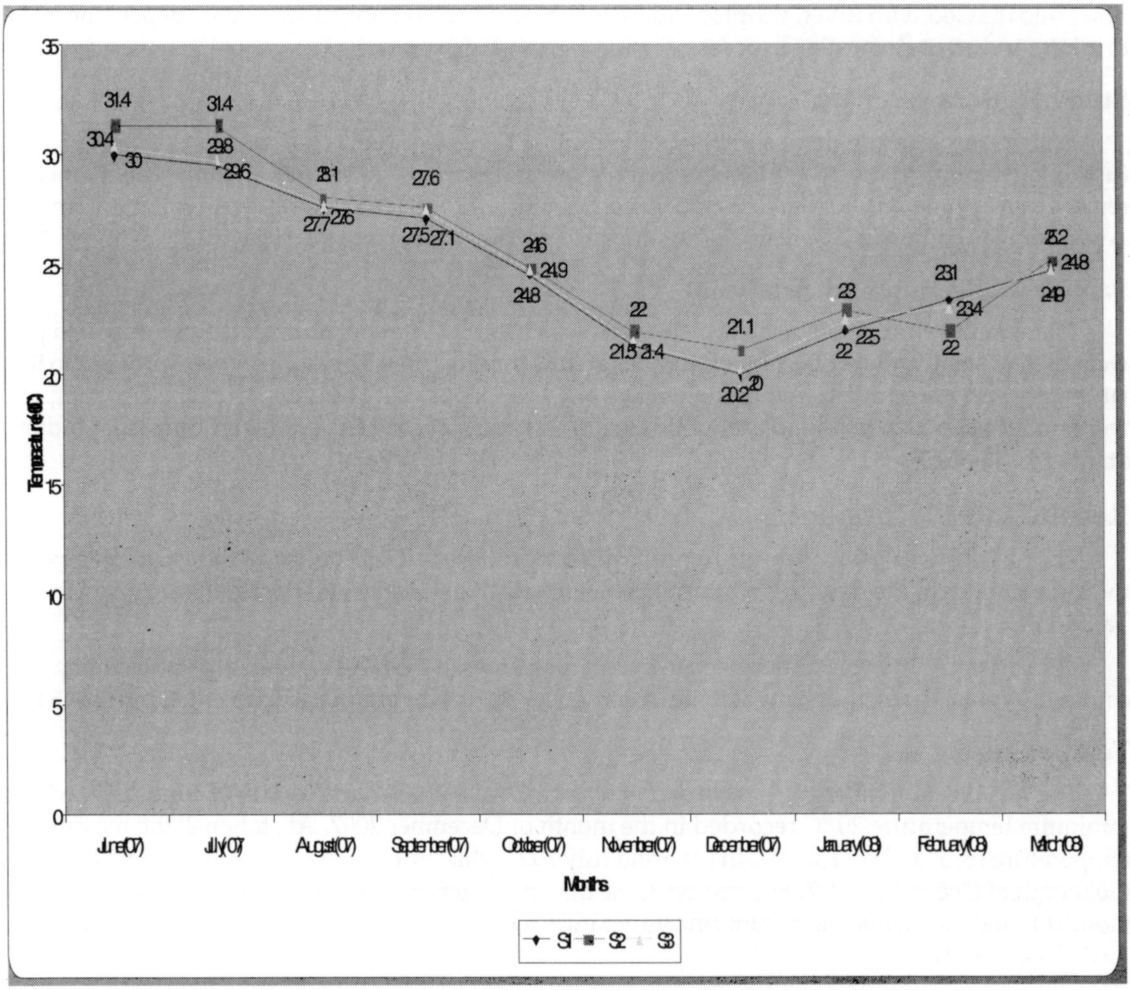

Figure 63.3: Temperature Recorded at Different Stations Along the Course of Wunna River During the Period of Study

Table 63.1: The Mean and SD Values Obtained for Different Parameters

Parameter	*WHO Standards*	*Mean and SD Value*		
		Station 1	*Station 2*	*Station 3*
Temperature (ºC)	–	25.05±3.45	25.67±3.82	25.24±3.50
Electrical conductivity (µS/cm)	750–2000	426.82±91.94	2739.40±827.02	499.70±97.12
Turbidity (NTU)	5–10	50.70±4.52	46.10±2.38	51.70±3.16
pH	6.5–8.5	7.79±0.25	7.95±0.37	7.85±0.15
Odor	UO	UO	O	PO

The rise in temperature at effluent mixed region may be attributed to the heat generated by the effluents.

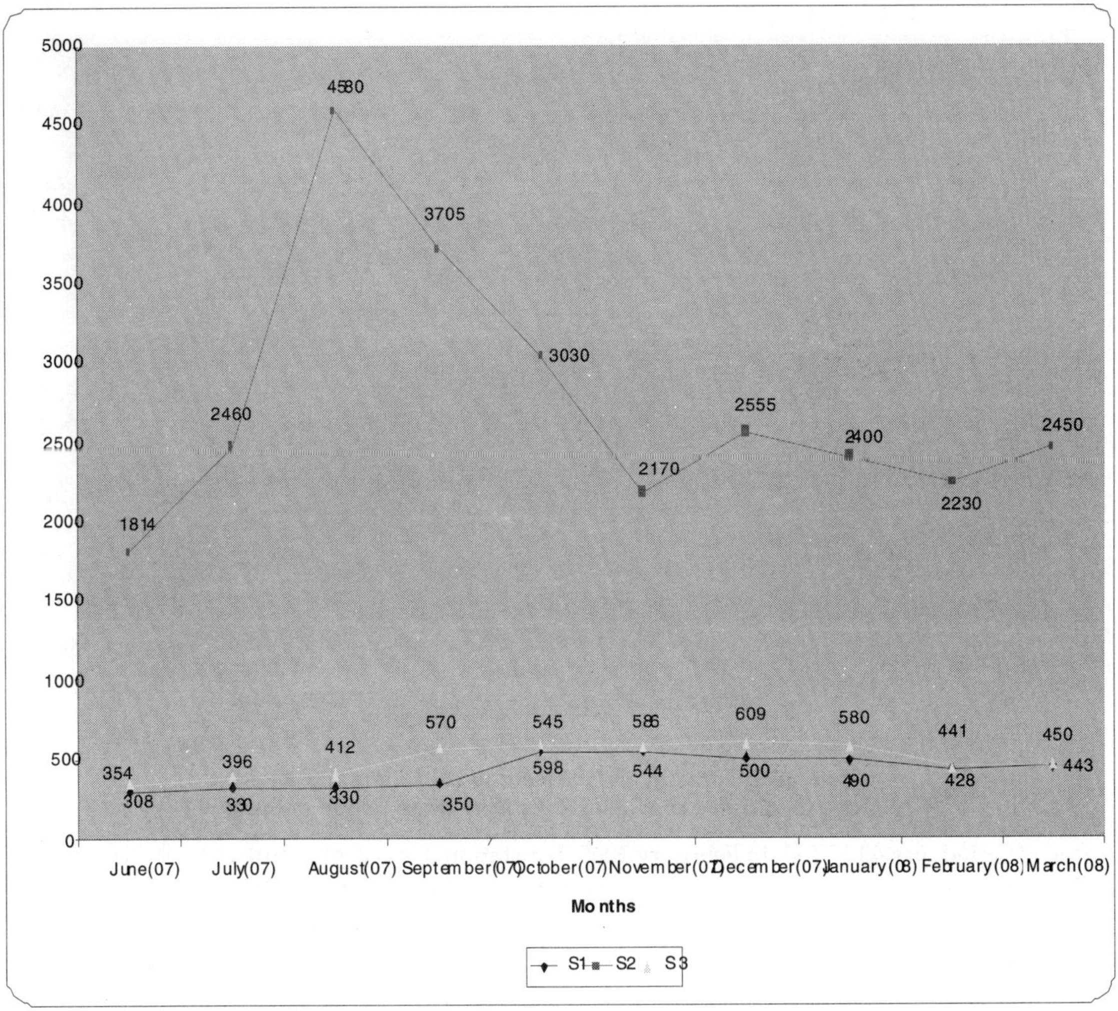

Figure 63.4: Electrical Conductivity (μS/cm) Recorded at Different Stations Along the Course of Wunna River During the Period of Study

Electrical Conductivity

Table 63.1 provides variations in EC at all the stations. At station 1, maximum conductivity recorded was 545 μS/cm in the month of October 2007, while minimum value recorded was 308 μS/cm in the month of June 2007. At station 2, maximum value recorded was 4580 μS/cm in the month of August 2007, while minimum value recorded was 1814 μS/cm in June 2007. At station 3, maximum value recorded was 609 μS/cm in the month of December 2007 and minimum value recorded was 354 μS/cm in June 2007 (Figure 63.4). The maximum value recorded at station-2 is well above the maximum permissible limit stated by WHO.

Turbidity

The variations in the turbidity (NTU) levels estimated at all stations were presented in Table 63.1. At station 1, the maximum value of turbidity recorded was 58 N.T.U. in July 2007, while minimum

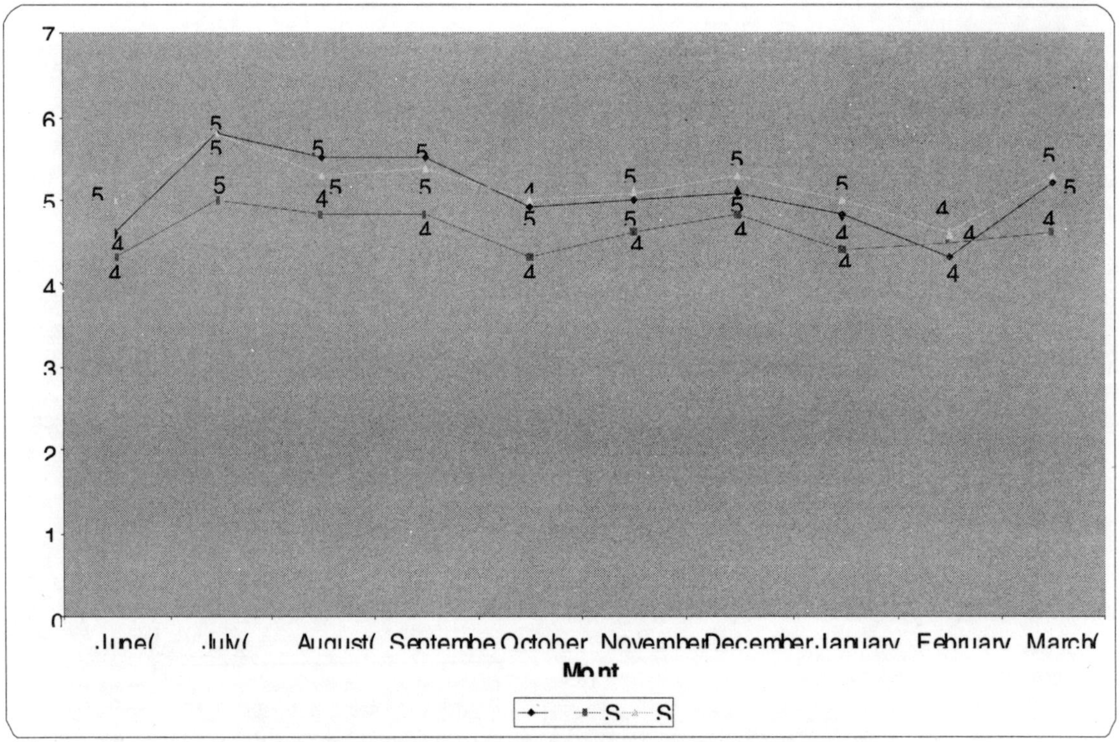

Figure 63.5: Turbidity (NTU) Recorded at Different Stations Along the Course of Wunna River During the Period of Study

value recorded was 43 N.T.U. in February 2008. For station 2, maximum turbidity observed was 50 N.T.U. in June and July 2007, while minimum turbidity observed was 43 N.T.U. in June 2007. For station 3, the maximum value of turbidity recorded was 53 N.T.U. in the month of August 2007 and December 2007 while minimum turbidity observed was 46 N.T.U. in February2008 (Figure 63.5).

pH

At station 1, the maximum pH value recorded was 8.4 recorded in the month of June 2007. While minimum value recorded were 7.5 in the month of January 2008. For station 2, the maximum pH value recorded was 8.68 in July 2007, while minimum value recorded was 7.46 in the month of March 2008. For station 3, the maximum value of pH recorded was 8.2 in the month of June 2007, while minimum recorded value of pH was 7.71 in the month of July 2007(Figure 63.6). The values of pH at the study stations clearly indicate the alkaline nature of water. Maximum pH value recorded at station–2 exceeds WHO permissible limits.

Conclusion

The discharge of textile industrial effluents into the Wunna river is known to alter the hydrographical parameters at different levels in the stations studied. Textile industry effluents influx in to the river may be considered as pollutant added to the river water thereby imparting bad odour along the course of the Wunna river at site-2 of the study area. As a result of this water of station-2 (S-2) has become unsuitable for drinking. These tremendous changes in the hydrography of the Wunna river could be expected to exert deleterious effects on the river ecosystem along the course of the down stream.

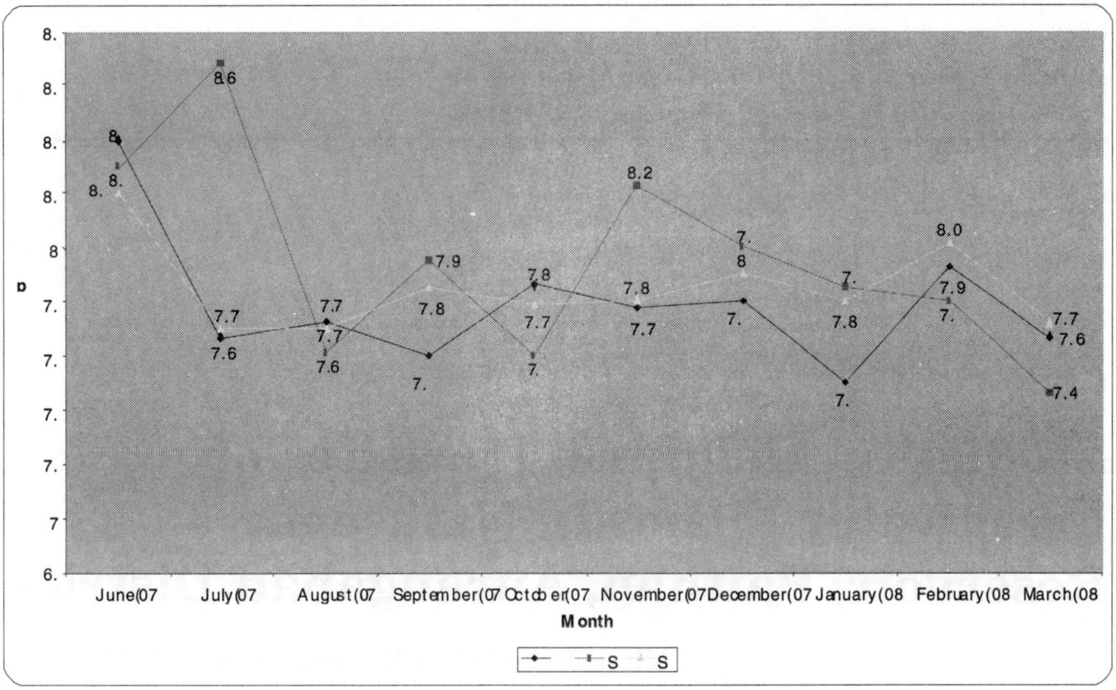

Figure 63.6: pH Recorded at Different Stations Along the Course of Wunna River During the Period of Study

Acknowledgements

The authors are thankful to Dr. Om Mahodaya, Principal, J.B. College of Science, Wardha, Dr. G. N. Paliwal, Prof. R. D.Nehete and P.V.Tekade for their help during this research work.

References

APHA, 1989. *Standard Methods for the Examination of Water and Wastewater*, 17th edn. APHA, New York, USA.

Bureau of Indian Standards–IS:3025. *Methods of Sampling and Test (Physical and Chemical) for Water and Wastewater.*

Chaudhari, N., 1982. *Water and Air Quality Control: The Indian Context for the Prevention and Control of Water Pollution*, New Delhi, India.

Godi, S.D., Barbudde, S.B., Hazel,D. and Dolley, C., 2003. Physico-chemical characteristics of drinking water at Velsao (Goa). *Jour. Ecotoxicol Env. Monitor.*, 13(3): 203–209.

Mayback, M., Chapman, D.V. and Helmer, R. (Eds.), 1989. *Global Environment Monitoring System: Global Freshwater Quality*. Basil Blackwell Ltd., Oxford, pp. 107–163.

Rangaraj, S., Elampooranan, T., Elango, C. and Ramlingam, V., 1996. Groundwater quality in Sub Urban regions of Madras city. *India Poll. Res.*, 15(4): 325–328.

Rao, C. Subba and Rao, N.V. Subba, 1995. Groundwater quality in residential colony. *Indian J. Environ. Hlth.*, 37(4): 295–300.

Todd, D.K., 1995. *Groundwater Hydrology*, 2nd edn. John Wiley and Sons, Inc., Singapore.

Water and Wastewater Management in 2 Vols (2011) *Pages* **440–444**
Editors: **D.R. Khanna, R. Bhutiani & Gagan Matta**
Published by: **DAYA PUBLISHING HOUSE, NEW DELHI**

Chapter 64

Studies on Helminth Parasites of Freshwater Fishes in Nathsagar Reservoir, Paithan, Aurangabad District

Nitin Podwal, B.V. Jadhav and D.B. Bhure
Department of Zoology, Dr. B.A.M.U., Aurangabad, M.S.

ABSTRACT

Freshwater fishes from Nathsagar reservoir, Paithan, Dist. Aurangabad (M.S.) were collected from June 2004 to May 2006. The 1296 fishes were observed from helminthic infections, out of them 761 fishes (58.71 per cent) were infected with helminth parasites. Thirty six species of helminthes were recovered throughout the investigation *i.e.*, 10 species of the trematodes include the genera *Genarchopsis* (Ozaki, 1925), *Orientochreadium* (Tubangui, 1931), *Oudhia* (Dayal *et al.*, Gupta, 1954), *Phyllodistomum* (Braun, 1899) and *Macrotrema* (Gupta, 1931). 20 species of the cestodes included six genera *i.e. Lytocestus* (Cohn, 1908), *Senga* (Dollfus, 1934; Shinde, 1968), *Proteocehalus* (Weinlend, 1858), *Gangesia* (Woodland, 1924) and *Silurotaenia* (Nybelin, 1942). And six species of the nematodes included two genera. *Rhabdochona* (Railliet, 1961) and *Camalanus* (Yeh, 1960). The high prevalence (75.69 per cent) occur in summer season where low prevalence (37.5 per cent) in monsoon season. The *Mastacembellus armatus* is highly infected (64.58 per cent) with helminth parasites.

Keywords: *Helminth parasites, Prevalence, Nathsagar reservoir, Freshwater fishes.*

Introduction

Nathsagar reservoir is one of the major irrigated project in Maharashtra State. It has been constructed across the river Godavari. The catchment area of this dam is 21,750 km^2. Helminth parasites (cestodes, trematodes and nematodes) were collected from Nathsagar reservoir, Paithan. Parasitic infection may occur with man and animals. From the present investigations, the results will be the key for identifying the helminthes and controlling of helminthes infecting fishes.

Materials and Method

Freshwater fishes were collected from different sites of Nathsagar dam during June 2004 to May 2006. The helminthes were collected, preserved, processed to a permanent slide and identified under a compound microscope, drawings are made-up with the aid of camera lucida and identified by Prof. B.V. Jadhav. Parasites distribution, host specificity, prevalence of helminthic infections were studied and recorded.

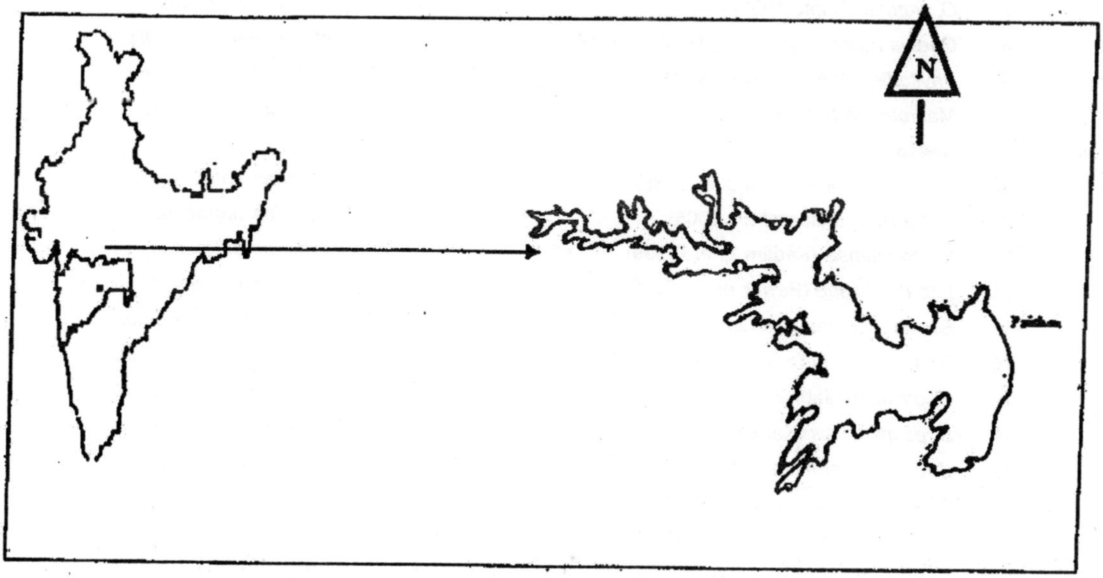

Figure 64.1: Location Map of Nathsagar Reservoir, Paithan, Dist. Aurangabad, M.S., India

Results and Discussion

There were 1296 fishes of 09 species observed with helminthic infection. 36 species of helminthes were recovered. The ten species of monogenea trematodes were *Genarchopsis piscicola, G. ozakii, Orientocradium dayali, P. phillppai, O. clariae, O. mahendrai, O. vermai, Oudhia horai, Phyllodistomum singhiai* and *Macrotrema macroni.* The twenty species of cestodes were *Lytocestus indicus, L. clarie, L. teranesnsis, L. bartrachusae, Senga maharashtrii, S. mohekarae, S. gachuae, S. paithanesis, Circumnobothrium aurangabadensis, C. alii, C. yamaguti, C. ophiocephali, C. khami, Proteocephalus vitellaris, P. gobiorum, gangeshia maharashtrii, G. dharurensis, G. mastacembalai, Silurotaenia macroni* and *S. godavari.* Six species of nematodes were *Rhobdochona singhi, R. Mazeedi, R. alii, R. sailuensis, Camallanus anafantis and C. unispicutus.* All worms of present investigation are shown in Table 64.1 with hosts.

Table 64.1: Helminths and their Fish Host

Sl.No.	Helminths	Hosts
	Trematoda	
1.	*Genarchopsis piscicoal* (Srivastava, 1933)	*Channa punctatus*
2.	*G. ozakii* (Bashirullah *et al.*, 1972)	*Channa punctatus*
3.	*Orientocreodiium dayali* (Dayal, 1949)	*Channa punctatus*
4.	*O. philippai* (Gupta, 1957)	*Channa punctatus*
5.	*O. clariae* (Chatterji, 1933)	*Clarias batrachus*
6.	*O. mahendrai* (Gupta, 1951)	*Clarias batrachus*
7.	*O. vermai* (Gupta,1951)	*Clarias batrachus*
8.	*Oudhia horai* (Dayal *et al.*, Gupta, 1951)	*Heteropneustes fossilis*
9.	*Phyllodistomum singhiai* (Gupta, 1951)	*Mastacembellus armatus*
10.	*Macrotrema macroin* (Gupta, 1951)	*Macronus cavasius*
	Cestoda	
11.	*Lytocestus indicus* (Moghe, 1925)	*Clarias batrachus*
12.	*L. clarias* (Tandon *et al.*, 2005)	*Clarias batrachus*
13.	*L. teranaensis* (Kadam *et al.*, 1999)	*Clarias batrachus*
14.	*L. batruchusae* (Pawar *et al.*, 2002)	*Clarias batrachus*
15.	*Senga maharashtyrii* (Jadhav *et al.*, 1991)	*Mastacembellus armatus*
16.	*S. mohekarae* (Tat *et al.*,1997)	*Mastacembellus armatus*
17.	*S. gachuae* (Jadhav *et al.*, 1991)	*Mastacembellus armatus*
18.	*S. paithanensis* (Kadam *et al.*,1981)	*Mastacembellus armatus*
19.	*Circumonobothrium aurangabadenis* (Jadhav *et al.*,1979)	*Mastacembellus armatus*
20.	*C.yamaguti* (Jadhav *et al.*, 1990)	*Mastacembellus armatus*
21.	*C. ophiocephali* (Shinde, 1968)	*Mastacembellus armatus*
22.	*C. atii* (Shinde *et al.*,1994)	*Mastacembellus armatus*
23.	*C. Khami*, (Shinde, 1977)	*Mastacembellus armatus*
24.	*Proteocephalus vitellaris* (Verma, 1929)	*Mystus senghala*
25.	*P. gobiorum* (Dogell *et al.*, 1939)	*Mystus senghala*
26.	*Gangesia maharashtrii* (Jadhav *et al.*,1995)	*Wallago attu*
27.	*G. dharurensis* (Jadhav *et al.*, 1997)	*Wallago attu*
28.	*G. mastacembali* (Wankhede, 2004)	*Wallago attu*
29.	*Silurotaenia macroni* (Shinde *et al.*, 1984)	*Mystus seenghala*
30.	*S. gondavari* (Wankhede *et al.*, 2002)	*Mystus seenghala*
	Nematoda	
31.	*Rhabdochona singhi* (Ali, 1956)	*Labeo rohita*
32.	*R. mazeedi* (Parsad *et al.*,1965)	*Labeo rohita*
33.	*R. alii* (Kalyankar, 1972)	*Labeo rohita*
34.	*R. sailuensis* (Khadap *et al.*, 2004)	*Labeo rohita*
35.	*Camalanus anabantis* (Pearse, 1933)	*Channa gachua*
36.	*C. unlspictus* (Khera, 1956)	*Channa gachua*

Prevalence

The prevalence results are shown in Table 64.2. The higher prevalence occur in summer season (75.69 per cent) followed by winter season (62.96 per cent) and rainy season (37.5 per cent). Because high temperature and sufficient moisture needed for the development of parasites. The higher incidence occur in host *Mastacembellus armatus* (64.58 per cent), whereas lower in *Wallago attu* (57.38 per cent). Because these infections are host specific and morphological, physiological and ecological factors affect the host specificity. The valuable information pertaining to the influence of season on the helminth parasites was contributed by served workers like Tornquist (1931) who described about the systematic method of occurrence of certain fish parasites *Camallanus lacustrs* that the infective stages invade the host during summer, the growth and maturation take place during autumn and winter and release of their infective progeny occurs during summer. According to Gibbons (1976), heavier incidence of nematodes occurs during late spring or summer month. Kennedy (1968) reported that the temperature is the major factor in controlling the seasonal distribution of many parasites.

Table 64.2: Total Number and Total Prevalence (Per cent) of Infected Fishes in Nathsagar Reservoir, Paithan

Name of Host	*No. of Host Dissected*				*No. of Host Infected*				*Prevalence (Per cent)*			
	M	*W*	*S*	*T*	*M*	*W*	*S*	*T*	*M*	*W*	*S*	*T*
Channa punctatus	48	48	48	144	18	26	34	78	37.30	54.16	70.83	54.86
Heteropneustus fossilis	48	48	48	144	17	28	37	82	37.41	58.33	77.08	56.94
Mastacembellus armatus	48	48	48	144	20	33	40	93	41.66	68.75	83.33	64.58
Clarias batrachus	48	48	48	144	20	37	35	92	41.66	77.68	72.91	63.88
Macronus cavasius	48	48	48	144	19	34	39	92	93.58	70.83	81.25	63.88
Wallago attu	48	48	48	144	16	24	34	74	33.33	50.00	70.83	51.38
Labeo rohita	48	48	48	144	17	29	36	82	35.41	60.41	75.00	56.94
Mystus seenghala	48	48	48	144	18	31	35	84	37.50	64.58	72.91	58.33
Channa gachua	48	48	48	144	17	30	37	84	35.41	62.25	58.33	58.33
Total	432	432	432	1296	162	272	327	761	37.5	62.96	75.69	58.79

M: Monsoon season; W: Winter season; S: Summer season; T: Total.

The present investigations also occur such type of results *i.e.* high infection as well as prevalence occur in summer months. The population investigations can prove data for the prediction of integrated methods to achieve the regulation of number of harmful parasites, because it has been stated that a single method of control or co-ordination acitivities are of little value since they ameliorate the infection (Kennedy, 1975 and 1978). This type of result indicates the morphological, physiological and ecological factor (seasons) affecting the distribution of parasites.

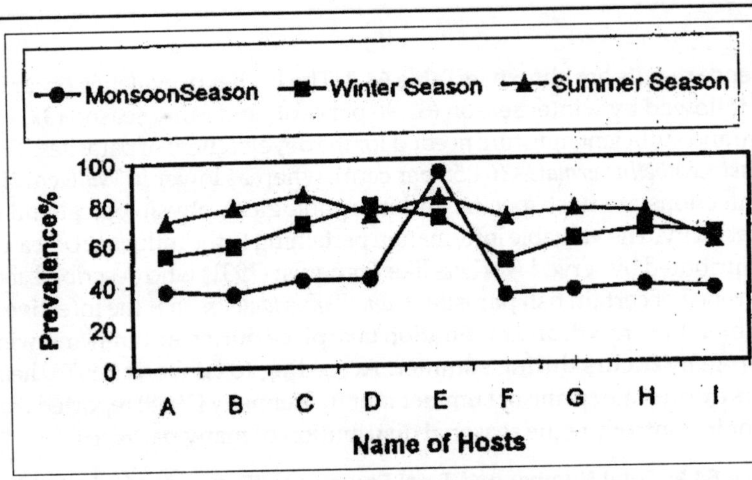

Figure 64.2: Prevalence of Helminth Parasites of Freshwater Fishes

A: *Channa punctatus;* B: *Heteropneustus fossilis;* C: *Mastacembellus armatus* D: *Clarias batrachus;* E: *Macronus cavasius;* F: *Wallago attu;* G: *Labeo rohita;* H: *Mystus seenghala;* I: *Channa gachua.*

Acknowledgements

The authors are sincerely acknowledged, to the UGC, New Delhi for sanctioning the major research project no-30-193/2004 (SR), dated 10-11-2004 for financial assistance.

References

Anderson, R.M., 1976. Seasonal variation in the population dynamics of *Caryophyllacus laticeps. Parasitology*, 72: 281–395.

Dobson, A.P. and Roberts, M.G., 1999. The population dynamics of parasite helminthic communities. *Parasitology*, 102 (Supl.): 507–510.

Gupta, 1951. On a new trematode *Phyllodistomium singhii* sp. of the subfamily, Allocreadinae Looss. 1899 from the intestine of a freshwater fish *Mastacembellus armatus. Ind. J. Helm.*, 3(1): 21–28.

Hiware, C.J., 1999. Population dynamics of the Coryophylloeid cestode parasitizing freshwater air breathing predatory fish *Clarias batrachus* (Linnaeus). *Ravi. Di. Para*, 19(63): IV–I.

Kennedy, C.R., 1976. *Ecological Aspects of Parasitology*. North Holland Publishing Company, Amsterdam Oxford.

Lawrence, J.L., 1970. Effect of season, host age, on endohelminthes of *Catastomus commersoni. J. Para.*, 56(3): 567–571.

Jadhav, B.V., Bhure, D.B. and Padwal, Nitin, 2005. A survey of cestode parasites of freshwater fishes from Pune and Ahmednagar district (M.S.) India. In: *Proc. Recent Trends in Para.*, at Secunderabad 29th–30th Sept, p. 48–51.

Wongsawad, 2004. Helminths of vertebrates in Mae-sa-stream, Thailand. *Southeast Asian J. Trop. Med. Publ. Health*, 35(Suppl): 140–146.

Water and Wastewater Management in 2 Vols (2011) *Pages* **445–451**
Editors: **D.R. Khanna, R. Bhutiani & Gagan Matta**
Published by: **DAYA PUBLISHING HOUSE, NEW DELHI**

Chapter 65

Sustainable Integrated Fish-cum-Livestock and Poultry Farming

S.K. Garg[1], R.P. Singh[2] and R.K. Gupta[1]
[1]Department of Zoology and Aquaculture, [2]Department of Animal Breeding,
CCS Haryana Agricultural University, Hissar, India

ABSTRACT

To maintain the ecological balance, a sustainable approach is very much important. So in this chapter fish fauna has been assessed in their no. by using the poultry waste.

Keywords: Fish, Sustainable, Poultry, Manure, Organic.

Introduction

At the world scenario, China is the first country who had leaded this world for the use of night soils in the fish ponds for the first time, for fish production and thus became a pioneer in integrated fish farming system by using other livestock wastes. This system in China was established in the early 19th century, estimated annual production of excreta of cow dung 900 kg, pig dung 2200 kg, chicken excreta 40-50 kg and the night soil of human beings 790 kg.The excessive use of natural resources exert pressure on the land and on the water body, therefore, it requires to maintain more productive yet sustainable aquaculture system. To maximize aquacultural production from a single unit area, through optimum utilization of resources, it involves crop farming, livestock farming and fish culture to maximize production through recycling of organic wastes. In this system, practically, nothing is wasted and simultaneously, ecological balance is maintained and thus, varieties of products are received. But if the system is poorly managed, accumulation of these waste materials in the aquatic system would lead to health hazards. Therefore, availability of different resources in various agroclimatic conditions, following kind of integrated fish farming systems can be established. Increased production is being achieved by the expansion of areas of land and water under culture, and the use of more intensive and modern farming technologies that involve higher usage of inputs such as water, feeds, fertilizers and chemicals. FAO (1990) defined the sustainable development as the management and conservation of the natural resources base, and the orientation of technological and institutional change in such a manner as to ensure the attainment and continued satisfaction of human needs for present and future generations. Such sustainable development (in agriculture, forestry and fisheries sectors) conserves land, water, plant and animal resources, is environmentally non-degrading, technically appropriate, economically viable and socially acceptable.

Most village farmers have few numbers of cattle, goats, sheep, birds, land and water areas, which can be exploited for food production. These components offer great scope for uplifting their socioeconomic conditions by following diversified activities. It is well known that the pond bed provides enough humus for increasing soil productivity, crop production, consequently, helps in fish culture, animal husbandry thereby acting as an integrated web cycle in the farm making it an independent multi-structural unit. The most integrated systems of aquaculture are duck and pig farming and poultry production. In some places cattle and fish culture are also partly integrated. Smaller fragmentation of agriculture land and population pressure calls for the diversification and integration of these enterprises.

Characteristics of Integrated Farming

The concept of unitary culture either of fish, crop, or animal husbandry is gradually changing to the integrated culture system with the view of producing fish, meat, egg, milk, vegetables and other allied products within a farm itself on an economical scale. Accumulation of wastes/byproducts can generate pollution to the environment. Such wastes/byproducts can be recycled for food production through integrated system of fish culture. The characteristics of integration is the development of structural network in line with local conditions and utilization of wastes by various ways of recycling as manure, bio-gas, crop production, feed *etc.*

Importance of Manure

Animal wastes are used to fertilize fish pond and are considered superior to inorganic fertilizers in producing and maintaining desirable species of planktonic and benthic food organisms in fresh and brackish water ponds. Yields obtained with organic fertilizers are usually higher than those obtained with inorganic compounds. Moav *et al.* (1977) reported that fish fed on plankton produced in manured ponds have lipid contents (6.2 per cent) less than those fed pellets of high protein content (15 per cent lipids) or those fed cereals (20 per cent lipids). The cost of manure is lower than that of inorganic fertilizers.

Organic fertilizers contain nitrogen and phosphorus, but also decaying organic matter. Nutrients benefit phytoplankton (and phytobenthos), while the organic matter promotes the development of bacteria in suspension in the water and in the sediment. The heterotrophic production cycle is thus stimulated leading to an increase in zooplankton and zoobenthos. Manure also supplies organic colloids, which increase the sorption capacities of soil and seston. Phytoplankton supplies organic matter for decomposition thereby regenerate the nutrients to the benefit of phytoplankton and themselves are consumed by zooplankton and fish and zooplankton in turn are consumed by fish.

Comparative Value of Manure

Some fish especially tilapia, feed directly on fresh excrements. They can be observed, for instance, below pigsties or at the entrance of feeder canals bringing sewage from stables, feeding on excreta floating in the water. This has led to the idea of using dung directly in the preparation of artificial feed pellets. Analysis of cow dung and chicken manure indicated that this contains an amount of metabolizable energy, in the order of 2510 to 3347 kJ kg and 3766 to 5021 kJ kg and 10 to 15 per cent and up to 25 per cent gross N-proteins, respectively. The composition of excreta of common animals is presented in Table 65.1. Amount of excrement expressed in fresh weight and dry weight per day per 100 kg of live weight produced by common animals are presented in Table 65.2.

Table 65.1: Composition of Excreta of Common Livestock Animals (Little and Muir, 1987)

Animal	*Faeces to Urine Ratio*	*Humidity*	*Nitrogen (Per cent)*	*Phosphorus (Per cent)*	*Potassium (Per cent of K_2O)*
Cattle	75:25	85	4.0	2.7	3.4
Pig	53:47	85	4.7	4.3	2.7
Sheep	67:33	70	3.9	2.4	3.5
Horse	75:25	75	2.3	1.3	1.4
Chicken	100	65	6.0	7.9	3.7

Table 65.2: Amount of Excrements of Common Animals in Fresh and Dry Weight Basis per Day (Little and Muir, 1987)

Animal	*Fresh Excreta/ Animal/Day (kg)*	*Dry Weight/ Animal/Day (kg)*	*Fresh Excreta/ 100 kg wt./ Day (kg)*	*Dry Wt. per 100 kg Live Wt/Day (kg)*
Cattle	30.0	4.50	6.0	0.90
Chicken	0.7	0.27	35.0	13.40
Ducks	1.0	0.38	33.3	12.70
Pigs	8.0	1.20	14.5	2.17
Sheep	2.1	0.71	7.0	2.40
Horse	24.0	8.20	6.3	2.15

Studies have indicated the superiority of chicken manure over cow dung for fish production. As compared to the control, chicken manure increases fish production by 21 per cent, while this value was 11 per cent for cow dung. The feed conversion rate of pellets decreased by 12 and 6 per cent, respectively, for chicken manure and cow dung, as compared to the control.

Types of Fish Cultured in Manured Ponds

Fertilization with manure can be applied to ponds where fish with various diets are cultured in polyculture, making use of different feed resources of the pond. Primary productivity is used by phytoplanktivorous fish, secondary productivity by zooplanktivorous fish, and detritus by detritivorous fish. Some fish especially tilapias have wide feeding regimes, capable of using a broad range of food items.

Disadvantage of the Use of Manure

1. Excess of organic fertilizer causes depletion in dissolved oxygen particularly during the hottest hours of the day because of high microbial activity.
2. There is release of ammonia especially when pH is alkaline which is toxic.
3. Risk of transmission of parasitic diseases to humans with fish as intermediate hosts, however, no such disease has yet been detected.

Precautions for Use of Manure

1. Spread manure in water in small amounts and several times a week.
2. Spreading should be homogenous over the whole surface.
3. Manure in an advanced stage of decomposition fertilizes better than less decomposed matter as the nutrients diffuse more easily to the water.
4. In dried ponds, spread dung/compost on the bottom of ponds before impoundment.
5. Ponds should be shallow, ideally with a mean depth of 1 m.
6. In stocked ponds, manuring is increased concurrently with fish growth @ 2 to 4 per cent (dry weight) of the standing fish biomass.
7. Application rates are 100 to 150 kg/ha/day (dry weight) in intensive production.

Number of Livestock Heads Needed for Integrated Fish Farming

In integrated livestock-cum-fish farming, the animals are raised near or on ponds so that the manure and other waste materials can be discharged directly into the ponds. The number of animals is regulated to ensure that the wastes discharged into the ponds do not exceed the desired quantities. This number varies with intensity of production, and with size and weight of the animals. For pigs, the numbers vary between 30 and 85 pigs/ha and for ducks, between 1000 and 3500/ha. Pigsties and duck pens can be built directly over ponds, on stakes, or nearby. Pigs are raised for 5 to 6 months, until they reach a commercial weight (60 to 100 kg). Chicken and ducks are raised for 1 to 3 months, till they reach their market size.

Types of Fish-cum-Livestock Integrated Farming

Fish-cum-Duck Integration

1. Ducks are popularly known as "Living Manure Machine'.
2. Ducks use both land and water as habitat.
3. Ducks feed on the predators (tadpoles, dragonfly, weed fishes) thus help fingerlings to grow.
4. Duck raising in fishpond reduce the demand for protein on duck feed.
5. Duck droppings go directly in to water bodies providing C, N and P.
6. Daily waste of duck feed can be utilized as fish feed in pond.
7. Manuring by ducks is homogeneously distributed without any heaping of duck droppings.
8. Ducks oxygenates pond water by their frequent movements and dubbing habits in water.

Table 65.3: Financial Considerations of An Integrated Cattle-Poultry-Fish System

Assumptions: The owner has two hectare of land, *i.e.* one hectare for ponds, remaining land for livestock and poultry sheds, stores and fodder cultivation *etc*. There are five family members to support the system in addition to one more hired labour. Five buffaloes and 200 layer chicks are considered to run the unit.

Financial Details

Sl.No.		*Amount (Rs.)*
A.	**Initial Capital Investment**	
1.	Layout and leveling charges for fish ponds (1 ha)	5000
2.	Cost of construction of 8 nursery ponds (40'x20'x5'), 4 rearing ponds (60'x40'x6'), 2 stocking ponds (200'x80'x8')	95000
3.	Cost of 5 buffaloes	100000
4.	Cost of buffaloes sheds (4 sq.m/head, @ Rs. 800/sq.m) and for open area of 8 sq.m/animal (@ Rs. 200/sq.m)	50000
5.	Poultry sheds of 500 sq.ft area (@ Rs. 70/sq.ft)	35000
6.	Office-cum store (100 sq.m, @ Rs. 1500/sq.m)	150000
7.	Construction of tubewell/guard shed *etc.*	100000
8.	Cost of dairy equipment, fish nets and other farm accessories	80000
9.	Miscellaneous expenses	70000
	Total Capital Cost	**6,85,000**
B.	**Expenses**	
1.	Interest on bank loan @ 10 per cent/annum on 80 per cent of capital	54800
2.	Depreciation ponds, building @ 5 per cent/annum	16750
3.	Depreciation on equipment @ 10 per cent/annum	18000
4.	Insurance premium of buffaloes @ 4 per cent/annum	4000
5.	Cost of 10000 fish seed @ Rs. 200/1000	2500
6.	Cost of 30 qtl. Fish feed @ Rs. 1000/qtl	30000
7.	Pond monitoring, liming charges *etc.*	3000
8.	Cost of 547.5 qtl green fodder (30kg/animal/day @ Rs. 55/qtl)	30112
9.	Cost of 73 qtl dry fodder (4kg/animal/day @ Rs. 150/qtl)	10950
10.	Cost of 55 qtl. Concentrates (3 kg/animal/day @ RS. 700/qtl)	38500
11.	Labour charges @ Rs. 3000/month	36000
12.	Cost of 200 chicks @ Rs. 14/chick	2800
13.	Cost of feeding chicks @ 6.5 kg/chick/day up to 18 weeks	71974
14.	Miscellaneous charges *etc*	55000
	Total expenses	**374386**
C.	**Total Receipts**	
1.	Sale of milk 10/lt/animal for 300 days (@ Rs. 14/lt)	210000
2.	Appreciation of heifers (@ Rs. 2500/head/annum)	12500
3.	Sale of 4500 kg fish (@ Rs. 40/kg)	180000
4.	Sale of eggs of 190 hens with 280 eggs/year @ Rs. 1.65/egg	87780
5.	Sale of 170 spent hens @ Rs. 45/hen	7650
	Total Receipts	**486310**
	Net Income per annum	**123544**
	Net income/month	**10295**

Fish-cum-Chicken Integration

1. Fishmeal is used as a protein source in poultry feed.
2. Poultry droppings are rich source of N and P.
3. Chicken manure is a very efficient fertilizer for fish production.
4. Chickens can be housed over the ponds and the droppings can fall directly into them.
5. For one acre of fishpond about 250-300 birds are required.
6. Droppings can also be utilized as a component in fish feed pellets.

Fish-cum-Pig Integration

1. Left over, residues of kitchen, aquatic plants, and wastes are used as feed for pigs.
2. Pig dung contains 70-75 per cent easily digestible nutrients for fish.
3. Feed cost for fishes can be reduced by 40-45 per cent.
4. Pig manure is consumed by bottom feeders and is beneficial for the growth of plankton.
5. About 25-30 growing pigs can fertilize one acre of pond.
6. Pig's pens can be constructed over the bank of the pond.
7. Three tier system (poultry in upper stall, pigs in lower stall and below it the fish ponds) is more profitable as pigs consume poultry droppings and fish consume pig's dropping.
8. Duration of the culture of fish and pigs is about one year.

Fish-cum-Cattle Integration

1. Cattle dung and urine also have super manure value.
2. Waste fodder and washings are also utilized as a fish feed.
3. Cow manure is rich in N and P levels, which are congenial for plankton growth.
4. Cow manure is very fine due to repeated digestion in stomach of cow and it can suspend longer in the water.
5. Such suspensibility of dung enables fish to get more feeds, reduces oxygen consumption and avoids the formation of harmful gases.
6. For manuring an acre of fish pond, 4-5 cattle heads are sufficient.
7. BOD of cow manure is relatively lower than other livestock manure because the cow forage has already been decomposed by micro-organisms in cow's body.

Fish-cum-Sheep/Goat Integration

1. Sheep and goats are very prolific small ruminants.
2. Droppings can be used as a component of fish feed thus reducing the cost of artificial feed.
3. Pens having slotted floors are constructed on the banks of pond such that the faeces are dropped directly into the pond water.
4. A unit of 25-30 animals is sufficient to manure an acre of fish pond.

Mathematical Management of Farming Systems

In integrated system, there is need of integrated management. Some of the management considerations to be adopted are given below.

1. Feed Demand

$$M = YF$$

M = Feed demand

Y = Target yield of fish

F = Food conversion rate

2. Number of Individuals/Water Area

$$N = (Y1-nY2)C/m$$

m = Amount of excreta per individual

C = Conversion factor of manure

Y1 = Net yield of filter feeder and omnivorous

Y2 = Gross yield of fish

N = Ratio of herbivorous fish faecal to filter feeder and omnivorous fish production (0.2–0.6).

3. Manure Requirement

$$M = (Y1-nY2)C$$

4. Construction Area of Animal House

$$S = N^{*}s/C$$

S = Standard area

N = Number of animals raised in whole year.

s = Average construction area per animal *i.e.* Cow–7m^2, grazing area 15-20m^2/cow

Duck–Dry run 45 individuals/m^2, wet run 3-4 individuals/m^2

Pig–2m^2/individual

5. Number of Cycles per Year for Different Species

Cow and chick–One cycle/year

Pig–Two cycles/year

Ducks–Four cycles/year

References

Little, D. and Muir, S., 1987. *A Guide to Integrated Warm Water Aquaculture*. Univ. of Sterling, U.K.

Moav *et al.*, 1977. *Aquaculture*, 10: 25–43.

Water and Wastewater Management in 2 Vols (2011) *Pages* 452–461
Editors: D.R. Khanna, R. Bhutiani & Gagan Matta
Published by: DAYA PUBLISHING HOUSE, NEW DELHI

Chapter 66

Maturation Biology of a Rare Hill-Stream Fish *Botia dayi* Hora from Garhwal Himalaya, Uttarakhand

Kishor Kumar[1], K.L. Bisht[2], Anoop K. Dobriyal[3], H.K. Joshil[3] and P.K. Bahuguna[3]

[1]*Department of Zoology, Government Degree College, Dakpathar, Dehradun*
[2]*Department of Zoology, Govt. P.G. College, Kotdwara, Garhwal*
[3]*Department of Zoology, HNB Garhwal University Campus, Pauri Garhwal*

ABSTRACT

The paper deals with maturation biology and spawning behaviour of a hillstream fish *Botia dayi* (Hora) collected from river Khoh in the foothills of Garhwal Himalaya, Uttaranchal. Seven maturity stages were determined for the fish. Based on GSI, Dobriyal index and availability of different maturity stages during months, it was inferred that the fish spawns during August-September as one time protracted spawner.

Keywords: *Maturation biology, Fish, Dobriyal index.*

Introduction

The spawning tendency of fish studied by the examination of its ova diameter round the year. A majority of teleost fishes all over the world are seasonal breeders. The Indian fishes are mostly monsoon breeder however in hill stream fishes six different spawning behaviours have been reported by Dobriyal (2005).

According to Lagler *et al.* (1962), the first attainment of ability to reproduce is termed as sexual maturity which is affected by several factors as difference in species, age, size and several environmental factor (temperature, pH of the water *etc.*). The maturation and spawning biology of fish has been extensively studied by various workers (Sobhna and Nair, 1974;Pathani, 1979; Nautiyal, 1984) Singh *et al.*, 1985; Dobriyal and Singh, 89,93; Uniyal *et al.*, 2004) Present contribution is related to one of the most rare species of Garhwal Himalaya.

Materials and Method

The fish were collected monthly from the two different sampling sites during 2002 and 2003 from the Khoh stream between Kotdwara and Dogadda. Morphometric measurements were taken in the fresh condition and the fish were then preserved in 5 per cent formalin solution for further examination. Prior to preservation the colour of the gonads was recorded.

The present study is based on the macroscopic as well as the microscopic examination of gonads in *B.dayi* (Hora). For the macroscopic studies the gonads were removed and sex, physical appearance, length and weight of the gonads were recorded. For microscopoic observation, only ovaries were taken in to consideration.

Gonado Somatic Index (G.S.I.) was calculated for each fish by the following formula :-

$$GSI = \frac{\text{Wt. of gonad}}{\text{Wt. of fish}} \times 100$$

Dobriyal Index (Dobriyal *et al.*, 1999): A newly designed and well in use index was found to be more accurate for the estimation of maturity, month and frequency of spawning. It was calculated as;

$$\text{Dobriyal Index (DI)} = \sqrt[3]{GW}$$

where,

D.I = Dobriyal Index, G.W.= Average gonad weight calculated for each months for male and female fish separately.

For the determination of maturity stages, the sample of ova were collected from different regions of ovary and were microscopically studied. The ova diameters were measured by means of an ocular micrometer and at least 100 ova from each ovary were measured. The measurements were taken in omd (ocular micrometer division) on the basis of developmental pattern of ova round the year, following stages of the maturity were determined for *Botia dayi* (Hora), which were almost in accordance with the ICES scale (Wood, 1930).

- ☆ Stage I: Immature I
- ☆ Stage II: Immature II
- ☆ Stage III: Maturing I
- ☆ Stage IV: Maturing II
- ☆ Stage V: Immature
- ☆ Stage VI: Spawning
- ☆ Stage VII: Spent

Month and the frequency of spawning were determined on the basis of percentage occurrence of fish of different maturity stages during different months of the year. Observations of mature, spent stage and occurrence of Ist immature stage are decisive factors of the month and the frequency of spawning. The value of GSI and Dl were also used for the determination of the month and frequency of spawning. Month with maximum values of these indices showed month of highest maturity, whereas fall is the indicator of first attempt of spawning.

Percentage occurrence of mature fishes during pre-spawning and spawning seasons were taken for the study of determination of size at first maturity when arrange size wise (Thakre and Bapat, 1981; Dobriyal and Singh, 1987).

Observations

The ovary of *Botia dayi* Hora was elongated and sac like structure. It had varied shape, size and colour during the different months of the year due to different maturity condition. A fully mature ovary was observed as 2.5-3.0 cm long and 4.5-5.0 gm in weight. When mature it was light yellow in colour and light pink during winter to spring season. The weight of ovary gradually increased up to the spawning period. Testes situated in the posterior region of the abdominal cavity were observed as elongated and thin structure with light pink colour which becomes dull and creamy on the advancement of maturity. The size and weight of the testes also increased along with the advancement of maturity (May to August). From October to January the testes were observed as thread like structures.

Gonado-Somatic Index

The gonado-somatic index (an important macroscopic observation) was calculated for each fish in each month for male and female fish separately. The monthly average values are presented in Figure 66.1. The maximum values were observed in the month of August for both sexes (2.59±0.38 and

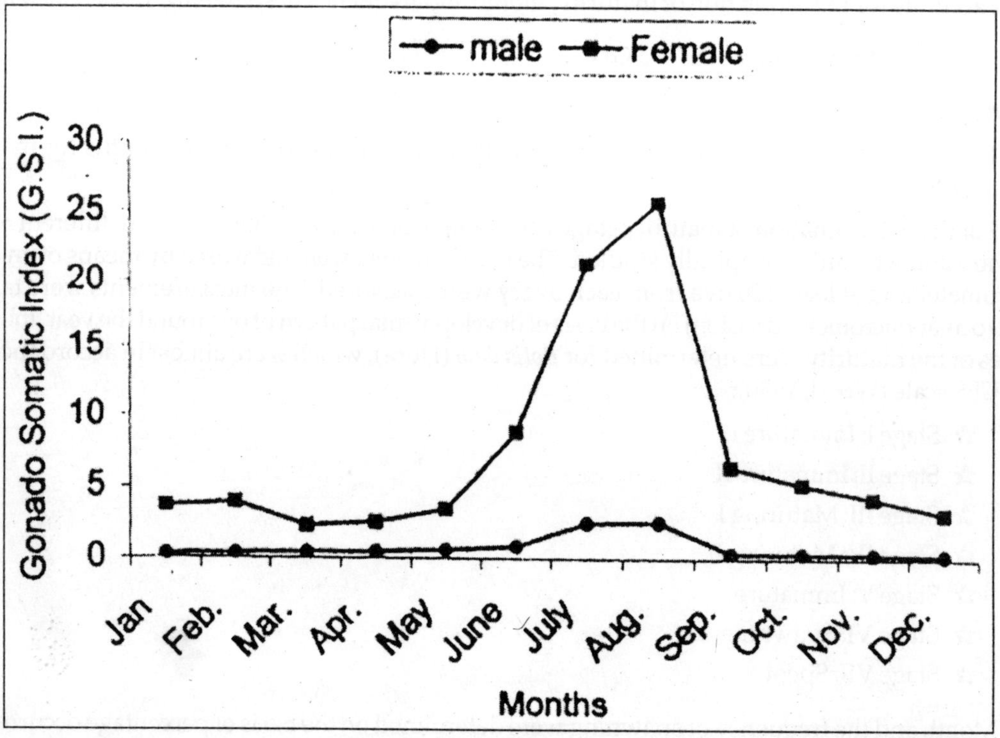

Figure 66.1: Monthly Variation in Gonado-Somatic Index (GSI) in *Botia dayi* Hora during 2002–2003

25.81.35 ± 8.91 for male and female fish respectively). The minimum values of GSI were recorded in the month of December (0.26 ±0.03 and 3.2 ± 1.80) for male and female fishes respectively.

Dobriyal Index

In the present investigation Dobriyal Index (DI) was used and found very effective to determine the start of spawning in *Botia dayi* Hora. The calculated values of DI were presented in Figure 66.2. The maximum DI was noted during the month of July for both the male and female fishes (0.67 and 1.51 respectively). A fall in the values of DI was observed in the month of August, which indicated that spawning has started in the month of August. DI was minimum (0.31 and 0.32) in the month of December for male and female fish respectively.

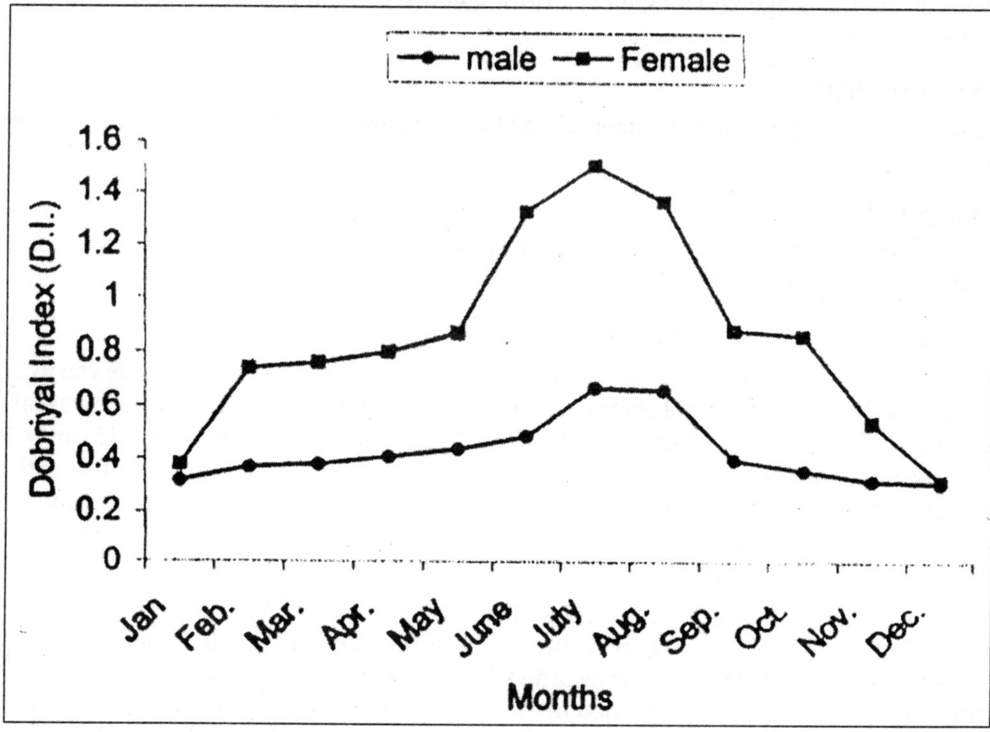

Figure 66.2: Monthly Variation in Dobriyal Index (DI) in *Botia dayi* Hora during 2002–2003

Maturity Stages

Maturity stages were determined on the basis of microscopic study of the ovaries of *Botia dayi* Hora in which ova diameters of each fish round the year were recorded. Maturity stages were determined based on the development of ova to maturity. Following stages were determined:

Stage I (Immature I)

This is the first stage of maturity and it is known as Immature I. In this stage the ova diameter *Botia dayi* Hora varied from 5 to 15 omd with peak value at 10 omd.

Stage II (Immature II):

In this stage the peak values of ova diameter was noted as 15 omd, ranging from 5 to 25 omd, at this stage it appeared in to distinct batches of maturity.

Stage III (Maturing I)

In this stage *Botia dayi* Hora represented a range of 10 to 40 omd ova diameter with a peak of 30 omd from this stage ova could be seen clearly without removing the membrane from the ovary.

Stage IV (Maturing II)

The maximum ova-diameter in this stage was 60 while the minimum was 15 omd. Ova diameter frequency polygon whom the peak values in this stage are 45 omd.

Stage V (Mature)

Ova diameter varied from 25 to 80 omd with a peak value of 65 omd. A rapid growth of ova were observed during this stage and single peak during entire growth period was reported during the course of study.

Stage VI (Spawning)

During spawning stage ova diameter ranged from 20 to 60 omd and the peak value was observed as 50 omd.

Stage VII (Spent)

The spent ovary was flaccid and contain the unspawned ova with a ranging diameter of 10-30 omd and peak values of 10 omd.

Frequency and Season of Spawning

For the study of frequency and season of spawning the percentage of fishes observed in different stages of maturity during different stages of maturity and different month of the year was taken into considerations (Table 66.1). The fish belonging to the 1st maturity stage were observed from October to January, the maximum being 100 per cent in October. The minimum percentage of fish at this stage was recorded in the month of January (20 per cent). The fishes placed in the second maturity stage were recorded from the month of November to March. Their maximum and minimum percentages were observed as 80 per cent in January and 25 per cent in March respectively.

Fishes belonging to stage third were observed from February to May, highest being 75 per cent during March-April. The fourth stage of maturity was recorded during April (25 per cent), May(40 per cent) and June (20 per cent). Stage V was recorded during May-June (20 per cent). The fish of sixth stage maturity (spawning stage) were obtained in the months of June, July and August. Their maximum number was recorded in the month of July (100 per cent) and the minimum in June (60 per cent). The spent fish (stage VII)were recorded in August and September month with the percentage of 33 and 100 respectively. This observation indicates protracted spawning with single frequency.

Gonado somatic Index (G.S.I.) and Dobriyal Index (D.I.) were also used as indicators of spawning and maturation biology of the fish. In present study the value of GSI was noted maximum in the month of August for both male and female fishes of *Botia dayi* Hora. The value was noted as 2.59±0.18 and 25.81 ± 8.91 for male and female fish respectively.

Sudden and heavy fall in the value of GSI was observed in the month of September indicated the spawning attempt in fish. Dobriyal Index (D.I.) was much closer to indicate spawning as also observed by other methods.

It was at its peak for male (0.67) in the month of July with a slight fall in August and sharp fall in September. The peak value of DI for female fish was noted as 1.51 in the month of July with a fall (1.37) in August. During the present investigation it has been noted that Dobriyal Index was much accurate for the study of spawning season and frequency of spawning in fish *Botia dayi* Hora.

Table 66.1: Percentage Occurrence of Different Maturity Stages of *Botia dayi* Hora (Female) During 2002–2003 from River Khoh

Sl.No.	*Months*	*Maturity Stage in per cent*							*No. of Fish Examined*
		Stage I	*Stage II*	*Stage III*	*Stage IV*	*Stage V*	*Stage VI*	*Stage VII*	
1.	Jan.	20	80	–	–	–	–	–	05
2.	Feb.	–	60	40	–	–	–	–	05
3.	Mar.	–	25	75	–	–	–	–	04
4.	Apr	–	–	75	25	–	–	–	04
5.	May.	–	–	40	40	20	–	–	05
6.	Jun.	–	–	–	20	20	60	–	05
7.	Jul.	–	–	–	–	–	100	–	04
8.	Aug.	–	–	–	–	–	67	33	06
9.	Sep.	–	–	–	–	–	–	100	06
10.	Oct.	100	–	–	–	–	–	–	04
11.	Nov.	67	33	–	–	–	–	–	03
12.	Dec.	33	67	–	–	–	–	–	03

Size at First Maturity

The percentage occurrence of mature fish in various size groups during prespawning and spawning period are presented in Figure 66.3 and 66.4 for male and female fish respectively. 40 per cent male mature fish were observed in the size group 9.5-10.4 cm while 62.5 per cent in size group 10.5-11.4 cm. This percentage increased up to 87.5 per cent in the size group 11.5-12.4 cm and to 100 per cent in size group of 12.5-13.4 cm. In female fish 44.5 per cent fishes were observed mature in size group 10.1cm, 87.5 per cent in size group 11.1-12.0 cm and 100 per cent in the size group 12.1-13.0 cm. Fifty per cent level of maturity calculated after interpolation was considered as size at first maturity and was found to be 10.4 cm for male and 10.6 cm for the female.

Ecology of Spawning Niche

During present study, the observed breeding months for the fish *Botia dayi* Hora were July and August. In these months the rising water level, low values of pH and high turbidity influenced the spawning. The habitat preference of fish substratum heterogeneity in the spawning site indicated that the fish spawns in the side pool under the boulders, stones and pebbles well covered or protected by the macrophytic vegetation. It was noticed that usually all the monsoon breeders in Khoh stream prefer the similar spawning ground. Physico-chemistry of these spot indicated the qualitative habitat as velocity of water current 0.90-1.0 meter/second, Dissolved oxygen 7.8-9.5 ppm, water temperature 26-28°C and pH 7.1-7.5.

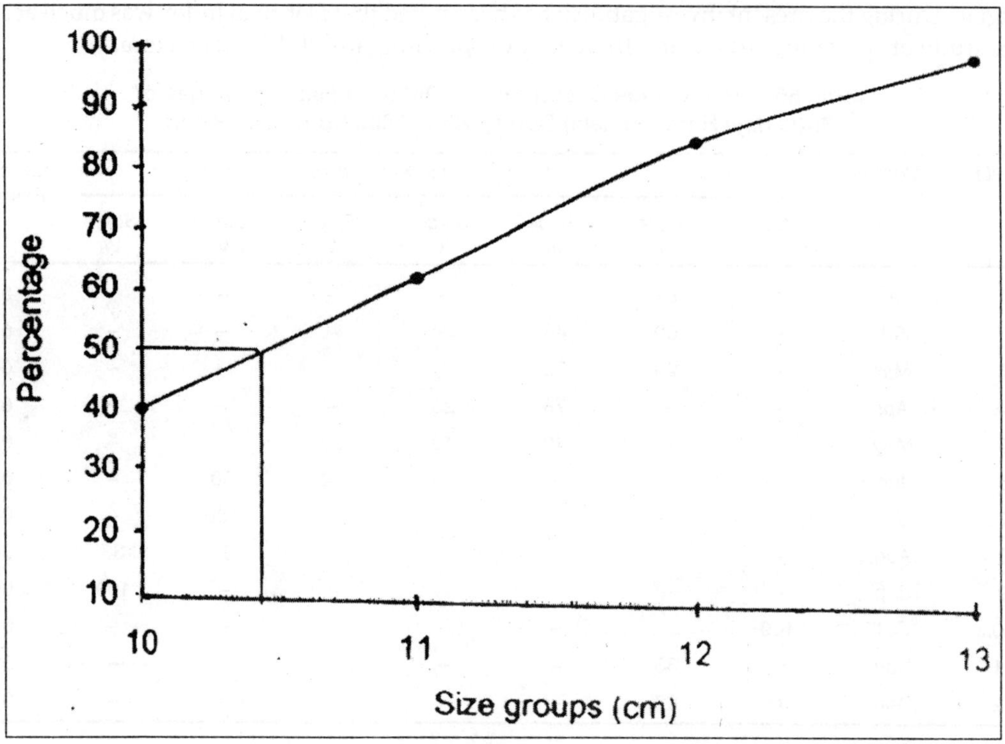

Figure 66.3: Size at First Maturity in Male *Botia dayi*, Hora during 2002–2003

Discussion

In life cyle of fish the spawning is an important act. There is a great diversity in spawning behaviour of fishes and the duration of spawning is variable from species to species. In each species the time of breeding is so naturally managed that the chances of survival for the egg and fry are maximal. The majority of freshwater fishes in India breed around monsoon months of heavy rainfall. The time of spawning has a protective significance because the developing eggs grow when the environmental conditions are most congenial and predators are least active.

According to Hickling and Rutenberg (1936) the spawning behaviour of fishes can be studied by the size of egg present in the ovary. The size of ova diameters were used as an important tool for identification of spawning stages of *Botia dayi* Hora. Seven different maturity stages were identified based on the pattern of ova diameters were used as an important tool for identification of spawning stages of *Botia dayi* Hora. Seven different maturity stages were identified based on the pattern of ova development.

Many workers have studied the ova development in the hill stream fishes of Garhwal Himalaya (Singh *et al.*, 1985; Dobriyal and Singh, 1989; Negi and Dobriyal, 1997; Uniyal *et al.*, 2004).

Generally the fish biologists follow ICES scale (Wood, 1930) and report seven maturity stages as done in the present case also. However there are reports who described less than seven stages of maturity depending upon slow growth rate of ova at a particular period or season (Malhotra, 1967 for *Schizothorax niger* and Bahuguna, 2004 for *Tor putitora*).

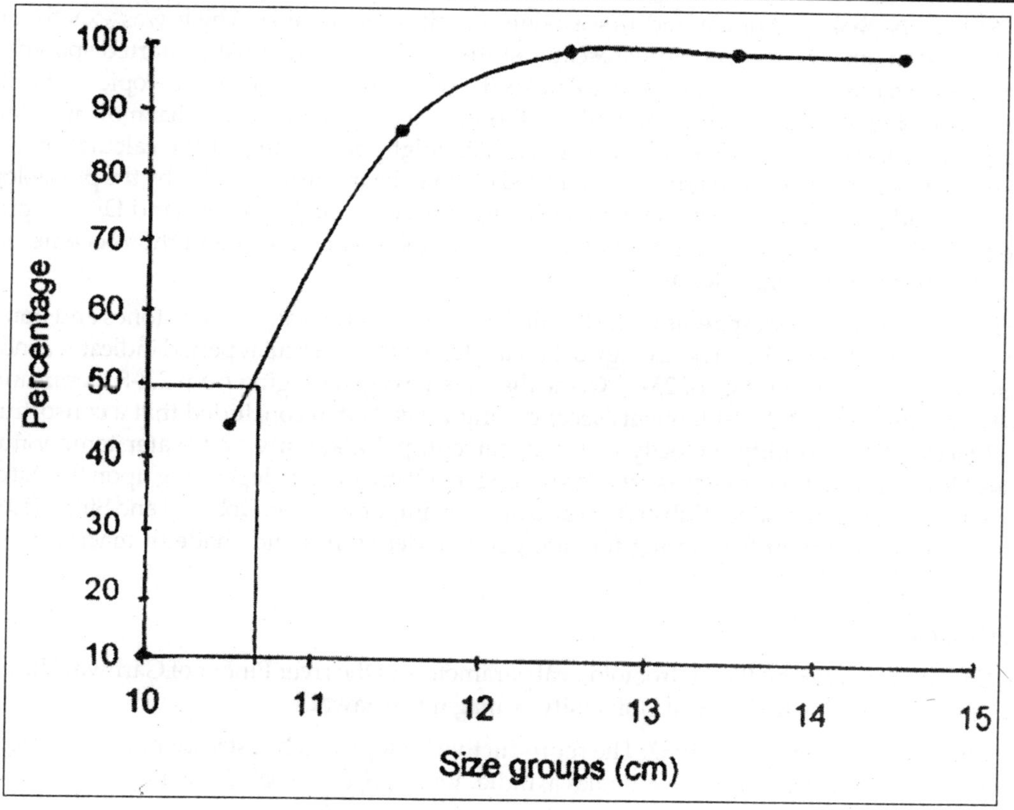

Figure 66.4: Size at First Maturity in Female *Botia dayi* Hora during 2002–2003

The present study indicates that the fish *Botia dayi* Hora spawn for a limited period (August-September) with a single batch of development eggs showing a single spawning frequency. On the other hand some species have a prolonged breeding period as *Tor chelynoides* (Uniyal *et al.*, 2004) or they spawn twice a year as *S. plagiostomus* of the river Pinder (Bisht, 1985) and *B. bendelisis* of river Nayar (Dobriyal and Singh, 1987).

On the basis of collected fish during pre-spawning and spawning time, a regression line was drawn for the percentage of occurrence of mature fish with different size groups. It was then interpolated with 50 per cent level in maturing for the determination of size at first maturity. It was observed that the male fish (*B.dayi*) matures upto a size of about 10.4 cm and the female 10.6 cm. Some other studies conducted on hill stream fishes indicated that *Glyptothorax pectinopterus* mature upto a size of 125 mm (male) and 121 mm (female) as reported by Dobriyal and Singh (1989). Dobriyal and Singh (1987) reported size at first maturity for *Barillius bendelisis* as 83 mm (female) and 104 mm (male). Uniyal *et al.* (2004) reported 96 mm and 148 mm for male and female *T. chelynoides*.

The tabulation of percentage occurrence of fish of various maturity stages in different months is the best indicator of frequency and season of spawning. Dobriyal and Singh (1987, 1989) and Negi and Dobriyal (1997) studied the tabulation of percentage occurence of fish of different maturity stages for determination of frequency and season of spawning. The macroscopic indicator like Gonado-somatic index (GSI) and Dobriyal index (DI) were studied in present investigation. It was observed that GSI had its peak in August and first spawning fall in September which was exactly not in accordance with other observation (microscopic studies) that fish had already started spawning in the late August month. However, the Dobriyal index (DI) was found closer to microscopic confirmation which showed highest value in July and first fall, though very slight, in August that indicates starting of spawning somewhere at last week of August. This might be due to fact the calculation of GSI involves body weight also which is not only affected by gonadal maturity but also by the physiological body stress and ecologically unstable habitat during monsoon. On the other hand DI is improved version of GSI in which body weight is not involved for calculating index and thus becomes much susceptible to any sort of egg release.

The fish *Botia dayi* Hora spawns in the flooded river in the side water under stones and boulders during August and September. The ecology of habitat during that particular period indicates condition of water temperature in a range of 25-28°C, the dissolved oxygen ranging from 7.8-8.0 ppm and the velocity of water current 0.90-1.0 meter/second. Singh *et al.* (1985) concluded that a consortium of several factors like turbidity, velocity of water current, pH alkalinity and water temperature is responsible for stimulation as well as act of spawning in hillstream fish depending upon the nature of their habitat and genetic instinct. Substratum also plays an important role. Dobriyal and Singh (1987,89, 93) reported the nature of fish opting for side pools under stones and shade of macrophytes for spawning.

References

Bisht, K.L., 1985. Studies on the hydrobiological parameters of the river Pinder of Garhwal Himalaya. *D. Phil. Thesis,* H.N.B. Garhwal University, Srinagar Garhawal.

Dobriyal, A.K. and Singh, H.R., 1987. The reproductive biology of a hill stream minor carp *Barillius bendelisis* (Ham) from Garhwal Himalaya, India. *Vest cs. Spolec. Zool.,* 51: 1–10.

Dobriyal, A.K. and Singh, H.R., 1989a. Ecology of rhithrofauna in the torential waters of Garhwal Himalaya, India: Fecundity and sex ratio of *Glyptothorax pectinopterus. Vest. Cs. Spole. Zool.,* 53: 17–25.

Dobriyal, A.K. and Singh, H.R., 1993. Reproductive biology of a hill stream catifish *Glyptothorax madraspatanum* (Day) from Garhwal, Central Himalaya, India. *Aquaculture and Fisheries Management,* U.K., 24: 699–706.

Dobriyal, A.K., Rautela, K.K. and Rautela, A.S., 1999 : Invention of a new index for the determination of sexual maturity in fishes. *Uttar Pradesh J. Zool.,* 19(1).

Hickling, C.F. and Rutnberb, E., 1936. The ovary as an indicator of spawning period of fishes. *J. Mar. Biol. Ass.,* U.K., 21: 311–317.

Legler, K.F., 1977. *Freshwater Fishery Biology.* WmC Brown Company Dubuque, Iowa, USA.

Nautiyal, P., 1984. Natural history of the Garhwal Himalaya Mahseer (*Tor putitora*). II. Breeding biology. In: *Proc. Indian Acad. Sci. (Anim. Sci.),* 93: 97–106.

Negi, K.S. and Dobriyal, A.K. 1997. Sexual maturity and spawning ecology of a hillstream carp *Crossocheilus latius latius* (Ham) from the glacier fed river Mandakini, Garhwal, Central Himalaya. *J. Inland Fish. Soc., India,* 29(2): 26–33.

Pathani, S.S., 1979. Studies on the ecology and biology of Kumaon Mahsheer *Tor tor* (Ham) and *Tor putitora* (Ham). *Ph.D. Thesis,* Kumaon University, Nainital.

Singh, H.R., Dobriyal, A.K. and Nautiyal, B.P., 1985. Spawning patterns and environmental regulation of spawning in hillstream fishes. In: *The Endocrine System and the Environment*, (Eds.) B.K. Follett *et al.* Japan Sci. Soc. Press, Tokyo, Springer-Verlag, Berlin, pp. 1–11.

Sobhana, B. and Nair, Balakrishnan, 1974. Observation on the maturation and spawning of *Puntius sarana* subanastus (Val.). *Indian J. Fish.*, 21: 357–368.

Thakre, V.Y. and Bapat, S.S. 1981. Maturation and spawning of *Rasbora daniconius* (Ham-Buch.). *J. Bombay Nat. Hist. Soc.*, 78(1): 38–45.

Wood, H., 1930. Scottish herring shoal: Pre-spawning and spawning movements. *Scotland Fish Bd S. Invest.*, 1 :1–71.

Water and Wastewater Management in 2 Vols (2011) *Pages* **462–466**
Editors: **D.R. Khanna, R. Bhutiani & Gagan Matta**
Published by: **DAYA PUBLISHING HOUSE, NEW DELHI**

Chapter 67

Water Quality of River Ganga in Respect of Physico-chemical Characteristics at Shyampur, District Haridwar

Unnati Vishnoi[1], Jagjeet Kaur[2] and S.K. Pathak[3]
[1]Department of Chemistry, K.L.D.A.V. (P.G.) College, Roorkee
[2]M.S. (P.G.) College, Saharanpur
[3]Department of Zoology, Govt. P.G. College, Mhow, M.P.

ABSTRACT

The water quality status of River Ganga at Shyampur was studied with respect to physico-chemical characteristics. Parameters studied were pit, total solids, total dissolved solid, TSS, conductivity, alkalinity, hardness, DO, BOD and COD. The nature of water quality of Ganga with respect to pit is alkaline.

Keywords: Water, Quality, River, DO, BOD, COD.

Introduction

Shyampur (District Haridwar), is situated at the east bank of river Ganga where number of people and cattle take bath daily. Shyampur is connected by the Haridwar-Najibabad road. Ganga river pass through the Haridwar, Kangri and Gendi Khata villages. The wastewater of nearby village ultimately mixes into the river. The washing and bathing activity by village may also affect the physico-chemical parameters of the river. Number of workers have carried out their investigations on water quality of river Ganga with respect to different physico-chemical characteristics from its origin to merging point

Table 67.1: Variation in Physico-chemical Characteristics of River Ganga at Sampling Station A of Shyampur (Haridwar), during January to December 2005

Parameters	*Jan*	*Feb*	*Mar*	*Apr*	*May*	*Jun*	*Jul*	*Aug*	*Sep*	*Oct*	*Nov*	*Dec*
Water Temp. (°C)	12.2±0.2	14.0±0.63	19.4±0.4	19.6±0.3	21.3±0.3	22.0±0.4	19.0±0.2	18.7±0.1	18.6±0.12	18.3±0.12	20.0±0.12	14.8±0.07
TS (mg/l)	125±2.0	120.9±1.0	130.0±2.8	160.0±2.1	179.1±7.6	213.0±10	224.2±18	232±.020	222±0.20	223.9±0.12	200.0±8	187.0±4.6
TDS (mg/l)	94±1.4	90±0.9	92±2.0	12±3.4	129±2.9	164.4.5	172±3.0	180±3.4	179±3.0	182±3.0	160±3.4	148±1.5
TSS (mg/l)	31.0±0.6	30.9±1.0	38.0±1.0	40.0±0.8	50.0±0.9	52.0±1.3	52.2±1.2	52±1.2	43.0±1.4	41.9±1.3	40.0±1.6	39.0±1.0
Conductivity (mmhos/cm)	0.11±0.0	0.16±0.0	0.21±0.0	0.20±0.02	0.28±0.02	0.30±0.03	0.60±0.02	0.48±0.4	0.50±0.00	0.40±0.01	0.3±0.01	0.22±0.0
pH (mg/l)	7.3±0.07	7.6±0.11	7.2±0.07	8.0±0.05	8.0±0.04	8.1±0.2	8.8±0.3	8.2±0.2	8.6±0.10	8.0±0.01	7.0±0.12	7.6±0.0
DO (mg/l)	12.5±0.2	11.8±0.02	11.0±0.1	11.0±0.7	11.2±0.5	10.0±0.4	9.5±0.7	10.5±0.6	10.5±0.1	11.0±0.04	11±0.01	12.0±0.0
Free CO_2 (mg/l)	1.8±0.3	2.0±0.09	2.0±0.10	2.5±0.7	2.2±0.8	2.6±1.2	3.8±1.0	3.2±1.4	2.6±1.2	2.5±0.02	2.0±0.7	1.8±0.3
COD (mg/l)	11.0±0.02	12.4±0.11	12.5±0.11	13.0±0.8	17.0±0.9	21.0±1.2	25.0±1.0	30.0±1.5	24.0±1.2	12.2±0.02	19.0±0.8	14.5±0.3
BOD (mg/l)	0.60±0.01	0.90±0.01	1.4±0.01	0.8±0.1	1.0±0.1	1.5±0.2	1.5±0.2	1.2±0.2	1.2±0.2	1.0±0.0	0.8±0.04	0.50±0.3
Alkalinity (mg/l)	55±1.1	59±0.6	60±1.0	64±0.3	72±1.0	74±1.2	75±0.7	80±1.0	72±0.8	98±3.0	89±0.07	75±1.1
Total hardness (mg/l)	79.0±1.2	66.0±1.0	72.0±1.3	72.0±0.4	76.0±1.3	89.0±1.3	11.6±1.6	76.0±0.6	81.0±1.0	82.0±2.0	81.0±1.2	74.0±0.9
Chloride (mg/l)	6.9±0.1	7.2±0.02	7.7±0.02	11.9±0.5	18.0±0.7	20.0±0.0	27.4±0.3	24.0±0.3	17.6±0.2	9.0±0.1	7.0±0.03	7.0±0.02

Table 67.2: Variation in Physico-chemical Characteristics of River Ganga at Sampling Station B of Shyampur (Haridwar), during January to December 2005

Parameters	*Jan*	*Feb*	*Mar*	*Apr*	*May*	*Jun*	*Jul*	*Aug*	*Sep*	*Oct*	*Nov*	*Dec*
Water Temp. (°C)	13.8±0.2	13.0±0.4	19.3±0.2	20.0±0.1	21.0±0.09	21.5±0.25	21.0±0.10	19.6±0.1	18.7±0.1	20.4±0.3	19.1±0.1	15.0±0.10
TS (mg/l)	133±2.0	146±1.7	148±2.0	175±10.2	188±10.2	231±17.5	261±14	266±20.2	255±2.5	233±3.12	225±3.8	198±3.0
TDS (mg/l)	101±3.2	110±2.7	98±1.5	138±2	130±3.2	179±3.2	189±4.1	185±7.2	181±3.9	188±2.5	161±3.8	137±4.9
TSS (mg/l)	32±1.5	36±1.6	50±1.3	37±1.0	58±1.3	52±1.3	72±1.3	81±1.5	74±2.0	45±2.1	64±1.3	61±1.8
Conductivity (mmhos/cm)	0.15±0.1	0.20±0.1	0.21±0.1	0.15±0.0	0.16±0.0	0.38±0.0	0.68±0.02	0.61±0.02	0.60±0.01	0.58±0.01	0.51±0.0	0.324±0.02
pH (mg/l)	7.0±0.5	7.0±0.3	7.1±0.5	7.1±0.10	7.2±0.025	7.3±0.2	7.9±0.03	7.8±0.25	7.6±0.2	7.7±0.45	8.6±0.2	7.6±0.15
DO (mg/l)	11±0.15	10.1±0.15	8.5±0.2	9.8±0.01	9.1±0.02	8.8±0.1	9.8±0.15	10.6±0.11	10.2±0.11	9.9±0.1	10.2±0.15	11.0±0.20
Free CO_2 (mg/l)	1.5±0.2	3.0±0.2	2.3±0.12	3.0±0.1	2.5±1.2	3.9±0.02	4.1±0.01	2.9±0.02	2.8±0.03	3.4±0.1	2.9±0.7	2.7±0.09
COD (mg/l)	9.5±0.4	11.0±0.03	11.5±0.04	12.0±0.15	15.5±0.14	20.2±0.15	23.5±0.15	28.2±0.40	23.4±0.24	10.2±0.20	17.5±0.07	13.5±0.05
BOD (mg/l)	2.8±0.02	0.8±0.03	1.9±0.02	2.2±0.05	2.9±0.02	3±0.03	3.1±0.02	1.6±0.01	1.8±0.0	1.5±0.1	2.1±0.01	2.9±0.20
Alkalinity (mg/l)	60±1.0	64±1.5	61±1.4	80±1.1	98±1.4	101±1.0	107±1.1	79±0.75	81±0.8	99±0.0	95±0.0	80±2.00
Total hardness (mg/l)	62±1.2	65±1.5	79±1.2	81±1.3	93±1.0	98±1.1	99±1.6	80±1.7	95±1.0	90±1.5	98±1.6	89±1.6
Chloride (mg/l)	7.4±0.11	7.9±0.13	9.5±0.15	13.0±0.02	22.0±0.2	34.0±0.15	59.1±0.16	25.5±0.14	21.0±0.18	17.0±0.17	15.0±0.12	13.0±0.10

(Singh *et al.*, 1988, 91; Khanna *et al.*, 2003; Gautam and Sati, 1987). But no study has been conducted to assess the water quality of river Ganga at Shyampur. Therefore it was proposed to conduct a study on water quality of river Ganga at Shyampur. The present chapter includes one year study (January–December, 2005).

Materials and Method

Water samples of river Ganga were collected in neat and clean two litre white plastic Jericanes for physico-chemical parameters. Samples for Dissolved oxygen (DO) were collected in a neat and clean 300 ml. Borosil glass stoppered DO bottles, DO was fixed by using 1 ml. of each reagent $MnSO_4$ and alkaline azide on the spot. Grab sample collection methods was adopted through out the study. Sample preservation and analytical methods were adopted as per APHA, AWWA, WPCF (1992), Khanna and Bhutiani (2004).

Two sampling stations were selected as Upstream Shyampur denoted as sampling station 'A' and Downstream Shyampur denoted as sampling station 'B'. Sampling station A is approximate 500 meter in Upstream of Shyampur and sampling station 'B' is situated 1 Km in Downstream of Shyampur.

Results and Discussion

Results of samples analyses for physico-chemical parameters are given in Tables 67.1–67.3. The values of studied parameters pH, TS, TSS, TDS, alkalinity, total hardness, chloride, free CO_2, conductivity, water temperature varied between 7.0–8.8, 120.9–232 mg/l, 30.9–52.2 mg/l,55–98 mg/l, 11.6–89 mg/l, 6.9–116 mg/l, 6.9–27.4 mg/l, 1.11–3.8 mg/l, 0.11–0.60 mmhos/cm, 12.2–22 mg/l; at sampling station A where as at sampling station B the values of these parameters ranged from 7.0–8.6,

Table 67.3: Average Value of Physico-chemical Parameters at Sampling Station A and B at Shyampur during Period January to December 2005

Parameters	*Jan–Dec 2005*			*Jan–Dec 2005*		
	Sampling A			*Sampling B*		
	Min.	*Max.*	*Average*	*Min.*	*Max.*	*Average*
Water Temp. (°C)	12.2	21.3	18.1	13.0	21.5	18.5
TS (mg/l)	187.0	232.0	184.7	133.0	226.0	204.9
TDS (mg/l)	90.0	182.0	142.2	98.0	189.0	142.3
TSS (mg/l)	31.0	52.2	42.5	32.0	81.0	55.1
Conductivity (mmhos/cm)	0.11	0.6	0.31	0.15	0.68	0.38
pH (mg/l)	7.0	8.1	7.8	7.0	8.6	7.5
DO (mg/l)	9.5	12.5	11.0	8.5	11.0	9.9
Free CO_2 (mg/l)	1.8	3.8	2.4	1.5	4.1	2.9
COD (mg/l)	11.0	30.0	17.6	9.5	28.2	16.3
BOD (mg/l)	0.5	1.5	1.03	0.8	3.1	2.2
Alkalinity (mg/l)	55.0	98.0	72.7	60.0	107.0	83.7
Total Hardness (mg/l)	66.0	116.0	80.3	62.0	99.0	85.7
Chloride (mg/l)	6.9	27.4	13.6	7.4	59.1	20.3

133–266 mg/l, 98–189 mg/l, 32–81 mg/l, 60–107 mg/l, 6299 mg/l, 7.4–59.1 mg/l, 1.5–4.1 mg/l, 0.1–5–0.68 mg/l, 13–21.5 mg/l respectively; and of DO, COD, BOD, ranged between 9.5–12.5 mg/l, 11.0–30 mg/l, 0.5–1.5 mg/l at sampling station A where as at sampling station B the values of these parameters ranged from 8.5–11.0 mg/l, 9.5–28.2 mg/l, 0.8–3.1 mg/l respectively. The average value of pH, TS, TDS, TSS, conductivity DO, COD, alkalinity, total hardness and chloride were obtained within the tolerance limit of drinking purpose where as average values of BOD was found 1.03 mg/l at sampling station-A and 2.2 mg/l at sampling station-B (Table 67.3).

Maximum values of the studied parameters were observed during monsoon season except DO and minimum values of these parameters were obtained during winter, lower values of DO were observed during rainy period at both sampling station which may be due to dilution in rainy season and super saturation of Oxygen at lower temperature and less human activity like bathing/washing *etc.* in winter. After winter period as temperature of water rises free CO_2, COD, BOD was increased and DO decreased of river Ganga, Khanna *et al.* (2003) has also found the similar trend. Water quality of river Ganga with respect to pH is alkaline similar trend of water quality of river Ganga was also found by Singh (1988) and Khanna (1993).

References

APHA, AWWA, WPCF, 1992. *Standard Methods for the Examination of Water and Wastewater*, 18th edn. American Public Health Association, Washington DC.

Khanna, D.R., 2003. *Ecology and Pollution of River Ganga.* Ashish Publishing House, Delhi, pp 1–241.

Khanna, D.R., Singh, Shakun, Gautam, Ashutosh and Singh, J.P., 2003. Assessment of water quality of River Ganga in district Bulandshahar (UP) India. *J. Nat. Con.*, 15(1): 167–175.

Khanna, D.R. and Bhutiani, R., 2004. *Water Analysis at a Glance.* Pub. ASEA, pp. 1–116.

Khanna, D.R. and Chugh, Tarun, 2004. *Microbial Ecology.* Discovery Publishing House, Delhi, pp. 1–277.

Singh, J.P., Yadav, P.K. and Singh, L., 1988. Pollution status of Sangam and its adjoining river before the Kumbh Mela at Allahabad. *Indian J. Environment Protection*, 8(11): 839–842.

Singh, J.P., Yadav, P.K., Singh, Shakun and Prasad, S.C., 1991. BOD Contamination in Kali river at Sadhu Asharam in Aligarh. *I.J.E.P.*, 11(5): 325–326.

Water and Wastewater Management in 2 Vols (2011) *Pages* 467–471
Editors: D.R. Khanna, R. Bhutiani & Gagan Matta
Published by: DAYA PUBLISHING HOUSE, NEW DELHI

Chapter 68

Studies on the Ecology of River Beas with Reference to Benthic Fauna

Ashok Kumar[1] and V.K. Saxena[2]
[1]*AD Hydro Power Ltd. V. and P.O. Prini, Manali, District Kullu, H.P.*
[2]*Fisheries Research Laboratory, Department of Zoology, Narain College, Shikohabad, U.P.*

ABSTRACT

This chapter incorporates the findings on the seasonal abundance and fluctuations of benthic fauna in relation to physico-chemical parameters of water in the selective stretches of river Beas. Considerable variations of benthic fauna in quality and quantity were observed. The population of benthic fauna was high in the month of July (418.0 units/m^2) while minimum in December (38.0 units/m^2).

Keywords: *Ecology, River Beas, Benthic, Physico-chemical.*

Introduction

The river Beas is one of the major rivers of Himachal Pradesh. It originates from southern slope of Rohtang Pass at an elevation of 4062 mts. The river receives number of tributaries both on right and left banks during its downward drift of over 470 kms. Its principal tributaries are Solang, Manalsu, Sujjain, Fojal and Sarvari on the right bank and Alain, Duhagan, Chhaki, Parbati, Tirthan and Sainj on the left bank. The disposition of the river in Kullu valley is shown in Figure 68.1.

The water of Beas and its tributaries remains shallow, rapid cool and clear except during the rainy season. The bed comprises of mainly the boulders, stones and rubble. The vegetation along the banks consists mainly of the alnus, willow, rubenia and conifers. The ecology of the river is constantly degrading. Once teeming with large number of trout the catches are declining each year. The major

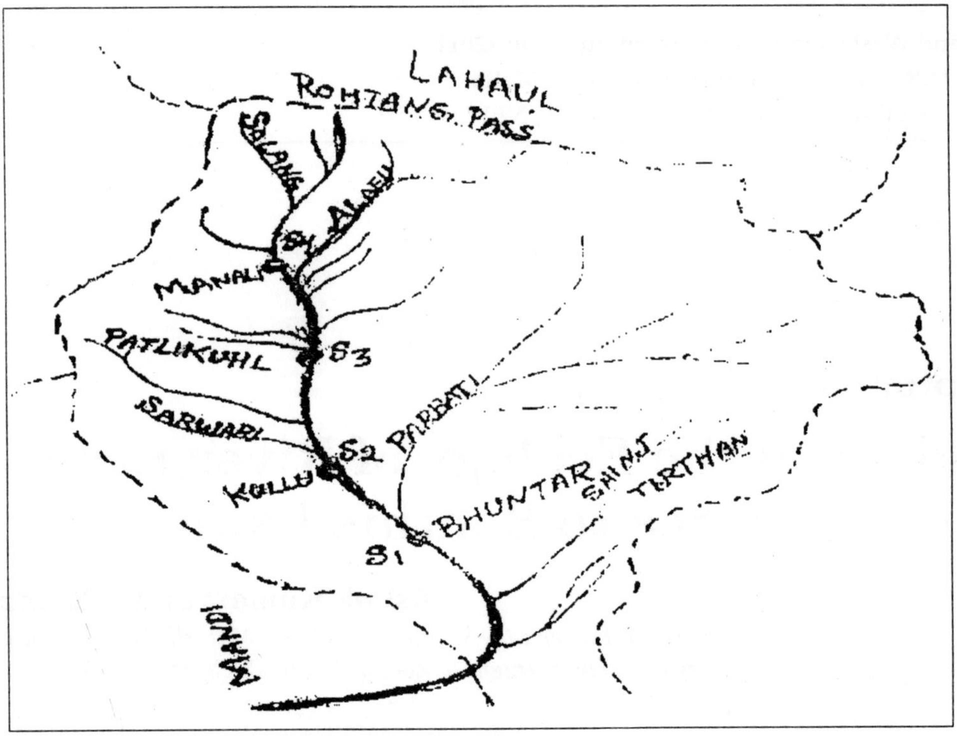

Figure 68.1: Disposition of the Experimental Sites of River Beas in Kullu District

reasons are attributed to fall in catches are ecological erosion, heavy silt deposition and shrinkage of feeding and breeding ground in the river. Observations were therefore made in the river Beas at Kullu District to study the abundance, seasonal fluctuations of benthic fauna.

Materials and Method

The physico-chemical parameters of water of river Beas like water temperature, pH, transparency, dissolved oxygen, total alkalinity were analysed fortnightly following APHA (1998) manual. Due to shallow depth, stony bottom and fast current, the transparency was measured by bright pin head method (Saba *et al.*, 1971). Hydrogen ion concentration of water was determined by pH digital meter. For dissolved oxygen 'Winkler's Method' (Welch, 1948) was adopted and total alkalinity was determined by 'titration' method. The aquatic insects and other bottom fauna were collected monthly from pre-selected sites of the river, by enclosing one square meter of river bed with square-meshed cloth. The bottom stones, gravel and sand were upturned to dislodge the aquatic life. This resulted in collection of all the benthic life in the square-meshed cloth. The organisms were hand-picked and preserved in 5 per cent formaline and were analysed quantitatively and qualitatively (Jhingran *et al.*, 1988). The species composition of benthic life was done group-wise, genera-wise and where ever possible, species wise.

Results and Discussion

The physico-chemical parameters studied in the river Beas includes water temperature, transparency, pH, dissolved oxygen and total alkalinity. Seasonal variations in physico-chemical parameters of water are shown in Table 68.1. Water temperature (average) ranged from 5.2°C to 17.7°C. The pH of the river water was always found to be in the alkaline range (7.0–7.7). River water was clear and transparent during winter and turbidity was found during monsoon (6.1 cm) which was due to fast current and greater inflow of muddy rainwater. Total alkalinity was high and the value was maximum in the month of February (85.0 ppm) and minimum was (53.7 ppm) in the month of August. Dissolved oxygen was quite high and showed wide fluctuation. Its value was maximum in December (12.8 ppm) and minimum during august (9.2 ppm).

Table 68.1: Variation in Physico-chemical Parameters of River Beas from December, 02–November, 03

Months	*Water Temp. (°C)*	*Transparency (cm)*	*pH*	*DO (ppm)*	*Alkalinity (ppm)*
December	6.2	51.5	7.5	12.8	75.6
January	5.2	47.8	7.6	12.0	70.6
February	7.1	43.3	7.4	11.2	85.0
March	9.2	46.1	7.4	11.2	83.1
April	11.3	51.5	7.3	10.0	80.1
May	10.0	29.3	7.1	10.8	61.2
June	12.8	10.2	7.3	10.2	57.5
July	16.0	6.1	7.5	9.7	59.3
August	17.7	7.7	7.7	9.2	53.7
September	17.1	9.9	7.0	9.9	56.2
October	12.9	29.5	7.0	10.9	67.5
November	8.1	53.2	7.4	11.4	75.6

The analysed benthic fauna samples showed that the bulk of benthic organisms in different beats of river Beas comprised nymph, larvae and adults of different orders of class insects. The fluctuation of benthic fauna in the river during different months of the year is shown in Figure 68.2. Insects contributed 98 per cent in the selective stretches of Beas River. Their population is constituted by nymph of Ephemeroptera and Plecoptera, larvae of Trichoptera, Diptera, larvae and adults of Coleoptera contributing 79 per cent, 2.6 per cent, 4 per cent, 7 per cent and 5.8 per cent respectively. The remaining was constituted by molluscs, annelids and others. The average numbers of organisms encountered were 223 units/m^2. Throughout the sampling year, the main bulk of benthic fauna was formed by Caenis and Baetis. The other genera *viz.*, Iron, Hexagenia, Ephemerella and Triaenode were also dominated in the collections. Perla, Tabanus and Simulium showed their seasonal dominance while species like Antocha, Psychoda, Gomphus, Maurina, Elimidae, Mollusc and Annelids occurred only sporadically and in considerable small number.

The nymphs of Ephemeroptera were abundant in all the stretches ranging from 26–334 units/m^2. May-flies were available throughout the year in the collections. The maximum number 334 units/m^2 of may-flies were encountered during the November while lowest were encountered during December. The important genera in order of their abundance were Caenis, Baetis, Hexageia, Ephemeralla, Iron and Epeorus. The nymphs of Plecoptera (stone-flies) were less abundant in all the four stretches of the river. Stone-flies were available in the river in an average number of 6 units/m^2. The important genera

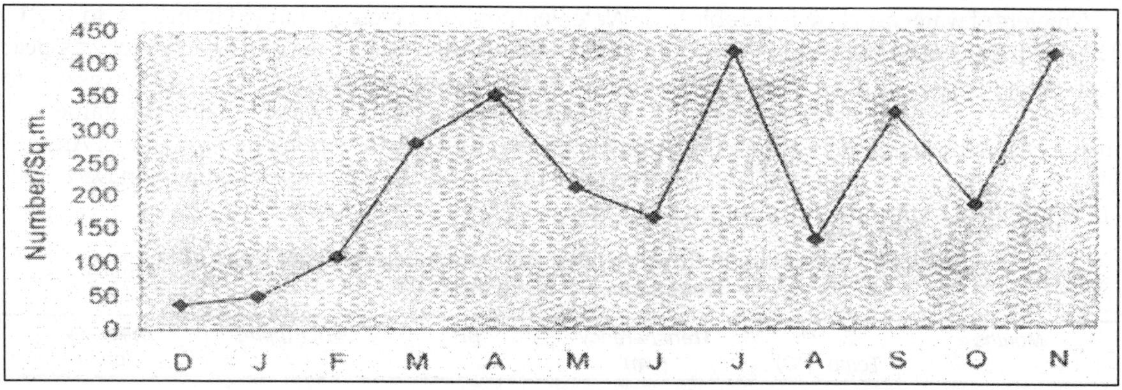

Figure 68.2: Monthly Variation of Benthic Fauna in the River Beas during December, 02 to November 03

in order of their abundance were Perla, Peltoperla, Nemoura. The larvae of Trichoptera (caddis-flies) both the cased and case less were available in the river in an average number of 9 units/m^2. The important genera in order of their abundance were Brachycentrus, Triaenode and Rhyacophila.

The important genera of order Coleoptera in order of their abundance were Elmis, Dytiscus, Hydrogobus and Hydrophilus in the river. The important genera of order Diptera in order of their abundance were Simulium, Tabanus, Altherix, Antocha, Maruina and Psychoda. The order Odonata represented by only one genus *viz.* Gompohus. The order Gnathobdellida was represented by Hirudinaria genus (Leeches) in the collections. The order Basommatophora was represented by mollusc in the collection. The aquatic insects and other benthic organism concentration were high in the river. The maximum number 418 units/m^2 of organisms were encountered during the July while minimum were 38 units/m^2 encountered during December. The genera belonging to order Ephemeroptera were recorded as the main biota of the river. The diversity and abundance of insect larvae which constitute the benthos are directly related to the substratum consisting of small sized boulders, cobbles which provide protection to the larvae (Johal, 2001). The insect population is numerically more in the months of higher alkalinity.

In the present investigation the relative abundance of major insect orders shows correlation with alkalinity and the carbonate alkalinity is more important than the other factor such as temperature, current, speed, volume of flow *etc.* in influencing the density of benthic fauna. The optimum temperature, high dissolved oxygen, slightly alkaline pH and reasonable high value of alkalinity were observed the main reasons of the high density in glacier-fed rivers (Dobriyal and Singh, 1990).

Acknowledgements

We are highly indebted to Dr. K. Kumar, Advisor of Fisheries, Himachal Pradesh for his keen interest, encouragement and critically going through the manuscript and offering valuable suggestions to undertake the study. We are also thankful to Mr. Rajan Sud, Assistant Director of Fisheries and Mr. Bijay Kumar Sahoo for their kind cooperation and help during course of investigation.

References

APHA, 1998. *Standard Methods for the Examination of Water and Wastewater*, 20th edn. American Public Health Association, Washington, D.C.

Dobriyal, A.K. and Singh, H.R., 1990. Ecological studies on the age and growth of *Barilius bendelisis* (Ham.) from India. *Arch. Hydrobiol.*, 118: 93–103.

Johal, M.S., 2001. *Final Report U.S. Fish and Wildlife Service*, 1–149 pp.

Saha, G.N., Sehgal, Eva Mitra and Nandy, A.C., 1971. Studies on the seasonal and diurnal variations in physico-chemical or biological conditions of a perennial freshwater pond. *J. Inland Fish. Soc., India*, 3: 79–102.

Jhingran, V.G., Natarajan, A.V., Banarjee, S.M. and David, A., 1988. *Methodology on Reservoir Fisheries Investigation in India.* Bull. 18; 36 pp.

Welch, P.S., 1948. *Limnology Method.* The Blakiston Company, Philadelphia Toranto, pp. 199–229.

Water and Wastewater Management in 2 Vols (2011) *Pages* **472–477**
Editors: **D.R. Khanna, R. Bhutiani & Gagan Matta**
Published by: **DAYA PUBLISHING HOUSE, NEW DELHI**

Chapter 69

Assessment of Water Quality of Phagwara Drain in Punjab

A. Polestya, S. Sharma and A.S. Reddy
Department of Biotechnology and Environmental Sciences, Thapar University, Patiala

ABSTRACT

An attempt has been made to study the pollution level of the Phagwara drain, by analyzing the physico-chemical and biological parameters. The duration of study was from January 2007 to December 2007. Water samples were collected monthly and analyzed to assess the quality of the drain. It receives wastes mainly from Phagwara and nearby areas. Domestic waste, industrial waste and agricultural runoff from surrounding areas is entering into the drain. The study revealed that the overall daily input of waste let into the drain has resulted in complete deterioration of water quality. Almost all the parameters were beyond the prescribed limits. The water of this drain is not suitable for any kind of use and it is poisonous in nature.

Keywords*: Water quality monitoring, Phagwara drain, Physico-chemical, Biological characteristics*

Introduction

Phagwara drain originated from Phagwara town and ultimately merged into river East Bein near Kangniwal. The main purpose of water analysis is to evaluate methods of treatment of wastewater to reuse or dispose, ascertain quality of water and aim at recovery of valuable products from waste effluents. In order to ascertain the above objectives, it is necessary to analyze various parameters, which would throw light on quality of water (Lokhande *et al.*, 2004). In comparison to water analysis, the analysis of wastewater is complicated; the reason being one does not encounter in wastewater the kind of impurities and deterious materials, which are dumped by different industries. Further the wastewater of different industries mixes together, before merging into the river or disposal point, to form very complex mixture. The study of water quality of different regions in India by Datar and Vashistha (1992), Athappan *et al.* (1992), Janaki Arunan *et al.* (2004), Ramamurthy *et al.* (2005), Dwivedi *et al.* (2006), Manohar *et al.* (2007) have shown remarkable pollution levels. Considering the present conditions of water quality of Phagwara drain, a quantitative study of physico-chemical and biological conditions of this drain was done.

Materials and Method

Samples were collected in precleaned and sterilized polythene or glass containers and preserved according to standard methods (APHA, 1998) at monthly intervals between 9 am to 5 pm from January 2007 to December 2007 and brought to laboratory for various physico-chemical and biological analysis. Temperature and Dissolved Oxygen (DO) were measured on the spot. The physico-chemical and biological analysis was carried out according to standard methods given by APHA (1998).

Table 69.1: Various Analytical Methods Used for Analysis

Parameter	*Analytical Method*
pH	pH metery
Temperature	Thermometry
Conductivity	Conductivity meter
Dissolved oxygen (DO)	Winkler's Iodometric with azide modification method
Biochemical oxygen demand (BOD)	5 day incubation method
Chemical oxygen demand (COD)	Dichromate reflux (closed) method
Chloride	Argentometric method
Total solids (TS)	Gravimetric method
Total dissolved solids (IDS)	Gravimetric method
Total suspended solids (TSS)	Difference of TS and TDS
Ammonical and organic nitrogen	Kjeldahl method
Nitrate and nitrite nitrogen	Spectrophotometric and Cd-reduction column method
Phosphate	Colorimetric (SnCh) method
Total coliform	Mac-Conkey broth
Faecal coliform	A-1 broth

Results and Discussion

The results for the various physico-chemical and bacteriological parameters determined in the water samples of the drain are presented in Table 69.2. Table 69.2 summarizes the maximum, minimum, mean values and standard deviation. The temperature values were lower in winter season as compared to summer and rainy season. The ranges of temperature of water samples were similar as observed by Muduli and Dhal (2006) in case of river Baitarani at Anandpur. Conductivity qualitatively reflects the status of inorganic pollution and is a measure of total dissolved solids and ionized species in the waters. The conductivity values ranged from 1,122 to 2,920 µmho/cm. These values indicate higher quantity of inorganic acid, base and salt in wastewater samples. Ram and Singh (2007) observed the same conditions in Ganga river.

Table 69.2: Various Physico-chemical and Biological Parameters at Phagwara Drain

Parameter	*Max*	*Min*	*Avg*	*Std*
pH	9.42	7.12	7.57	0.60
Temp. (°C)	32.40	20.30	26.56	4.56
E.C (μ mho/cm)	2920.00	1122.00	1893.92	522.49
TS (mg/l)	2050.00	1220.00	1587.08	283.71
TDS (mg/l)	1630.00	900.00	1312.42	250.24
TSS (mg/l)	504.00	99.00	274.67	114.97
DO (mg/l)	0.00	0.00	0.00	0.00
BOD (mg/l)	237.50	62.50	161.99	54.84
COD (mg/l)	575.00	275.00	410.33	107.24
NH_3-N (mg/l)	40.04	22.12	31.34	5.94
Org-N (mg/l)	2.74	1.90	2.37	0.24
NO_2+NO_3-N (mg/l)	4.17	0.02	0.61	1.20
Org-N (mg/l)	2.74	1.90	2.37	0.24
Total-P (mg/l)	3.54	0.79	2.60	0.94
Chloride (mg/l)	519.10	195.66	343.58	104.58
Total MPN/100 ml	460000000	1359779	82500761	139898204
Faecal MPN/100 ml	75894664	314484	20345803	24585711

The total solids (TS) values were ranged from 1,220 to 2,050 mg/l. These higher values of total solids indicate higher pollution load in the drain. The total dissolved solids (TDS) values correlated well with the conductivity values. In agreement with the high conductivity values, TDS levels were high. TDS values ranged from 900 to 1,630 mg/l. Ahmad *et al.* (2003) observed same results for river Damodar, where solids come from fly ash and fine coal particles from thermal plants and coal washeries.

The suspended solids represent the floating material (bacteria, algae) and undissolved particles, which ranged from 99 to 504 mg/l. No fix trend was observed in various solids seasonally this may be due to different discharges at different time. The concentration of dissolved oxygen in water depends on temperature, pressure and the concentrations of various ions. Due to pollution load the concentration of DO depletes and possess thrust on the aquatic life. Low oxygen in water can be detrimental to fishes and many other organisms in the aquatic system. The DO values were nil throughout the year showing higher pollution load. A minimum value of 4.0 mg/l of dissolved oxygen in water is essential for the survival of aquatic life. BOD is the empirical test to determine the relative oxygen requirement of water mostly due to organic ingredients. Its application is to calculate the pollution load.

The BOD values were from 62.5 to 238.0 mg/l (Figure 69.1). Higher values of BOD were also observed by Rajurkar *et al.* (2003) in case of river Unkhrah at Shillong. The level of BOD depends on temperature, density of plankton, concentration of organic matter and related factors. In the present study it was observed that the BOD in rainy season was lesser than winter and summer seasons, which may be due to dilution of drain water with rain. In this study the COD values ranged from 275 to 575 mg/l (Figure 69.1). These results were showing higher organic pollution load in the drain. The COD values were also minimum in rainy season; this again may be due to dilution of river water with rain. Higher COD values were also observed by Verghese *et al.* (2005) in the river Yamuna at Agra. Ammonical nitrogen imparts an adverse impact on aquatic life. The values of ammonical nitrogen were very much higher, which ranged from 22.1 to 40.0 mg/l. Ammonical nitrogen values were highest in summer season and lowest in rainy season at all stations. Ammonical nitrogen is extremely soluble and is readily transported by run off from cultivated lands and also a major component of raw sewage. It occurs in water as a break down product of nitrogenous material. Such higher concentrations of ammonical nitrogen were also observed by Rajurkar *et al.* (2003) in case of river Unkhrah at Shillong. The presence of ammonical nitrogen may be perhaps due to frequent flow of sewage into water systems, since sewage has large quantities of nitrogenous matter which tend to increase the ammonia content in water. The values of organic nitrogen were in the range of 1.90 to 2.40 mg/l (Figure 69.2). In case of organic nitrogen there was not much difference in different seasons.

Nitrate, the most highly oxidized form of nitrogen compounds is commonly present in surface waters because it is the end product of the aerobic decomposition of organic nitrogenous matter. The values of nitrate and nitrite nitrogen were in the range of 0.02 to 4.17 mg/l (Figure 69.2). Similar trend in nitrate and nitrite nitrogen was observed by Jaji *et al.* (2007) in Ogun river. Phosphate values were

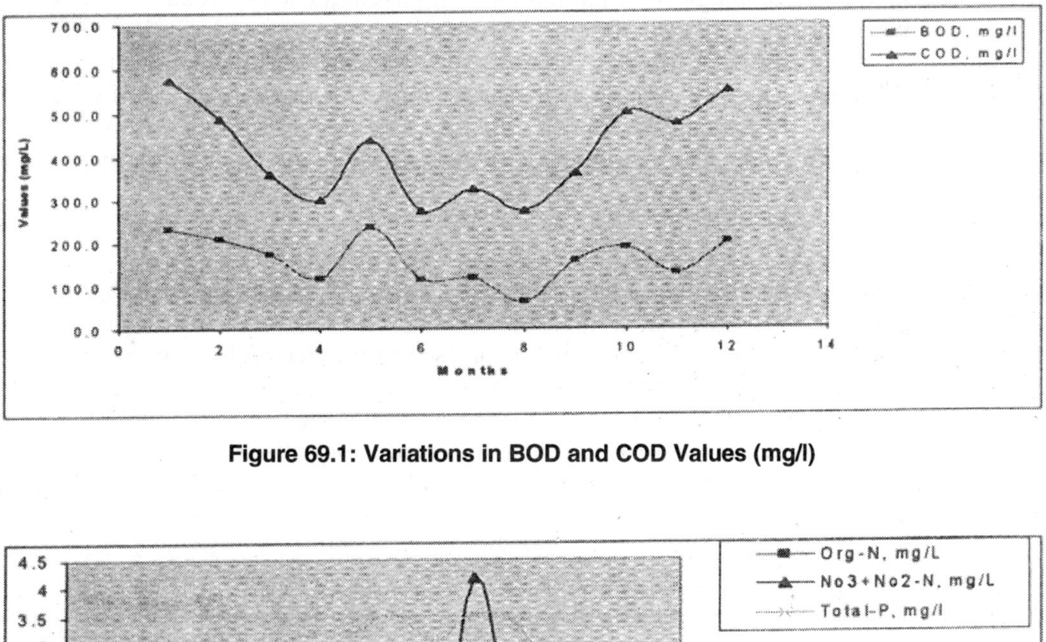

Figure 69.1: Variations in BOD and COD Values (mg/l)

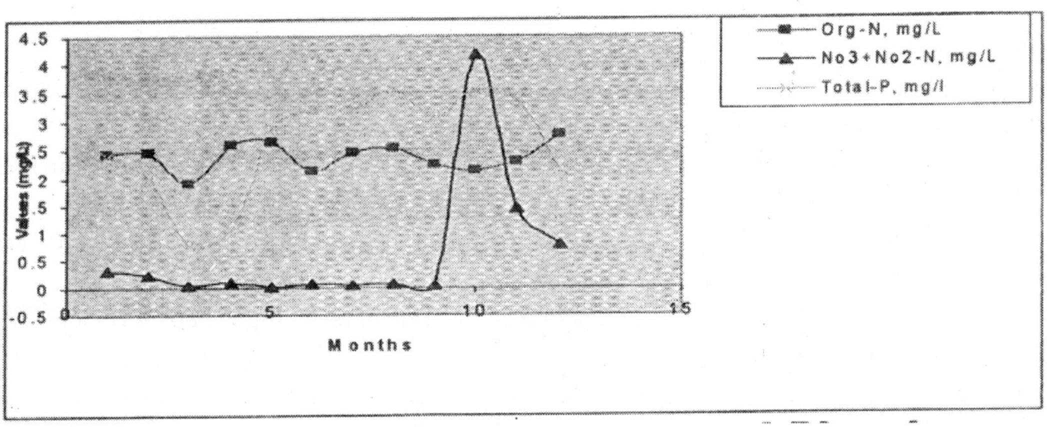

Figure 69.2: Variations in Organic, Nitrate+Nitrite and Phosphate Values (mg/l)

in the range of 0.79 to 3.54 mg/l (Figure 69.2). High levels of phosphorous increase the growth of vegetation in water systems and increase the oxygen demand. Very high phosphate values were observed throughout the year. Photosynthesis and respiration play an important role in the self-purification of natural water. The disturbance of the stationary state between photosynthesis and respiration leads to chemical and biological changes reflecting pollution. High level of phosphate in the drain indicates the wastewater discharges from Phagwara. Fertilizer industries were the most likely source of phosphorous in the waters. The high phosphate values obtained from the samples could be attributed to phosphorous in runoff from domestic, municipal and agricultural wastes (non-point sources) flowing into drain as well as washing along the side with detergents (Correll, 1998).

Chlorides values were in the range of 195.66 to 519.10 mg/l. Higher values of chlorides were also observed by Verghese *et al.* (2005) in the study of river Yamuna at Agra city. High chloride concentrations were due to excess dumping of domestic and industrial effluent into drainwater. Coliform bacteria are common indicators of overall water quality and although they are ubiquitous in the environment and are generally not considered harmful to humans, their presence in high concentrations often coincides with more dangerous bacteria (WHO, 1996). Higher total coliform values indicate severe microbiological contamination, originated due to the wastewater discharges and sewage contamination. The constant presence of total coliform through all the study period clearly indicates a pathogenic risk caused by the accidental consumption of water during recreation (WHO, 1996). The total and faecal coliforms were not following any trend with seasons. Total coliform were ranged from 13,59,779 to 46,00,00,000 MPN/100 ml. Faecal coliform were from 3,14,484 to 7,58,94,664 MPN/100 ml (Table 69.2). From the above study it was concluded that the drain was highly polluted through out the year.

Conclusion

The parameters analyzed revealed higher pollution load in the drain. The colour of the drainwater was black throughout the year. This state result from discharges coming from various industries such as distilleries, leather, fertilizer, textile, sugar industries and sewage discharges from the town Phagwara and nearby areas. Almost all the parameters were found to be beyond permissible limits. The high levels of pollution parameters are definitely harmful to aquatic living beings. Significant pollution was observed at sampling station because of huge quantity of industrial effluents and sewage discharge. A continuous monitoring of the drainwater for various parameters is required because it merge into river East Bein.

References

Ahmad, M., Chakraborty, M.K., Tiwari, R.K. and Gupta, A., 2003. Impact of mining and other industries on the major rivers with special reference to river Damodar. *Indian J. Env. Prot.*, 23(11): 1208–1216.

APHA, 1998. *Standard Methods for the Examination of Water and Wastewater*, 20th edn. American Public Health Association, Washington, D.C.

Athappan, P.R., Sethuraman, K. and Kannan, N., 1992. A study on the pollution of river Vaigai at Madurai. *Indian J. Env. Prot.*, 12(11): 818–823.

Correll, D.L., 1998. The role of phosphorus and eutrophication of receiving waters: A review. *J. Env. Quality*, 27: 261–266.

Datar, M.D. and Vashistha, R.P., 1992. Physico-chemical aspects of pollution in Betwa. *Indian J. Env. Prot.*, 12(8): 577–580.

Dwivedi, A., Kotiyal, A., Sharma, S., Bisht, G.R.S. and Thakur, R.L., 2006. Assessment of bacteriological quality and aspect of pollution along the stretch of river Ganga in Garhwal Himalayas. *Indian J. Env. Prot.*, 26(2): 112–115.

Jaji, M.O., Bamgbose, O., Odukoya, O.O. and Arowolo, T.A., 2007. Water quality assessment of Ogun river, South West Nigeria. *Env. Monit. Assess.*, 133: 473–482.

Janaki, A., Lalitha, S., Kasthuri, R., Banumathi, K. and Agila, A., 2004. Study on quality of drinking water at pilgrim centers in Tiruchirapalli. *Indian J. Env. Prot.*, 24(3): 193–198.

Lokhande, P.B, Gawas, A.D. and Mujawar, H.A., 2005. Study of water quality parameters of river water in Konkan region. *Indian J. Env. Prot.*, 25(3): 212–217.

Manjappa, S. and Naik, V.K., 2007. Physico-chemical properties of Malaprabha river. *J. Env. Sci. Engg.*, 49(1): 1–6.

Manohar, S., Jadia, C. and Fulekar, M.H., 2007. Impact of Ganesh idol immersion on water quality. *Indian J. Env. Prot.*, 27(3): 216–220.

Muduli, S.D. and Dhal, N.K., 2006. Classification of water quality of Baitarani at Anandpur. *Indian J. Env. Prot.*, 26(2): 175–177.

Ramamurthy, N., Subashini, J. and Raju, S., 2005. Physico-chemical properties of Palar river in Tamil Nadu. *Indian J. Env. Prot.*, 25(10): 925–928.

Rajurkar, N.S., Nongbri, B. and Patwardhan, A.M., 2003. Water quality status of river Unkhrah at Shillong. *Indian J. Env. Prot.*, 23(9): 990–998.

Ram, P. and Singh, A.K., 2007. Ganga water quality at Patna with reference to physico-chemical and bacteriological parameters. *J. Env. Sci. Engg.*, 49(1): 28–32.

Verghese, P.S., Singh, M. and Mishra, A., 2005. An assessment of water quality of river Yamuna during monsoon at Agra city. *Indian J. Env. Prot.*, 25(10): 893–898.

WHO, 1996. *Health Criteria and Other Supporting Information, 2nd edn, Vol. 2: Guidelines for Drinking Water Quality*. World Health Organization, Geneva, Switzerland.

Water and Wastewater Management in 2 Vols (2011) *Pages* 478–482
Editors: D.R. Khanna, R. Bhutiani & Gagan Matta
Published by: DAYA PUBLISHING HOUSE, NEW DELHI

Chapter 70

Suitability of Groundwater for Irrigation in Lakhanpur Block of Jharsuguda District, Orissa

Atish Kumar Pattanaik
Department of Earth Sciences,
Sambalpur University, Jyoti Vihar, Sambalpur

ABSTRACT

Lakhanpur block of Jharsuguda district rests predominantly on crystalline and sedimentary rocks of Precambrian and Permo-carbonifeous age. In most of the part of the study area cultivation sustain on groundwater. The area is characterized by shallow to moderate deep phreatic aquifers system. The present study has been carried out to evaluate the suitability of groundwater for irrigation. On the basis of various criteria like salinity, magnesium hazard, RSC *etc.* The majority of water samples are found suitable for irrigation.

Keywords: Groundwater quality, SAR, Per cent sodium, Magnesium hazard.

Introduction

Groundwater is a valuable as well as a dependable source to meet both the domestic and agricultural need. In groundwater evaluation quality is as important as quantity. The quality of groundwater depends on the various chemical constituents and their concentration, which decides its suitability for any particular use. Water used for irrigation can vary greatly in quality depending upon type and quantity of dissolved solid present. The soil problems most commonly encountered and used as the basis of evaluation of water quality are those related to salinity, water infiltration *etc.*

Water used for irrigation should therefore meet the demands of both soil as well as crops for better results.

Lakhanpur block comprises 602.71 sq.km. The livelihood of this area largely depends upon agriculture. The study area rests on crystalline and sedimentary rock of precambrian and permo-carboniferrous age. Geomorphologically it is pedeplain with nos scattered structural and denudation hill. The drainage of the area is controlled by the Mahanadi River. Loamy soil is the most common soil and is of medium depth. The study area enjoys a humid tropical climate with hot summer (maximum 43°C) and chilled winter (minimum 12°C). The average rainfall is 1300 mm. The irrigated area with respect to Khariff and Rabi crops are 3376 hectares and 1648 hectares respectively. The total groundwater resources available in the block are 3297 ham, out of which utilizable groundwater for irrigation purpose is 1929 ham. The purpose of this chapter is the evaluation of groundwater quality of shallow aquifer and finds its suitability for irrigation purposes. Use of groundwater for irrigation if not judged from angle of suitability, may cause serious problems.

Materials and Method

A total of 20 water sample; representing shallow groundwater of the study area were collected as per standard procedures laid down by Brown *et al.* (1974). Environmental sensitive parameters such as pH, electrical conductivity (EC) and temperature were measured on the spot at the time of sampling with water analysis kit (CK711 of Century make). For chemical analysis, the groundwater samples were collected in precleaned, dried one liter polyethylene bottles after rinsing. The chemical analysis for major cations (Na^+, K^+, Ca^{+2}, Mg^{+2}) and anions (HCO^{-3}, Cl^-, SO_4^{-2}, F^-) of water samples were done at the laboratory, according to standard procedure (APHA, 1989; Jain *et al.*, 1987). The sodium adsorption ratio (SAR), residual sodium carbonate (RSC), per cent sodium (per cent Na), magnesium hazard were calculated, which are used to evaluate water quality for irrigation use. The results of chemical analysis are given in Table 70.1.

Results

Water Quality Evaluation for Irrigation

Groundwater quality depends largely on both, type and quantity of dissolved solids. The important characteristic properties of groundwater used for determine its suitability for irrigation in present study area are:

Salinity

Salinity of irrigation water is of major importance. The evaluation of salinity within the root zone is the controlling factor for plant growth and crop yield. The total salt concentration is usually measured as EC in irrigation work. Out of 20 groundwaters, no water falls under salinity hazard. Most of the water samples fall in good to permissible zone.

Sodium Hazard

Owing to its effect on both soil and plant, sodium is one of the governing specific ion. A number of indices have been proposed to assess sodium effects and the equilibrium between soil chemistry and soil water chemistry. per cent sodium (per cent Na) is calculated as

$$\text{Per cent Na} = (Na^+ + K^+) \times 100/(Ca^{+2}+Mg^{+2}+Na^++K^+)$$

where,

all values are in epm.

Table 70.1. Chemical Quality of Groundwater Samples of Lakhanpur Block

Sl.No.	Location	pH	EC	Na	K	Ca	Mg	HCO_3	Cl	SO_4	SAR	Per cent NA	RSC	Mg Hazard
1.	Charpali	7.6	402	43	12	27	6	150	50	17	1.94	54.06	0.61	27.03
2.	Kanaktura	7.9	546	36	22	34	20	137	75	16	1.21	38.74	–1.12	49.50
3.	Kandijharan	8.2	401	30	16	24	14	120	60	12	1.20	42.01	–0.40	49.30
4.	Bikampali	7.1	460	43	8	42	18	200	46	30	1.39	36.56	–0.32	41.67
5.	Rengali	7.5	289	29	5	20	7	50	53	27	1.42	46.73	–0.76	36.84
6.	Machida	8.3	951	90	65	60	30	315	195	56	2.36	50.36	–0.34	45.45
7.	Jamgaon	7.3	640	60	25	52	9	150	110	38	2.02	49.24	–0.89	22.39
8.	Kuremal	7.1	1390	118	4	114	41	330	305	15	2.40	36.47	–3.71	37.48
9.	Kodapad	8	680	68	5	40	22	260	85	30	2.14	44.59	0.43	47.83
10.	Durga	8.4	378	19	14	34	12	150	32	16	0.71	30.50	–0.24	37.04
11.	Bagmunda	8.8	812	20	11	26	32	215	113	8	0.62	22.50	–0.44	67.23
12.	Pandra	7.3	360	34	6	20	8	120	30	18	1.62	49.48	0.30	40.00
13.	Lakhanpur	.8.4	300	25	6	21	to	200	25	5	1.12	39.72	1.40	44.25
14.	Kudloi	7.9	515	29	3	34	14	165	64	18	1.05	31.82	–0.16	40.70
15.	Ubda	8.1	390	43	16	20	8	110	35	30	2.05	57.77	0.14	40.00
16.	Kursoloi	8.6	195	16	4	28	1	100	.23	4	0.81	34.99	0.16	5.62
17.	Kumarbandha	7.5	220	14	6	20	6	90	25	9	0.70	33.70	–0.02	33.33
18.	Piplikani	8	425	30	3	38	12	200	30	7	1.08	32.26	0.38	34.48
19.	Sarandamal	7.1	656	47	10	50	22	225	45	12	1.39	34.67	–0.64	42.31
20.	Karpabahal	7.2	467	36	6	45	8	120	30	30	1.30	37.08	–0.95	22.86

All concentrations are in mg/l, except pH, EC (μS/cm), SAR and RSC (meq/l).

According to the classification depending on per cent Na as proposed by Wilcox (1967), the groundwater of the study area is of good quality.

Sodium adsorption ratio (SAR) is an important parameter for determination of suitability irrigation water and is calculated as

$$SAR = Na^{+}1\{(Ca^{+2}+Mg^{+2})/2\}^{1/2}$$

where,

all the values are in epm.

The SAR values in groundwater of study area ranges from 0.62 to 2.4 which implies that no alkali hazard is anticipated in the study area. According to classification for irrigation based on SAR the water is of excellent quality (Raghunath, 1987).

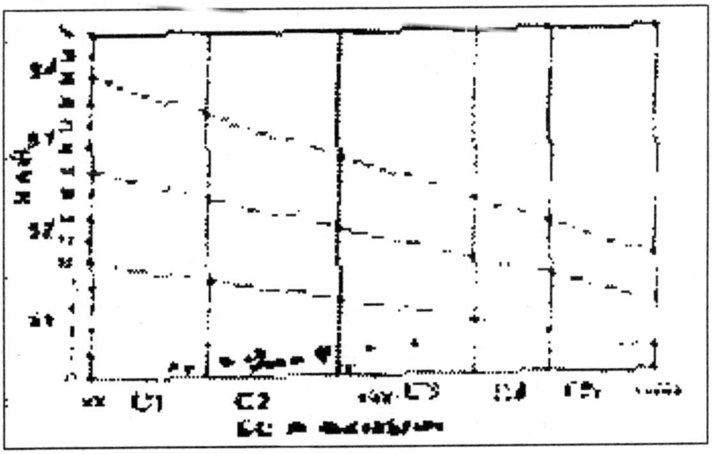

Figure 70.1: U.S. Salinity Diagram for Classification of Irrigational Water (After Richards, 1954)

Magnesium Hazard

A ratio of (Mg×100)/(Ca + Mg), is used as magnesium hazard of irrigational water. When this ratio is less than 50, no magnesium hazard is found. According to this scheme all of the groundwater samples except one are found to be within 50.

Alkalinity Hazard

When the sum of CO_3^{-2} and HCO_3^{-1} is in excess of Ca and Mg, there occurs a complete precipitation of the later. It is judged by the RSC value (Eaton, 1950).

$$RSC = (HCO_3^{-1} + CO_3^{-2}) - (Ca^{+2} + Mg^{+2}), \text{ all values are in epm.}$$

The water of low RSC values (< 1.25) is safe for irrigation where RSC value above 2.5 is unsuitable and may cause hardening of soil which may lead to infertile soil. Except one (1.6), all the water samples are found to be within safe limit.

Integrated Effect of Electrical Conductance and SAR

The SAR (Sodium Hazard) and EC (Salinity Hazard) values of the groundwater of the study area were plotted in the graphical diagram of irrigational water (Richards, 1954) (Figure 70.1). Most of the samples (16) fall in the category of good quality with low alkali hazard and moderate salinity hazard (C_2S_1). Only 3 sample fall under class C_3S_1, show medium to high salinity hazard.

Toxicity

Certain ions like sodium and chloride in the soil and water are taken up by plant and accumulate to concentration high enough to cause crop damage or reduce yield. The degree of damage depends upon the exposure time, concentration, sensitivity of the crop and volume of water transpired by crop. Sodium and chloride can also cause toxicity by absorbing through wetted leaves by over head sprinklers. No sodium, chloride and bicarbonate toxicity have been encountered in the water samples of the study area.

Conclusion

On the basis of various criteria and guidelines the groundwater quality of the phreatic aquifers of Lakhanpur block of Jharsuguda District is found suitable for agricultural purpose. There is no harm in using groundwater as source of irrigation in present soil condition.

References

APHA, 1985. *Standard Methods for the Examination of Water and Wastewater.* American Public Health Association, 1134 pp.

Brown, E., Skougstod, M.W. and Fishman, F.J., 1974. *Methods for Collection and Analysis of Water Samples for Dissolved Minerals and Gases.* U.S. Deptt. of interior. Book No.5.

Eaton, EM., 1950. Significance of carbonates in irrigation water. *Soil Science,* 69: 123–133.

Jain, C.K. and Bhatia, K.K.S., 1987. *Physico-chemical Analysis of Water and Wastewater.* User's Manual, U.M.–26, National Institute of Hydrology, Roorkee.

Raghunath, H.M., 1987. *Groundwater.* Wiley Eastern Limited, Delhi, India.

Richards, L.A., 1954. *Diagnosis and Improvement of Saline and Alkaline Soils.* U.S. Deptt. Agril. Handbook, 60 pp.

Wilcox, L.V., 1967. *Classification and Use of Irrigation Water.* USDA Circ. 969, Washington, D.C., 19 pp.

Water and Wastewater Management in 2 Vols (2011) *Pages 483–489*
Editors: **D.R. Khanna, R. Bhutiani & Gagan Matta**
Published by: **DAYA PUBLISHING HOUSE, NEW DELHI**

Chapter 71

A Study of Plankton Population in the Moghat Reservoir at Khandwa District, M.P.

Vivek Keshre[1], L.K. Mudgal[2], D.R. Khanna[3], Gagan Matta[3], Dheeraj Kumar[3] and Vikas Singh[3]

[1]Department of Zoology, Government M.L.C. Girls P.G. College, Khandwa, M.P.
[2]Department of Zoology, Government Girl's P.G. College, Indore, M.P.
[3]Department of Zoology and Environmental Science, Gurukula Kangri University, Haridwar, U.K.

ABSTRACT

Khandwa District is situated in the south west part of the Madhya Pradesh, along the border of state Maharashtra. With increasing water demand in 1897, Moghat reservoir was constructed specially to store the rainwater so that water requirement of the nearby area can be fulfilled. This reservoir is very important for local people but no much information on quality of water is available. This study was carried out with the purpose to know the planktonic characteristics of the reservoir.

Keywords: Khandwa, Reservoir, Planktonic, Rainwater.

Introduction

With the technological advancement man made several new things which either had multiple uses or specific use; one of such example is of reservoir, which have multiple uses. Reservoir construction needs very high techniques and technological knowledge. They are used to store water to solve the water problem, control flood, generate electricity, pisciculture and sometimes for recreational activity and to conserve endangered freshwater organisms. Water quality and productivity of any aquatic (reservoir) ecosystem and its health can be evaluated by studying the planktonic status of that water body. Plankton in an aquatic system is an essential link in the food chain, capable in affecting the entire aquatic biota. Freshwater sources are degrading fastly due to men's increased activity (Rao and Durve, 1989) and ignorance towards the health of water source, Moghat reservoir of Khandwa is one of such examples. Indian freshwater planktonic information was given by Chakraborty *et al.* (1959), Rao *et al.* (1988), Kumar (1995), Khanna *et al.* (1999), Khanna *et al.* (2000). District Khandwa is situated in the south west part of the Madhya Pradesh. The studied Moghat reservoir was constructed in the year of 1897 on a small river near village named Moghat. Moghat village is 3 km. away from the Khandwa Township. Catchment area of the reservoir was 23.30 sq. km. and storage area was 2.02 sq. km. This water reservoir is situated at a height of 324.54 meter above the sea mean level, height of the band is 12 meter and depth is 8.78 meter. To increase the water supply to the reservoir, a 6.4 km long canal was constructed. The studied reservoir has a long history but till now no much information on its planktonic ecology is available. This study was carried out with the purpose to develop planktonic abundance data and to study and understand their ecological characteristics.

Materials and Method

Study Site

Moghat reservoir is located at a distance of 3 km from the Khandwa city in northwest. Location of this reservoir is 21º49′ N, 76º20′ E and is situated 1017 ft. above the sea level.

Sampling

Monthly sampling was done for two years (2005–2007). Samples were collected between 7.00 am to 11.00 am on first Monday of every month. Plankton were studied following the standard methods of Welch (1948), Khanna (1993) and APHA-AWWA-WPCF (1998).

Figure 71.1: Moghat Reservoir at Khandwa

Results and Discussion

During the course of study of plankton of the Moghat reservoir it was observed that the number of total plankton in the reservoir increases from October to January and then started decreasing from February to July. The maximum number of total plankton was observed in the month of January (3535.13 unit/l ± 854.55) and minimum in month of July (661.25 unit/l ± 99.51). The similar trend of planktonic number was also observed by Eddy (1934) and Khanna (1993). The maximum value of zooplankton was observed (398.75 unit/l ± 113.95) in January and minimum number (48.50 unit/l ± 29.36) in the month of July. As the total plankton is the, sum of phytoplankton and zooplankton, both showed the same trend in their number similar to the total plankton.

The maximum number of phytoplankton (3175.63 unit/l ± 1190.54) in February and minimum (602.25 unit/l ± 98.97) in July. Roy (1955) reported similar results. The study showed a dominance of phytoplankton as also reported by Pahwa and Mehrotra (1966), Dobriyal and Singh (1987), Khanna and Singh (2000). The variation in the number of phytoplankton and zooplankton can be observed in the Table 71.1 and in Figure 71.2. Phytoplankton observed in the Moghat reservoir belongs to the classes diatoms, green algae and blue green algae. Variation in their number is tabulated in Table 71.2 and graphically represented in Figure 71.3. During the study maximum number (1429.87 unit/l ± 775.66) of plankton observed belongs to the class diatoms and minimum number (39.42 unit/l ± 28.61) of plankton belongs to the class blue green algae. Singh and Ahmed (1990), Khanna *et al.* (1999) reported the same trend. Diatoms observed to be more and blue green algae minimum in comparison to other class of phytoplankton in every month during the study. Maximum number of diatoms were found in February (2662.75 unit/l ± 1217.0 I) and minimum (590.63 unit/l ± 178.48) in July and its number showed a continuous decreasing trend from March to November and then started increasing from December to February. Khanna and Bhutiani (2003) worked on limnological status of Satikund pond at Haridwar and they observed the same phenomenon in relation to diatoms. The maximum number (363.13 unit/1±59 .15) of green algae was found in January and minimum number (39.88 unit/I±26.02) in August.

Table 71.1: Planktonic Number (unit/l) in Moghat Reservoir during 2005–2007

Month	*Phytoplankton*	*Zooplankton*	*Total Plankton*
October	1185.13±389.29	144.50±36.99	1330.00±400.71
November	1178.13±350.19	181.88±56.18	1360.63±340.51
December	1787.63±749.94	272.00±69.54	2160.63±799.94
January	3135.88±823.38	398.75±113.95	3535.13±854.55
February	3175.63±1190.54	355.25±52.89	3406.50±1209.62
March	2837.38±1100.69	459.13±195.82	3157.00±1194.49
April	2123.25±824.85	376.38±145.09	2500.13±911.81
May	1685.00±618.47	288.88±109.80	1974.13±601.73
June	894.00±185.83	118.88±59.16	1013.25±188.12
July	602.25±98.97	48.50±29.36	661.25±99.51
August	772.75±80.79	50.00±20.19	848.00±125.02
September	1090.13±362.93	97.88±24.74	1188.38±370.67
Average	1645.29±921.63	225.84±143.97	1858.26±1022.35

Table 71.2: Number (unit/l) of Different Group Among Phytoplankton of the Moghat Reservoir during 2005–2007

Month	*Total Diatoms*	*Total Green Algae*	*Total Blue Green Algae*
October	985.63±369.63	155.50±32.50	37.00±23.79
November	912.50±316.45	240.38±37.69	13.50±11.93
December	1423.75±772.45	275.75±59.80	68.25±9.71
January	2327.25±878.64	363.13±59.15	90.00±30.47
February	2662.75±1217.01	238.63±87.I3	69.38±26.46
March	2567.63±1046.17	229.88±85.87	54.38±18.33
April	1885.25±780.59	182.00±69.52	66.25±41.91
May	1584.25±649.63	124.00±60.71	33.75±11.88
June	882.75±368.26	85.00±17.27	17.00±21.59
July	590.63±178.48	41.75±29.71	5.00±2.73
August	747.8&±117.89	39.88±26.02	5.63±1.99
September	939.38±350.34	104.38±51.35	23.75±15.44
Average	1429.87±775.66	167.68±99.53	39.42±28.61

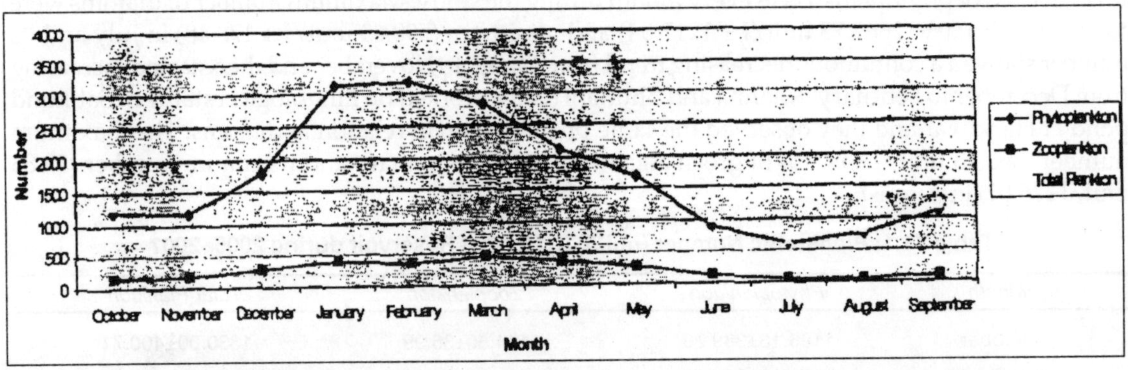

Figure 71.2: Average Monthly Variation in Plankton Number During 2005–2007

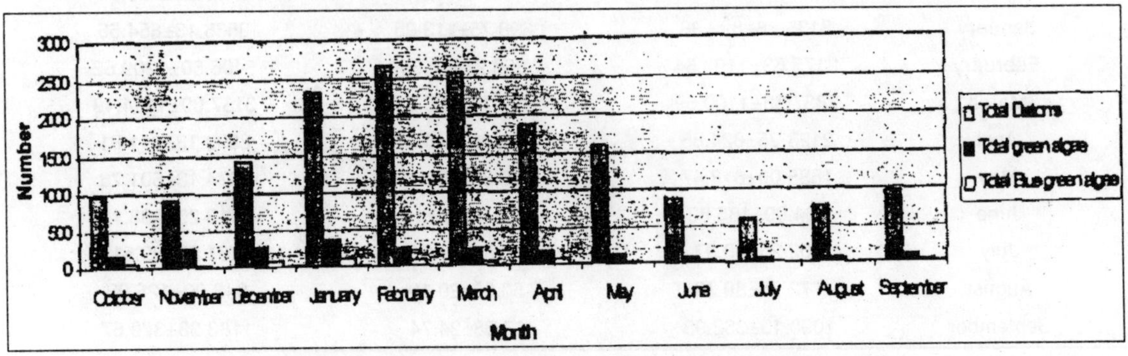

Figure 71.3: Average Monthly Variation in Different Classes of Phytoplankton During 2005–2007

The green algae showed increasing trend from November to March and then decreasing trend from April to August. Blue green algae is also known as cyanobacteria and during the study of Moghat reservoir the maximum number (90.00 unit/l ± 30.47) in the month of January and minimum number (5.00 unit/l ± 2.73) in the month of July was observed. Similar trend was observed by Prasad and Singh (2003) during their study of tropical water body.

In this study zooplankton of five different classes protozoa, rotifera, cladocera, copepod and ostracods of were identified and out of all the five class maximum average number (60.55 unit/l ± 39.34) was found for rotifers and minimum (42.40 unit/l ± 29.00) for ostracods. Results are tabulated in Table 71.3 and graphically presented in Figure 71.4.

Table 71.3: Number (unit/l) of Different Genera Among Zooplankton in the Moghat Reservoir During 2005–2007

Month	*Protozoa*	*Rotifera*	*Cladocera*	*Copepod*	*Ostracods*
October	31.88±14.34	38.25±14.47	36.25±18.84	22.00±7.21	33.63±13.09
November	47.63±11.94	52.13±12.00	49.00±31.14	32.67±5.77	44.25±13.47
December	75.00±20.11	75.38±17.05	30.88±21.33	62.00±10.00	64.63±17.89
January	137.6±37.19	104.00±60.52	76.75±48.73	75.33±17.62	90.88±43.34
February	100.63±40.17	87.00±59.01	83.00±59.29	85.33±21.01	72.75±35.98
March	112.25±41.97	104.75±46.74	76.63±56.90	109.67±22.50	80.00±36.13
April	91.88±39.49	132.00±115.56	84.38±59.03	83.67±15.04	55.88±30.14
May	63.13±23.17	70.88±14.60	37.75±19.34	55.67±5.51	27.6±13.59
June	32.25±21.39	26.50±12.12	12.63±7.17	12.00±2.00	13.00±5.13
July	12.75±6.65	14.38±9.29	7.38±4.27	5.67±2.52	5.88±1.81
August	16.88±9.09	14.63±7.74	10.13±6.59	9.3±3.79	9.50±5.83
September	23.50±1 1.01	27.50±7.21	19.63±12.33	15.33±1.53	24.2±10.09
Average	60.51±41.28	60.55±39.34	42.62±29.62	46.49±35.71	42.40±29.00

Protozoa found maximum (137.63 unit/l ± 37.19) and minimum (12.75 unit/l ± 6.65) in number during January and July respectively. Protozoa showed an increasing trend from October to January and started decreasing from March to July. Maximum and minimum number of rotifera observed was (132.00 unit/l ± 115.56) in April and (14.38 unit/l ± 9.29) in July respectively. Rotifera showed regular

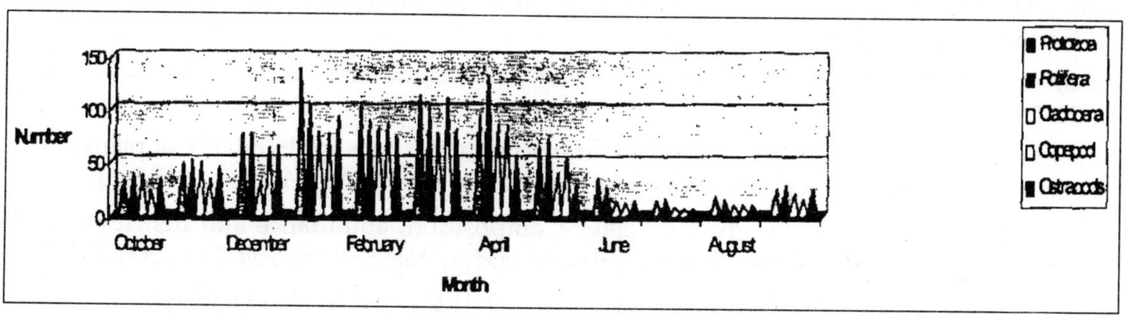

Figure 71.4: Average Monthly Variation in Different Class of Zooplankton During 2005–2007

increase in number from October to January. Das and Srivastava (1956), Das and Upaddhyaya (1979) found the same trend in the number of protozoa and rotifera. Cladocera didn't showed any specific trend in number, its maximum number was found in the month of April (84.38 unit/l ± 59.03) and minimum in the month of July (7.38 unit/l ± 4.27). Similar to cladocera, copepod also did not showed any trend in number during the study period. Copepod was found maximum (109.67 unit/l ± 22.50) in March and minimum (5.67 unit/l ± 2.52) in July. Ostracods found to be maximum in January (90.88 unit/l ± 43.34) and minimum in the month of July (5.88 unit/l ± 1.81).

References

APHA, AWWA, WPCF, 1998. *Standard Methods for the Examination of Water and Wastewater*, 20th edn. Washington D.C., New York.

Chakraborty, R.D., Ray, P. and Singh, S.B., 1959. A quantitative survey of plankton and physiological conditions of the river Yamuna at Allahabad. *Indian J. Fish*, 6(1): 186–203.

Das, S.M. and Srivastava, V.K., 1956. Quantitative studies on freshwater plankton. Part II. Correlation between plankton and hydrological factors. In: *Proc. Nat. Acad. Sci., India*, 26B: 243–254.

Das, S.M. and Upaddhyaya, J.C., 1979. Studies on qualitative and quantitative fluctuations of plankton in two Kumaon lakes, Nanital and Bhimtal (India). *Acta. Hydrobiol.*, 21(1): 9–17.

Dobriyal, A.K. and Singh, H.R., 1987. A case study on the origin of rhithroplankton in Garhwal Hill Streams. *Agri-Biol.*, 3(2): 104–106.

Eddy, S., 1934. A study of the freshwater plankton communities. *Illinois. Biol. Monoar*, 12(4): 1–93.

Khanna, D.R., 1993. *Ecology and Pollution of Ganga River*. Ashish Publishing House, Delhi, pp. 1–241.

Khanna, D.R., Badola, S.P. and Dobriyal, A.K., 1993. Plankton ecology of the river Ganga at Chandighat, Haridwar. In: *Advances in Limnology*, (Ed.) H.R. Singh. Narendra Publishing House, New Delhi, pp. 171–174.

Khanna, D.R., Malik, D.S., Seth, T.R. and Rupendra, 1999. Correlation between abiotic factors and planktonic population in river Ganga at Rishikesh (UP.). In: *Sustainable Ecology System and Environment*, (Ed.) D.R. Khanna and A. Gautam. ASEA, Rishikesh, pp. 69–76.

Khanna, D.R. and Singh, R.K., 2000. Seasonal fluctuations in the plankton of Suswa River at Raiwala (Dehradun). *Env. Cons. J.*, (2–3): 89–92.

Khanna, D.R., Gautam, Ashutosh, Chugh, Tarun and Sarkar, Praveen, 2000. Impact of abiotic factors on the phytoplanktonic population of a pond. *Env. Con. J.*, 1(1): 41–46.

Khanna, D.R. and Bhutiani, R., 2003. Limnological status of Satikund pond at Haridwar (UA). *India J. Env. Set.*, 7(2): 131–136.

Kumar, Arvind, 1995. Studies in pollution in river Mayurakshi in South Bihar. *Indian J. Env. Poll.*, 2(1): 21–26.

Pahwa, D.V. and Mehrotra, S.M., 1966. Observations on fluctuations in abundance of plankton in relation to certain hydrological conditions of river Ganga. In: *Proc. Nat. Acad. Sci.*, 36B(2): 157–158.

Prasad, Bijay Bhushan and Singh, R.B., 2003. Composition, abundance and distribution of phytoplankton and zoobenthos in a tropical water body. *J. Nat. Env. and Poll. Tech.*, 2(3): 255–258.

Rao, K.S., Kartha, K.N., Gupta, D.R., Pandya, S.S and Iyer, H.K., 1988. Studies on morphometry and hydrobiology of Gandhi Sagar Reservoir with special reference to its fisheries. *J. Fish Tech.*, 25(1): 21–28.

Rao, N.G. and Durve, V.S., 1989. Cultural eutrophication of the lake Rangasagar, Udaipur, Rajasthan. *J. Env. Biol.*, 10: 127–134.

Roy, H.K., 1955. Plankton ecology of the river Hoogly at Patna. *Ecology*, 36(2): 169–175.

Singh, A.K. and Ahmad, S.H., 1990. A comparative study of the phytoplankton of the river Ganga and pond of Patna (Bihar), India. *J. Indian Bot. Soc.*, 69: 153–158.

Trivedi, R.K. and Goel, P.K., 1984. *Chemical and Biological Methods for Water Pollution Studies.* Environmental Publications, Karad, pp. 1–250.

Welch, P.S., 1948. *Limnological Methods*. The Blakiston Co., Philadelphia, pp. 1–381.

Water and Wastewater Management in 2 Vols (2011) *Pages* ***490–493***
Editors: **D.R. Khanna, R. Bhutiani & Gagan Matta**
Published by: **DAYA PUBLISHING HOUSE, NEW DELHI**

Chapter 72

Study of Physico-chemical Characteristics of Treated Effluent of Sugar Industry

A. Polestya, S. Sharma and D.R. Khanna
Department of Zoology and Environmental Science,
Gurukula Kangri University, Haridwar, U.K.

ABSTRACT

An attempt has been made to study the characteristics of the treated effluent of Saraswati Sugar Mill in Yamuna Nagar, by analyzing various physico-chemical parameters (pH, temperature, TS, TDS, TSS, DO, BOD, COD and oil and grease). Saraswati Sugar Mill is India's second largest sugar industry having well equipped effluent treatment plant. The duration of study was from January 2001 to March 2001. Almost all the monitored parameters were within the prescribed limits. Thus this industry does not possess any threat to water quality.

Keywords: *Sugar industry, Physico-chemical characteristics.*

Introduction

All over the world, industrial development has caused indiscriminate exploitation of natural resources, without any regard to conservation aspect. Sugar industry is one of the major industries in India. On one hand sugar industry is playing a vital role in the economy of country on other hand it is a potent and problematic source of environmental pollution. Effluent treatment plant (E.T.P.) of Saraswati Sugar Mill was designed by Hindustan Dor Oliver. It contains equalization tank, primary clarifier, aeration tank, secondary clarifier and sludge drying beds. Process adopted for the treatment of effluent is activated sludge method, which involves primary treatment, secondary treatment and dewatering of sludge on sludge drying beds.

The main purpose of water analysis is to evaluate methods of treatment of wastewater; to reuse or dispose, ascertain quality of water and aim at recovery of valuable products from waste effluents. In order to ascertain the above objectives, it is necessary to analyze various parameters, which would throw light on quality of water (Lokhande *et al.*, 2005). Several workers have carried out studies on water of different sources in respect of physico-chemical parameters (Khanna, 1993 and Khanna *et al.*, 2003). Considering the present conditions a quantitative study of physico-chemical characteristics of treated effluent of Saraswati Sugar Mill was done.

Materials and Method

Sampling

Water samples were collected from the outlet of the treatment plant.

Analytical Methods

The physico-chemical analysis was carried out according to standard methods given by APHA (1998) and Trivedi and Goel (1986). The analytical methods for various parameters are given in Table 72.1.

Table 72.1: Various Analytical Methods Used for Analysis

Sl.No.	*Parameters*	*Analytical Method*
1.	pH	pH metery
2.	Temperature	Thermometry
3.	Dissolved Oxygen (DO)	Winkler's Iodometric with azide modification method
4.	Biochemical Oxygen Demand (BOD)	5 days incubation method
5.	Chemical Oxygen Demand (COD)	Dichromate reflux method
6.	Total Solids (TS)	Gravimetric method
7.	Total Dissolved Solids (TDS)	Gravimetric method
8.	Total Suspended Solids (TSS)	Difference of TS and TDS
9.	Oil and grease	Petroleum ether method

Results and Discussion

The results for the various physico-chemical parameters determined in the treated effluent sample are presented in Table 72.2. It summarizes the maximum, minimum, mean values and standard deviations of the parameters monitored. The graphical representations are given in Figure 72.1.

The pH values were in the range of 7.33 to 7.50 (Table 72.2). The pH range for the discharge of treated effluent is 5.5 to 9.0. pH of the samples were found to be in the safe range. The values obtained were similar as observed by Manjappa and Naik (2007) for Malaprabha river. It is apparent from results that temperature values ranged between 7.50 to 16.0°C and these values were within the prescribed limits (<45°C). The range of temperature of water samples was similar as observed by Muduli and Dhal (2006) in river Baitarani at Anandpur.

Table 72.2: Various Physico-chemical Parameters at the Outlet of Effluent Treatment Plant

Sl.No.	*Parameters*	*Maximum*	*Minimum*	*Average*	*Standard Deviation*
1.	pH	7.50	7.33	7.43	0.07
2.	Temp. (°C)	16.0	7.50	12.9	3.15
3.	TS (mg/l)	261.0	214.0	227.0	19.12
4.	TDS (mg/l)	250.0	204.0	215.2	19.59
5.	TSS (mg/l)	14.0	10.0	12.0	1.58
6.	DO (mg/l)	4.1	2.0	3.02	0.99
7.	BOD (mg/l)	10.0	5.0	6.8	2.17
8.	COD (mg/l)	50.0	20.0	32.0	10.95
9.	Oil and grease (mg/l)	0.2	0.0	0.06	0.09

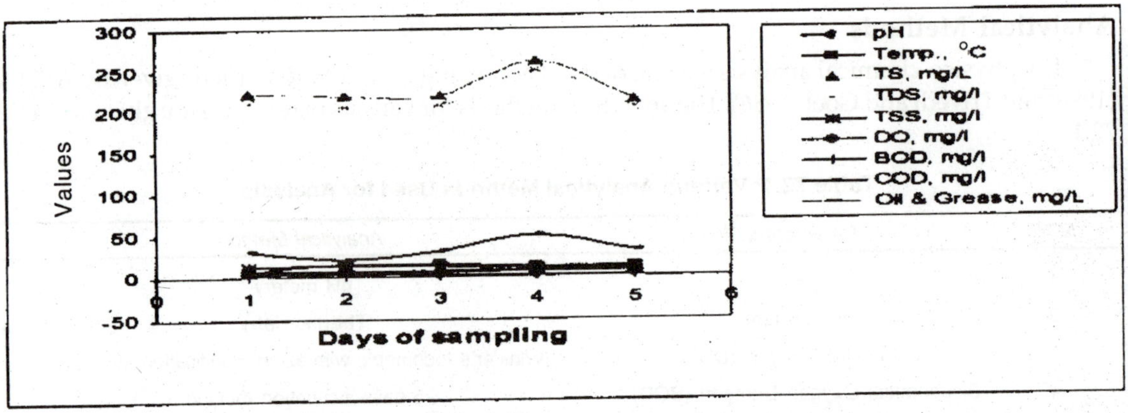

Figure 72.1: Variations in Different Physico-chemical Parameters

The total solids (TS) values ranged from 214 to 261 mg/l. TDS values ranged between 204 to 250 mg/l. The suspended solids (SS) represent the floating material (bacteria, algae) and undissolved particles, which ranged from 10 to 14 mg/l. All types of solids were found within tolerable limits.

The concentration of dissolved oxygen in water depends on temperature, pressure and the concentration of various ions. Due to pollution load the concentration of DO depletes and possess thrust on the aquatic life. Low oxygen in water can be detrimental to fishes and many other organisms in the aquatic system. The DO values ranged between 2.0 to 4.10 mg/l. BOD is the empirical test to determine the relative oxygen requirement of water mostly due to organic ingredients. Its application is to calculate the pollution load. The BOD values ranged from 5 to 10 mg/l. Maximum permissible limit for the discharge of effluent for BOD is 30 mg/l. Therefore, BOD is not possessing any threat to water quality.

COD is the amount of oxygen required for oxidation of organic constituents with strong oxidizing agent and it is also an important parameter for stream and industrial wastewater pollution studies. In the present study the COD ranges from 20 to 50 mg/l. Maximum permissible limit for the discharge of effluent for COD is 250 mg/l. Oil and grease values were from 0 to 0.2 mg/l. Permissible limits for oil and grease is 10 mg/l.

Conclusion

Saraswati sugar mill follows advanced process technology and system to achieve the environment management and performance. Therefore this industry is conscious of its responsibility and has invested heavily in reducing pollution. All the monitored parameters (pH, temperature, TS, TDS, TSS, DO, BOD, COD and Oil and grease) were within the prescribed limits and this industry is not possessing any threat to water quality.

References

APHA., 1998. *Standard Methods for the Examination of Water and Wastewater*, 20th edn. American Public Health Association, Washington, D.C.

Khanna, D.R., 1993. *Ecology and Pollution of Ganga River.* Ashish Publishing House, Delhi, p. 1–241.

Khanna, D.R., Singh, S., Gautam, A. and Singh, J.P., 2003. Assessment of water quality of River Ganga in District Bulandshahar (U.P.), India. *J. Nat Cons.*, 15(1): 167–175.

Lokhande, P.B., Gawas, A.D. and Mujawar, H.A., 2005. Study of water quality parameters of river water in Konkan region. *Indian J. Env. Prot.*, 25(3): 212–217.

Manjappa, S. and Naik, Y.K., 2007. Physico-chemical properties of Malaprabha river. *J. Env. Sci. Engg.*, 49(1): 1–6.

Muduli, S.D. and Dhal, N.K., 2006. Classification of water quality of Baitarani at Anandpur. *Indian J. Env. Prot.*, 26(2): 175–177.

Trivedy, R.K. and Goel, P.K., 1986. *Chemical and Biological Methods for Water Pollution Studies.* Env. Publications, Karad.

Water and Wastewater Management in 2 Vols (2011) *Pages* ***494–498***
Editors: **D.R. Khanna, R. Bhutiani & Gagan Matta**
Published by: **DAYA PUBLISHING HOUSE, NEW DELHI**

Chapter 73

Water Quality Assessment of Neri Nala at Durgapur, District Chandrapur, Maharashtra

P.M. Telkhade[1], N.R. Dahegaonkar[1] and S.B. Zade[2]
[1]*Department of Zoology, Arts, Commerce and Science College, Tukum, District Chandrapur, M.S.*
[2]*P.G. Department of Zoology, M.B. Patel Science College, Sakoli, District Bhandara, M.S.*

ABSTRACT

The Neri Nala is a small rivulet flowing between Chandrapur and Durgapur cities. Seasonal Studies were made to know the water quality of this small rivulet for assessing the water quality parameters *viz.* BOD, COD, TDS, Chloride, Sulphate, Phosphate along with other physico-chemical parameters. The additional burden to this small rivulet is received from washing, bathing and other domestic activities of adjoining slums. The studies indicate the polluted status of this rivulet due to inputs from adjoining localities and the degraded water quality, subsequently points out that it is unfit for bathing or any other activity. The presence of phosphate in large quantity originates due to surfactants utilized by local slum dwellers which might cause the eutrophication problem in long run.

Keywords: Neri Nala, Surfactant, Eutrophication

Introduction

The Neri Nala is a small rivulet which is situated about 2.5 km away from Chandrapur city. The slum area is situated on both the sides of this rivulet. The surrounding locality uses water for bathing, washing and other domestic purposes. The water after utilization is directly discharged into this rivulet which contain harmful detergents and domestic wastes. The study of water pollution of this rivulet has not been reported earlier, so keeping this in view, the present communication is an attempt to study the water quality. The samples were taken in month of March (summer) and August (monsoon) to study the physico-chemical variations during premonsoon and monsoon season. The present study deals with the assessment of different physico-chemical parameters like DO, BOD, COD, TDS, Chloride, Sulphate, Phosphate and other parameters of relevance. The assessment of these parameters is vital for knowing the water quality thoroughly.

Materials and Method

Water samples were collected from three sampling stations *viz.* station S_1, S_2 and S_3 once in a month of March and August for estimation of various physico-chemical parameters. Samples were collected and analysed according to APHA (1998). The samples were analysed within few hours of collection. The pH, Temp. and DO were measured on the spot. Determination of Chloride, Sulphate, Phosphate *etc.* were carried out by following standard methods in the laboratory.

Table 73.1: Parameters Estimated and Methods Used

Sl.No.	*Parameter*	*Method/Equipment Utilized*
1.	Temperature	Direct measurement on site by thermometer
2.	pH	pH meter (Model-EQ-610)
3.	Conductivity	Conductivity meter (Model-EQ-660)
4.	Alkalinity	Titrimetric method
5.	DO	Modified Winkler's iodide-azide method
6.	BOD	Modified Winkler's iodide (as per std. method)
7.	CO_2	Titrimetric method
8.	COD	Titrimetric method
9.	TDS	Gravimetric method
10.	Chlorides	Argentometric method
11.	Sulphate	Spectrophotometric (EQ-820)
12.	Phosphate	Spectrophotometric (EQ-820)

Results and Discussion

Variations of physico-chemical characteristics during March and August months are given in Table 73.2. The waters were discharged in the Neri Nala from surrounding village Durgapur. The chemical analysis of the Nala water showed that it is rich in organics as reflected by high BOD and COD values. Temperature is basically important for its effects on certain chemical and biological activities. In present study maximum temperature was recorded 32.22 °C in month of March at site S_3 and minimum temperature recorded 28.00°C in month of August at site S_2. The temperature showed an inverse relationship with the dissolved oxygen, which is consistent with result reported by Das and Shrivastava (1956) and Khanna (1997). pH is the scale of intensity of acidity and alkalinity of water and measures the concentration of hydrogen ions. Maximum pH values recorded in month of August was 7.62 at site S_2 and minimum pH value recorded in month of March was 7.02 at site S_1. Identical results were reported by Sangu and Sharma (1985) and Prapurna and Shashikanth (2002).

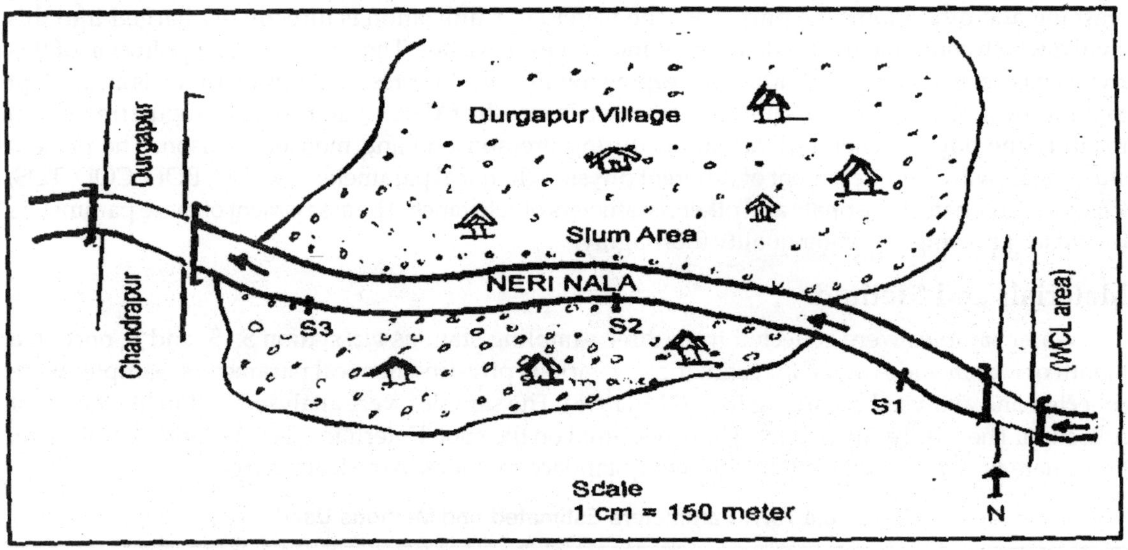

Figure 73.1: Sampling Station of Neri Nala

The high pH in rainy season is due to dissolved organic substances from the catchment area. During the study, the highest concentration was observed in month of March at site S_3 (102.00 mg/l) and lowest concentration was observed in month of August (42.90 mg/l) at site S_1. Alkalinity was mainly due to bicarbonates throughout the year, similar observations have also been reported by Holden and Green (1960) and Venkateshwarlu and Jayanti (1968).

Table 73.2: Physico-chemical Parameters of Neri Nala in Summer and Monsoon Month (March 2006 and August 2006)

Sl.No.	Parameter	March			Mean	August			Mean
		S_1	S_2	S_3		S_1	S_2	S_3	
1.	Temp. (°C)	32.1	32.17	32.22	32.16	28.5	28.0	28.02	28.17
2.	pH	7.02	7.06	7.12	7.06	7.3	7.62	7.46	7.46
3.	Alkalinity (mg/l)	89.0	96.0	102.0	95.66	42.9	45.77	46.96	45.21
4.	Conductivity (μmhos/cm)	530.0	580.0	610.0	573.33	410	422.0	489.0	437.0
5.	DO (mg/l)	3.8	4.1	4.6	4.16	5.8	6.7	7.1	6.35
6.	BOD (mg/l)	12.24	13.05	13.89	13.6	15.6	17.2	17.89	16.71
7.	Free CO_2 (mg/l)	8.02	9.11	12.1	9.74	5.21	6.29	7.62	6.37
8.	COD (mg/l)	10.9	11.18	11.89	11.32	8.2	9.67	10.2	9.29
9.	TDS (mg/l)	290.0	379.0	420.0	363.0	490	580.0	632.0	567.33
10.	Chlorides (mg/l)	21.22	22.89	24.22	22.72	34.0	35.5	36.2	35.23
11.	Sulphate (mg/l)	3.2	3.6	3.7	3.5	2.0	2.29	2.4	2.48
12.	Phosphate (mg/l)	5.0	5.2	5.8	5.4	3.2	3.7	3.85	3.58

On the basis of the data obtained from the water sampling, conductivity range in month of March was 573.33 µmhos/cm and in month of August was 437 µmhos/cm. The dissolved oxygen and free carbon dioxide are usually inversely related to one another because of the photosynthetic and respiratory activities of the biota (Hynes, 1970). Dissolved oxygen is one of the important parameter in water quality assessment. The DO was found to be maximum 7.1 mg/l in month of August at site S_3 and minimum 3.8 mg/l in month of March at site S_1. The high temperature and low dissolved oxygen during summer, was also reported by Badola and Singh (1981) in the river Alaknanda.

The free carbon-dioxide is released during the decomposition of certain substances and metabolic activities of living organism, since high temperature accelerate the decomposition of organic substances as well as respiratory activity of the biota. The carbon-dioxide maximum value recorded in month of March was 12.1 mg/l at site S_3 and minimum value recorded in month of August at site S_1 was 5.21 mg/l. This phenomenon was also reported by Quadri and Shah (1984). BOD determination is a most useful technique to assess the level of organic pollution in water. The maximum BOD was observed 17.89 mg/l at site S_3 in month of August and minimum 12.24 mg/l in the month of March at site S_1, the same trend was reported by Raina *et al.* (1984). Chemical oxygen demand is the amount of oxygen required to oxidize all the organic material. It was noted maximum in month of March *i.e.* 11.89 mg/l at site S_3 and minimum 8.20 mg/l in month of August at site S_1. Total Dissolved solids cause ecological imbalance in the aquatic ecosystem by mechanical abrasive action. In present study, it was noted that TDS was maximum 632.00 mg/l in month of August at site S_3 and minimum 290.00 mg/l in month of March at site S_1. Similar condition was also reported by Verma and Shukla (1969), Zingde *et al.* (1980). Chloride is one of the important chemical indicator of pollution in present study. The maximum value 36. 22 mg/l noted in month of August at site S_1 and minimum value 21.22 mg/l in month of March at site S_1. The similar result was also reported by Raina *et al.* (1984).

The major sources of phosphates are domestic sewage, agricultural run off and detergents and phosphate are normal constituent of human excreta. Phosphates in large quantity indicates pollution through sewage. In the present study, maximum phosphate concentration is 5.8 mg/l in month of March at site S_1 and minimum concentration recorded 3.20 mg/l in month of August at site S_1 which is due to dilution by rain. Sulphate is an important component in protein metabolism and plays important role in growth of plants. Maximum sulphate concentration recorded in present study is 3.7 mg/l in month of March at site S_1 and minimum sulphate concentration recorded 2.00 mg/l in month of August at site S_1, which is due to dilution.

In the present study values of physico-chemical parameters BOD and COD show excess values. Also phosphate and sulphate are noted to be high and it indicates that water is polluted. Pollution of water due to various agencies is one of the important cause to increase a number of blue green algae. These conditions promote eutrophication of water at a rapid rate due to which water is unsuitable for drinking, washing and bathing purpose (Wilhm and Dorris, 1968). From the present study it is noted that there is a definite impact on the water quality of Neri nala due to adjoining slums and their household discharges. Also there is a dilution effect due to rain in monsoon.

Acknowledgement

Authors are thankful to Dr. N.H. Khatri, Principal, Arts, Commerce and Science College, Tukum, Chandrapur (M.S.) for providing laboratory facilities.

References

APHA, AWWA, WPCF., 1998. *Standard Methods for Examination of Water and Wastewater*, 20th edn. American Public Health Association, Washington, DC.

Badola, S.P. and Singh, H.R., 1981. Hydrobiology of the river Alaknanda. *Poc. Nat. Sci.*, 15(B): 133–142.

Das, S.M. and Shrivastava, V.K., 1956. Some new observation on freshwater plankton. Part I. Plankton of fish tanks of Lucknow. *Sci. and Cult.*, 21(8): 466–467.

Holden, J.M. and Green, J., 1960. Hydrobiology and plankton of the river Sokota. *J. Anim. Eco. I*, 29(I): 65–84.

Khanna, D.R., 1997. Phytoplanktonic communities in relation to certain physico-chemical parameters of Ganga Canal at Haridwar. *Him. J. Env. and Zoo.*, 12: 193–197.

Prapuma, N. and Shashikanth, K., 2002. Pollution level in Hussain Sagar lake of Hyderabad: A case study. *Poll. Res.*, 21(2): 187–190.

Quadri, M.Y. and Shah, G.M., 1984. Hydrobiological features of Hoassar. A typical wetland of Kashmir–1. Biotope. *Indian. J. Ecol.*, 2(2): 203–206.

Raina, Y., Saha, A.R. and Ahmed, S.R., 1984. Pollution studies on river Jhelum–1: An assessment of water quality. *Indian J. Env. Hlth.*, 26: 187–210.

Sangu, R.P.S. and Sharma, K.D., 1985. Studies on water pollution on Yamuna River at Agra. *Indian J. Env. Hlth.*, 27(3): 257–261.

Verma, S.R. and Shukla, G.R., 1969. Pollution in a perennial stream Khala by the Sugar factory effluent near Laksar, U.P. *Indian Env. Hlth.*, 2: 145–162.

Venkateshwarlu, T. and Jayanti, T.V., 1968. Hydrobiological studies of the river Sabarmati to evaluate water quality. *Hydrobiologia*, 33(3–4): 422–448.

Wilhm, J.L. and Dorris, T.C., 1968. The biological parameter of water quality criteria. *Bio. Sci.*, 18: 477–481.

Zingde, M.D., Narvekar, P.V., Sharma, R.V. and Desai, B.N., 1980. Water quality of the River Damanganga (Gujarat). *Indian J. Mar. Sci.*, 9: 94–99.

Water and Wastewater Management in 2 Vols (2011) *Pages 499–502*
Editors: D.R. Khanna, R. Bhutiani & Gagan Matta
Published by: DAYA PUBLISHING HOUSE, NEW DELHI

Chapter 74

Quantitative Analysis of Phytoplankton and Zooplankton of Masala Lake, Masala, District Chandrapur, Maharashtra

P.M. Telkhade[1], N.R. Dahegaonkar[1], S.B. Zade[2] and A.N. Lonkar[3]
[1]A.C.S. College, Tukum, District Chandrapur, M.S.
[2]M.B. Patel Science College Sakoli, District Bhandara, M.S.
[3]Nutan A.A.C. and Smt. M.H. Wegad Science College, Umrer, District Nagpur, M.S.

ABSTRACT

Quantitative analysis of both phytoplankton and zooplankton have greater importance in limnological studies, Masala lake is situated at 6 km north east from Chandrapur city. The phytoplankton and zooplankton samples were collected from sampling site S_1, S_2 and S_3 in month of September 2006. At site S_1 the phytoplankton were found in larger number in comparison to sites S_1 and S_3, however at site S_3 the species variation is noted. About zooplankton the larger amount of zooplankton were reported at site S_2. Along with all these zooplankton some Nematode species were also noted, these species were *Diplogaster fictor* and *Rhabdolaimus minor*. In the present study the result of the quantitative analysis of both phytoplankton and zooplankton noted at different collection sites are discussed.

Keywords: *Quantitative, Phytoplankton, Zooplankton, Lake.*

Introduction

Amongst the aquatic ecosystems the lake ecosystem have unique characteristics and importance. The lake water at many places used for drinking as well as for domestic purposes. Besides studying the physico-chemical parameters of lake water, the study of phytoplankton and zooplankton have equal importance. Various workers contributed their studies on this aspect such as Hutchinson (1967), Prasadam (1977), Jyoti and Sehgal (1979), Balki *et al.* (1984), Kaur *et al.* (1999), Khanna and Bhutiani (2003), Sawane *et al.* (2006) and Veerendra *et al.* (2006). Masala lake is situated at 6 km north east from Chandrapur city, Maharashtra. This lake has greater importance for its domestic use, so the study of biological parameters as quantitative analysis of phytoplankton and zooplankton were attempted at 3 collection sites.

Materials and Method

Lake Masala having earthen embankment from one side and have a additional attached depression site. The samples were collected in the month of September 2006 from collection stations, S_1, S_2 and S_3. For studying the quantitative analysis of phytoplankton and zooplankton at each collection site separately, 70 liters of water was passed through the plankton net. The plankton samples were preserved in 4 per cent formaldehyde and brought in the laboratory for quantitative and qualitative analysis. With the help of broad mouth dropper the sample was transferred to the Sedgwick Rafter cell and the plankton were counted. The identification was made with the help of available current literature (Penak, 1978). The quantitative analysis are presented in Tables 74.1 and 74.2.

Observation

In the present study 12 species of phytoplankton and 7 species of zooplankton with their quantitative estimates were noted at collection station S_1, S_2 and S_3. The result of quantitative analysis of phytoplankton is presented in Table 74.1 and zooplankton is in Table 74.2.

Table 74.1: Quantitative Analysis of Phytoplankton at Collection Stations S_1, S_2 and S_3

Sl.No.	*Name of Phytoplankton*	*Number of Phytoplankton*		
		S_1	S_2	S_3
1.	*Spirogyra*	76	23	32
2.	*Acanthes lanciolate*	67	16	21
3.	*Anabaena*	69	74	88
4.	*Nostoc linekia*	102	98	83
5.	*Spirulina*	87	39	56
6.	*Diatom vulgare*	103	78	94
7.	*Phacus succica*	20	37	18
8.	*Geodinium montanum*	37	46	30
9.	*Closterium*	18	17	12
10.	*Chlorocloster pirenigera*	22	188	24
11.	*Dinobryon stipitatum*	–	–	7
12.	*Volvox*	–	3	11

Results and Discussion

The phytoplankton were found more in number at collection station S_1, however the species variation is more at station S_3, where *Dinobryon stipitatum* was noted as well as more number of *Volvox* were noted. Khanna and Bhutiani (2003) reported 3 genera of Cyanophycae, *Oscillatoria*, *Anabena* and

Nostoc. These were recorded highest in winter in Sitapur pond at Haridwar. Veerendra *et al.* (2006) identified 34 species of phytoplankton under 4 classes, among them maximum density was recorded under Bacillariophyceae, Chlorophyceae, Cyanophyceae and Euglenophyceae, in Mani reservoir, Hosangar, Karnataka. In the present investigation the phytoplankton *Nostoc* and *Anabena* were noted, the findings agrees with the findings of Khanna and Bhutiani (2003) and Veerendra *et al.* (2006). In the quantitative analysis of zooplankton 7 species of zooplankton were noted these includes *Cyclops, Brachionus forticula, Keratella tropica, Asplanchnopus, Lepadella oralis, Diplogaster fictor* (Nematode) and *Rhabdolaimus minor* (Nematode). Hutchinson (1967) and Prasadan (1977) noted presence of Cladocerons in lower profile in both annual cycle and such as no definite pattern of their variations were observed. However they are mostly abundant in winter and summer seasons.

Table 74.2: Quantitative Analysis of Zooplankton at Collection Stations S_1, S_2 and S_3

Sl.No.	*Name of Zooplankton*	*Number of Zooplankton*		
		S_1	S_2	S_3
1.	*Cyclops*	4	6	5
2.	*Brachionus forticula*	12	14	17
3.	*Keratella tropica*	9	11	7
4.	*Asplanchnopus*	4	3	4
5.	*Lepadella oralis*	1	4	3
6.	*Diplogaster fictor* (Nematode)	4	5	6
7.	*Rhabdolaimus minor* (Nematode)	3	6	2

Jyoti and Sehgal (1979) and Balki *et al.* (1984) observed Rotifera from the dominant zooplankton fauna in many aquatic habitats, Kaur *et al.* (1999) identified 6 taxas of Protozoa, 12 Rotifers, 11 Crustaceans, 14 Insects, 7 Annelids, 8 Mollusca, 2 Nematodes, 1 Nemartina, from 6 sites of Kanjali lake from November 1996 to March 1997. In the present work the Nematode species *Rhabdolaimus minor* and *Diplogaster fictor* are noted. Pathak and Mudgal (2002) reported 19 species which belonged to Protozoa, Cladocera, Ostracoda, Copepoda and Rotifera. Sawane *et al.* (2006) noted the 8 zooplankton species in Irai dam Chandrapur. These are *Difflugia, Cyclops, Diaptomus, Chydorus, Moina, Brachionus calciflorus, Brachionus fulcatus* and *Cypris*. The density of zooplankton were noted more in winter and less in summer. The present study showed that the phytoplankton *Nostoc linekia* and *Diatoms vulgare* are abundant in number and *Dinobryon stipitatum* are least. The species variation is more at collection station S_3. Amongst zooplankton *Cladoceras* are absent. *Brachionus* are found more in number and *Lepadella oralis* are least in number. It was noted that the Rotifers were present in larger number as well as the algae like *Anabena* and *Nostoc* were present, this shows that the lake water may be contaminated with domestic pollutants. The plankton both phytoplankton and zooplankton were considerable present in good amount, it may be due to favorable physico-chemical conditions in the month of September or approaching winter as many worker noted abundance of plankton in the winter season.

Acknowledgement

Authors are thankful to Dr. N.H. Khatri, Principal, Arts, Commerce and Science College, Tukum, Chandrapur, (M.S.) for providing laboratory facilities.

References

Balki, M.H., Yousuf, A.R. and Quadri, M.Y., 1984. Rotifera of Anchor lake during summer and winter. *Geobios News Reports*, 3: 163–165.

Hutchinson, G.E., 1967. *A Treatise on Limnology, Vol. II: Introduction to Lake, Biology and the Limnoplankton*. John Wiley and Sons, New York.

Jyoti, M.K. and Sehgal, M., 1979. Ecology of rotifera of surino 9-subtropical freshwater lake in Jammu (Jammu and Kashmir) India. *Hydrobiologia*, 27: 160–187.

Kaur, H.K.S., Bath, G.M. and Dhillon, S.S., 1999. Aquatic invertebrate diversity of Kanjali lake Punjab. *Indian J. Environ. and Ecoplan.*, 2(1): 37–41.

Khanna, D.R. and Bhutiani, R., 2003. Ecological status of Sitapur pond at Haridwar (Uttarakhand) India. *Indian J. Environ. and Ecoplan.*, 7(1): 175–178.

Penak, R.W., 1978. *Freshwater Invertebrates of United States*, 2nd edn. John Wiley and Sons, New York.

Prasadam, R.D., 1997. Observation of zooplankton population of some freshwater impoundments in Karnataka. In: *Symp. Warm Wat. Zooplankton*, N10, Goa, pp. 214–225.

Sawane, A.P., Puranik, P.G and Lonkar, A.N., 2006. Preliminary study on the seasonal distribution of plankton in Irai river at Irai dam site District Chandrapur, Maharashtra. *Indian J. Environ. and Ecoplan.*, 12(1): 207–212.

Veerendra, D.N., Manjappa, S. and Puttaiah, E.T., 2006. Diversity of phytoplankton in Mani reservoir Hosangar, Karnataka. *Indian J. Env. and Eco.*, 12(2): 335–338.

Water and Wastewater Management in 2 Vols (2011) *Pages* **503–507**
Editors: **D.R. Khanna, R. Bhutiani & Gagan Matta**
Published by: **DAYA PUBLISHING HOUSE, NEW DELHI**

Chapter 75

Impact of Improper Disposal of Solid Waste on Groundwater Quality: A Case Study

S.C. Gupta and Seema Sambyal
Department of Environmental Sciences,
University of Jammu, Jammu

ABSTRACT

The present study bas been made to evaluate the groundwater quality of Samba town of Jammu division of J&K state. The solid waste generated in urban area is increasing day-by-day. Many municipalities and corporations are facing the problem of managing large volumes of solid waste generated in their area. One of the alternatives for disposal of solid waste is filling the depressions, drains, nallah, open spaces *etc.* In the present study, a big nallah flowing through Samba town carries huge load of pollutants in the form of refuse and wastewater opens at many places into self-created garbage pits. Groundwater monitoring was carried out at two stations around garbage pit to assess the groundwater quality of hand pumps used by the residents of the area for domestic and other uses. Tile study revealed that the quality of groundwater, however, conform to the drinking water quality standards but there is need for periodic evaluation of various water quality parameters as well as adoption of proper water quality management practices as natural quality of groundwater is being degraded near the garbage pit.

Keywords: *Solid waste, Groundwater quality, Depressions, Drains, Nallahs, Open spaces.*

Introduction

Due to rapid growth of population, urbanization and industrialization, there is several fold increase in generation of waste and wastewater. Due to improper disposal of solid waste, there is a great threat to the quality of groundwater.

In the present study, a nallah flowing through Samba town, carries huge load of pollutants in the form of waste and wastewater was studied. This nallah opens at many places into self-created garbage pits before draining into the river Basantar. In garbage pits, solid waste stays on the top but water percolates deep inside into the surface. Therefore, it is important to monitor the surface and groundwater quality parameters. The monitoring programme was planned to measure groundwater quality with the objectives to evaluate, analyze and summarize existing groundwater quality, and to correlate the results of the data with permissible standards for drinking water quality.

Materials and Method

Study Area

The present study has been aimed to assess the quality of groundwater of Samba town of Jammu division of J&K state. The solid waste generated in Samba town is mostly dumped in open land fills in low lying areas, drains, nallah *etc.* Since there is no sewage treatment plant and recycling facilities in the area, the sewage and the solid waste is directed in the nallah which flows in the town. For this purpose, two different stations or study sites have been selected Station–I (hand pump located near the garbage pit) and Station–II (hand pump located at a distance of 300 meters from the garbage pit).

Since the town has developed in an unplanned manner, there is no underground drainage system in the area. Moreover, the nallah opens up at many places into self-created garbage pits. In these pits the solid waste stays at the top and the stored water seeps into soil cover. Once the soil cover is saturated, the infiltrated water moves to subsurface water through fissures and joints.

Methodology

A total of 6 representative samples were collected fortnightly from the study area during November 2005 to January 2006. The samples were collected from both hand pumps, which were being extensively used for drinking and other domestic purposes. The location of groundwater sampling stations (Hand pumps) is shown in Figure 75.1. Both physical and chemical analysis were carried out according to standard methods (APHA, 1998). The various analyzed parameters include pH, DO, COD, chloride, nitrate, calcium, magnesium and total hardness.

Results and Discussion

For studying the impact of refuse and wastewater of the garbage pit on groundwater quality, the water quality parameters were analyzed and are presented in Table 75.1. A comparison of groundwater quality of the study area with drinking water standards per guidelines laid down by WHO (1971) and BIS (1990) is presented in Table 75.2. A critical examination of the tables reveals that quality of water considerably vary from location to location. pH is a measure of intensity of acidity or alkalinity and the concentration of hydrogen ions in water. The pH of the water samples collected around the garbage pit ranged from 6.9 to 8.21, where as the samples collected from the hand pump located at a distance 300 meters away from the garbage pit showed a range of 7.42–8.61. The pH values of both samples are normally acceptable as per guidelines suggested by WHO (1971) and HIS (1990). These fubdings are in conformity with Murugesan *et al.* (2005), *i.e.* 6.62–8.07; Mor *et al.* (2003), *i.e.* 7.2–8.6 and

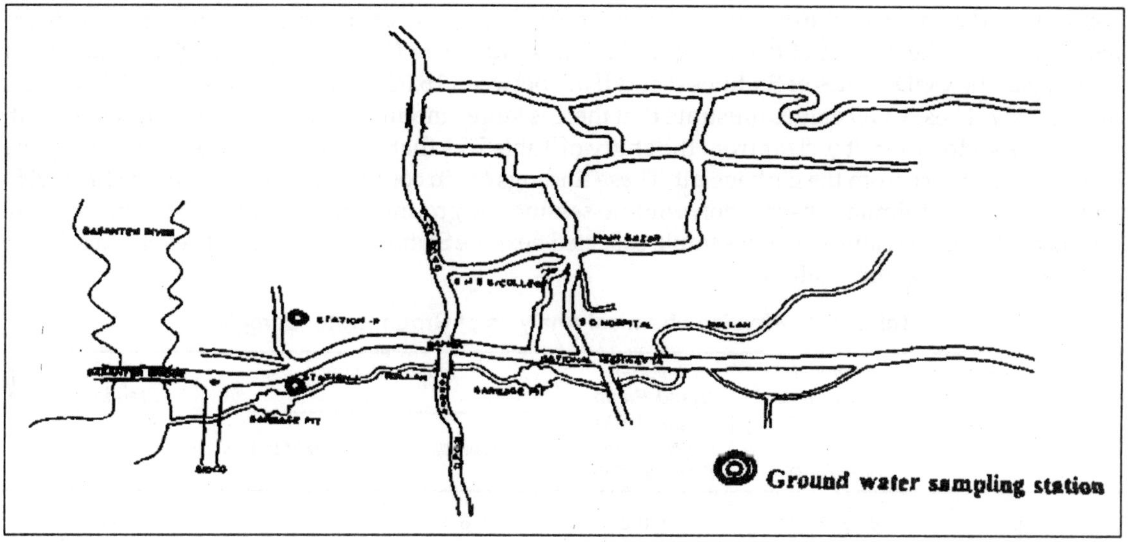

Figure 75.1: Location Map of Groundwater Sampling Stations

Aurangabadkar *et al.* (2000), *i.e.* 6.5–8.7 who have worked on the levels of groundwater pollution in the urban environment.

Table 75.1: Physico-chemical Analysis of Groundwater Sample

Parameter	*Station–I*						*Station–II*					
	Hand Pump Located 5 m Away from the Garbage Pit						*Hand Pump Located 300 m Away from the Garbage Pit*					
	I	*II*	*III*	*IV*	*V*	*VI*	*I*	*II*	*III*	*IV*	*V*	*VI*
pH	6.92	8.21	7.7	8.2	7.6	7.4	7.42	8.61	8.0	8.4	8.0	7.8
DO	1.48	3.2	3.2	4.0	4.0	4.0	1.65	3.5	4.0	5.3	5.3	4.2
COD	28.8	36:0	26.0	16.0	17.0	17.0	14.4	24.0	16.0	13.0	13.0	15.0
Cl^-	58.03	109.9	83.8	113.6	120.4	138.0	24.87	14.4	20.40	30.55	45.0	24.0
SO_4^{2-}	33.0	17.5	28.0	26.0	42.0	26.0	1.5	6.0	18.0	16.0	14.0	16.0
NO_3^-	9.5	3.5	0.7	6.6	3.8	16.8	0.25	1.1	2.2	9.0	2.8	1.2
Ca^{2+}	117.02	135.82	88.9	113.0	75.0	92.3	87.37	82.46	40.1	42.5	72.1	74.5
Mg^{2+}	23.16	29.39	35.1	38.5	25.0	31.7	17.96	13.71	7.3	18.0	21.9	25.3
TH	387.15	459.67	366.0	440.0	292.0	360.0	291.82	262.09	130.0	180.0	270.0	290.0

* All parameters are in ppm except pH.

Dissolved oxygen (DO) measured for all samples ranged between 1.48–4.00 ppm and 1.65–5.30 ppm at Stations–I and II, respectively. These values of DO are below 5.00 ppm which clearly indicate pollution which may be due to the contents of decomposable matter in the samples which is more in the water samples collected from Station–I as compared to those collected from Station–II. These findings are comparable with the findings of Ravinder *et al.* (2005) who recorded high DO values from wells situated away from the dumping site and low DO values in the nearby wells of Warangal town while studying the impact of dumping municipal solid waste (MSW) on groundwater quality of Warangal. The COD values of Stations–I and II are between 16.00–36.00 ppm and 13.00–24.00 ppm, respectively. These COD values illustrate that there is large amount of organic matter present in both the samples. However, it is clear from the study of Table 75.1 that the COD value decreases with the increase in distance from the garbage pit. These findings are in conformity with Ravinder *et al.* (2005) who has recorded similar observation while assessing the groundwater quality of Warangal town. The increase in COD value of water sample collected near the Station–I is due to high level of discharge of domestic sewage from Nallahas.

Table 75.2: Physico-chemical Analysis of Groundwater Sample

Parameter	*Station–I (Hand Pump Located at 5 m Away from Garbage Pit)*	*Station–II (Hand Pump Located at 300 m Away from Garbage Pit)*	*Indian Standard*		*WHO Standard*
			Acceptable	*Maximum*	
pH	6.92–8.21	7.42–8.61	7.0–8.5	6.5–9.2	6.5–9.2
DO	1.48–4.0	1.65–5.3	–	–	–
COD	16.0–36.0	13.0–24.0	–	–	10
Cl^-	58.03–138	14.4–45.0	200.0	1000.0	500.0
SO_4^{2-}	17.50–42.0	1.5–18.0	200.0	400.0	–
NO_3^-	0.7–16.8	0.25–9.0	45.0	100.0	–
Ca^{2+}	75.0–135.82	40.1–87.37	75.0	200.0	75.0
Mg^{2+}	23.16–38.50	7 .30–25.30	30.0	–	150.0
TH	292.0–459.67	130.0–291.82	200.0	600.0	500.0

*All parameters are in ppm except pH.

Chloride concentration in the water samples collected from the Station–I ranged from 58.03 to 138.00 ppm whereas it ranged from 14.40 to 45.00 ppm at Station–II. Both these concentrations or levels of chloride are well within the permissible limits of WHO (1971) and BIS (1990). But chloride content from the Station–I showed higher content as compared to the Station–II. Therefore, it can be explained that there is regular addition of large quantities of sewage from nearby localities to the garbage pit, it is absorbed by the soil or it leached into the soil and move with the groundwater. Moreover, food waste also contributes to the chloride content in water.

Sulphate content in the water samples of present study ranged from 17.50 to 42.00 ppm from the Station–I and 1.50 to 18.00 ppm from the Station–II. The higher concentration of sulphate in water samples of Station–I is due to the seepage of pollutants from the wastewater which is being regularly added to the garbage pit by the Nallah. But both the values of the sulphate are found to be within the prescribed limits of WHO (1971) and BIS (1990). The present values of sulphate content are in agreement with the values recorded by Mor *et al.* (2003) who have tried to work out the pollution status of groundwater of Jind city.

Nitrate content in both the samples collected from Stations–I and II ranged from 0.70 to 16.80 ppm and 0.25 to 9.00 ppm respectively. Table 75.2 shows that although both the values are within the permissible limits of 45–100 ppm as prescribed by BIS (1990), the Station–I values are higher as compared to the Station–II values which may be due to large addition of decayed vegetable and animal matter, sewage sludge, domestic effluents disposal to land, leachates from refuse dumps and atmospheric washout. Similar findings were recorded by Ravinder *et al.* (2005).

Calcium concentrations from Stations–I and II ranged 75.00 to 135.82 ppm and 40.10 to 87.37 ppm, respectively and this level/concentration of Ca^{++} content is found to be within the maximum limit prescribed (75–200 ppm) by the Indian Standards (1990) by above (75 ppm) WHO (1971) standards. Again Ca^{++} content is high in the water sample collected from the Station–I which is due to fact that domestic wastewater contributes ions to the groundwater. Magnesium concentration in the groundwater samples collected from Stations I and II ranged from 23.16 to 38.50 ppm and 7.30 to 25.30 ppm, respectively and is well within the prescribed limits of WHO (1971) and BIS (1990). Hardness values ranged from 292.00 to 459.67 ppm and 130.00 to 291.82 ppm in water samples collected from Stations–I and II, respectively.

These values of Ca^{++}, Mg^{++} and total hardness are in agreement with the values recorded by Mor *et al.* (2003), while assessing the pollution status of groundwater due to landfills and septic tanks in Jind city of Haryana. Ca^{++} and Mg^{++} ions in greater quantities may be present in groundwater either by leaching of soil deposits or through seepage of ions from domestic wastewater. The hard water causes ill-effects on digestive system and moreover, the possibilities of forming Ca^{++} oxalate crystals (leading to stone formations) in the urinary tracts.

Summary and Conclusion

From the above results, it is concluded that the groundwater of Samba town is not highly contaminated but there is an indication of increasing pollution due to discharge of sewage and solid waste into rivers and nallah along with other land sites which ultimately percolates into the ground and is thus responsible for groundwater pollution. Hence, there is an urgent need to take immediate necessary steps for the protection of this valuable natural resource in Samba town.

Acknowledgement

The authors are highly thankful to the Department of Environmental Sciences, University of Jammu, Jammu for providing necessary facilities to undertake the present investigation.

References

APHA, AWWA, WPCF, 1998. *Standard Methods for the Examination of Water and Wastewater*, 20th edn. American Public Health Association, Washington DC.

Aurangabadkar, K., Swaminath, K., Sandhya, S., Uma, T.S., Jyotikumar, N., and Paramasivam, R., 2000. Groundwater quality around a municipal solid waste dumpsite at Chennai, India. *J. Environ. Prot.*, 24(94): 323–327.

BIS, 1990. *Drinkng Water Specifications.* IS 10500. Bureau of Indian Standards, New Delhi.

Mor, S., Bishnoi, M.S. and Bishnoi, N.R., 2003. Assessment of groundwater quality of Jind city. *Indian J. Environ. Prot.*, 23(6): 673–679.

Murugesan, A., Ramu, A. and Kanan, N., 2005. Characterization of groundwater quality in Madurai region. *Indian J. Environ. Prot.*, 25(10): 885–892.

Ravinder, Y., Ravinder, Ch. and Rao, K.Y., 2005. Groundwater pollution due to dumping of municipal solid waste at Warangal. *Indian. J. Environ. Prot.*, 25(6): 523–526.

WHO, 1971. *International Standard for Drinking Water*, 3rd edn. World Health Organization, Geneva.

Water and Wastewater Management in 2 Vols (2011) *Pages 508–512*
Editors: **D.R. Khanna, R. Bhutiani & Gagan Matta**
Published by: **DAYA PUBLISHING HOUSE, NEW DELHI**

Chapter 76

Studies on Physico-chemical and Biological Parameters of Chorgaon Lake, District Chandrapur, M.S.

P.M. Telkhade[1], N.R. Dahegaonkar[1], S.B. Zade[2], C.J. Khune[2] and A.N. Lonkar[3]

[1]*A.C.S. College, Tukum, District Chandrapur, M.S.*
[2]*M.B. Patel Science College, Sakoli, District Bhandara, M.S.*
[3]*Nutan A.A.C. and Smt. M.H. Wegad Science College, Umrer, District Nagpur, M.S.*

ABSTRACT

Chorgaon lake is a man made freshwater reservoir at Chorgaon village, in District Chandrapur. The present study has been made to investigate the physico-chemical and biological status of the lake. The study involves the physico-chemical analysis like DO, COD, BOD, Alkalinity, Total Hardness, Chloride, Sulphate and Phosphate along with other parameters and quantitative and qualitative analysis of phytoplankton and zooplankton in biological parameters. Samples were collected from four sampling stations of the lake. The results revealed that site S_2 and S_3 shows more Nitrogen and Phosphate, which favoured the growth of phytoplankton and zooplankton. Phytoplankton shows dominance of Bacillariophyceae. In zooplankton abundance of Rotifers, Cladocerons and Copepods were observed. The average value revealed that Dioptomus shows dominance in zooplankton and Diatoms in phytoplankton. At other sites parameters remained in constant range showing no much variation thus indicating better quality of water, which was free from pollution.

Keywords: *Quantitative, Phytoplankton, Zooplankton.*

Introduction

Water in its various forms is a major element of all the components of biosphere and one of the most needed factor for the existence of living organisms. Besides studying the physico-chemical parameters of lake water, the study of biological parameters also have equal importance. Chorgaon lake is a man made reservoir situated in the North-Eastern part of District Chandrapur. The lake water is used for irrigation, aquaculture as well as for domestic purposes.

The present work was carried out during August 2006 to July 2007 in which focus was given on the study of water quality in terms of physico-chemical parameters and biological diversity of the lake.

Materials and Method

The water samples were collected from Chorgaon lake from four location sites S_1, S_2, S_3 and S_4. The water samples were collected in pre cleaned 5 liters plastic can. The sample collection was usually completed during morning hrs between 8.00 AM to 10.00 AM. The parameters like DO, Temp., pH were analyzed on the spot while Total solids, Total alkalinity, BOD, COD, were analyzed in the laboratory by standard methods given in APHA (1998). The plankton were collected by plankton net and were preserved in 4 per cent formaline.

Results and Discussion

In an aquatic ecosystem physico-chemical environment has profound influence on its biotic components. They exert their influence both individually and collectively. The values of physico-chemical parameters of water samples collected from various sites are shown in Table 76.1 and 76.2.

During the study, the temperature of water ranged between 25.6°C to 33.6°C minimum temperature *i.e.*, 25.6°C was observed in month of January at Site S_1 and maximum temperature was recorded 33.6°C in month of May at site S_4. Temperature variation is due to depth and inflow of water in catchment area. The permissible limit of pH for potable water ranges within 6.0 to 8.5. In the present investigation the pH value noted down minimum 7.1 in August at site S_4 and maximum 8.4 at site S_4 in April.

Total Alkalinity of water is a measure of weak acid present in it and of the cations balanced against them. The highest concentration is 24.2 mg/l in month of May at site S_2 and the lowest concentration *i.e.* 156 mg/l was noted down in month of August at site S_4. Throughout the investigation period, it was noted that the Dissolved Oxygen ranges between 3.6 to 6.95 mg/l. Similar variation in oxygen was reported by Khatavkar *et al.* (1989) and Bhosle *et al.* (1994). Free CO_2 value was observed maximum 4.89 mg/l at site S_2 in month of August, this may be due to higher turbidity. Minimum value 2.11 mg/l was noted down in month of September at site S_4. In winter season turbidity was lowest.

The value of Total hardness was maximum in month of May at site S_3 *i.e.*, 298 mg/l and minimum 106 mg/l at site S_1 in month of December. This may be due to presence of high content of Ca and Mg in addition to sulphates and nitrates. The maximum value of TDS was noted down 1684 mg/l in month of June at site S_3 and minimum value was 1058 mg/l in month of February at site S_4, Chloride concentration was recorded maximum 23.7 mg/l in month of May at site S_3 while the minimum was observed 14.05 mg/l in month of December at site S_4. The value are within permissible limit with respect to DIS, ICMR. Sulphate varied from minimum 0.89 mg/l to maximum 2.12 mg/l during the investigation period. Its value increased during month of May. Phosphate value was maximum in month of July at site S_3 *i.e.*, 3.01 mg/l and minimum value recorded in month of December at site S_1 *i.e.* 1.88 mg/l. The Nitrate shows the range of 2.9 mg/l to 8.15 mg/l. The Nitrate level was max in the month of August at site S_3 and minimum was recorded in month of February at site S_1.

Table 76.1: Physico-chemical Parameter of Chorgaon Lake during August 2006 to January 2007

Sl.No.	Months	August				September				October				November				December				January			
	Sites / Parameters	S_1	S_2	S_3	S_4	S_1	S_2	S_3	S_4	S_1	S_2	S_3	S_4	S_1	S_2	S_3	S_4	S_1	S_2	S_3	S_4	S_1	S_2	S_3	S_4
1.	Temp.	29.1	29.2	29.1	29.1	28.9	28.6	28.8	29.1	28.2	28.4	28.2	28.2	27.1	27.1	27.2	27.1	26.00	26.2	26.3	26.00	25.6	25.7	25.7	26.7
2.	pH	7.2	7.4	7.3	7.1	8.1	8.2	8.0	7.99	7.91	7.98	7.99	7.92	7.76	7.71	7.69	7.70	7.22	7.82	7.86	7.70	7.91	7.89	7.96	7.90
3.	Alkalinity	158	172	168	156	169	175	167	167	174	184	186	179	187	168	182	179	182	189	187	190	198	197	199	189
4.	DO	3.8	4.1	4.00	3.9	4.00	3.6	3.8	3.8	4.1	4.6	4.2	4.1	5.1	4.9	4.1	4.8	5.7	5.4	5.5	5.4	6.4	6.1	6.2	6.4
5.	Free CO_2	4.52	4.89	4.81	4.4	2.19	2.23	2.27	2.11	2.64	2.67	2.71	2.59	2.42	2.67	2.86	2.41	3.08	3.09	3.10	3.02	2.89	2.69	2.71	2.88
5.	Total Hardness	268	279	271	262	252	284	276	258	212	217	229	240	196	203	210	188	106	157	168	42	217	219	225	154
7.	Ca-Hardness	189	180	198	181	160	201	189	179	158	171	162	158	154	149	153	142	82	120	123	111	164	172	173	122
8.	Mg-Hardness	79	99	73	81	92	a3	87	79	54	46	67	82	42	54	5	46	24	37	45	31	53	47	52	32
9.	Chlorides	20.21	19.81	21.23	19.98	19.20	20.41	20.89	20.21	18.89	18.60	19.01	18.90	16.12	17.02	17.89	16.09	14.99	15.03	15.72	14.05	17.02	17.86	17.46	17.01
10.	TDS	1630	1612	1590	1628	1487	1510	1521	1460	1300	1380	1372	12.89	1287	1350	13.64	1288	1265	1298	1290	279	1120	1210	1219	1189
11.	Sulphates	1.60	1.40	1.78	1.58	1.10	1.15	1.25	1.49	1.02	1.20	1.38	1.10	0.89	1.13	1.13	0.92	0.90	1.02	1.14	0.98	0.98	1.06	1.08	0.99
12.	Phosphates	2.80	2.89	2.87	2.71	2.20	2.33	2.39	2.58	2.01	2.14	2.17	2.05	1.92	2.01	2.03	1.99	1.88	1.99	2.01	2.10	1.80	2.03	2.06	2.15
13.	Nitrate	7.10	8.12	8.15	7.12	7.13	8.01	8.04	7.13	6.15	6.23	6.27	6.69	5.02	6.01	5.98	5.12	4.20	4.68	4.98	4.18	3.53	3.89	3.60	3.83

* All parameters are in mg/l except pH and Temp. (°C).

Table 76.2: Physico-chemical Parameter of Chorgaon Lake during February 2007 to July 2007

Sl.No.	Months	February				March				April				May				June				July			
	Sites	S_1	S_2	S_3	S_4	S_1	S_2	S_3	S_4	S_1	S_2	S_3	S_4	S_1	S_2	S_3	S_4	S_1	S_2	S_3	S_4	S_1	S_2	S_3	S_4
	Parameters																								
1.	Temp.	27.00	27.3	27.8	27.4	29.00	28.8	28.9	29.4	32.00	31.6	32.4	32.6	33.00	32.18	33.1	33.6	31.00	31.4	31.6	31.4	30.00	30.4	30.4	30.1
2.	pH	8.1	7.9	8.1	8.3	7.9	7.9	7.6	7.7	8.3	8.1	8.4	8.3	7.7	7.3	7.9	7.9	7.6	7.5	7.3	7.9	7.3	7.4	7.5	7.6
3.	Alkalinity	178	187	186	172	269	204	211	214	213	210	212	236	231	242	236	241	189	160	179	180	168	169	170	165
4.	DO	6.17	6.12	5.9	6.1	6.95	6.90	6.80	6.70	5.70	5.2	5.5	5.2	4.1	3.9	4.3	4.1	3.9	3.6	3.8	3.9	4.1	4.2	4.6	4.5
5.	Free CO_2	2.20	2.50	2.40	2.44	3.74	3.60	3.71	3.50	3.9	4.11	4.18	4.12	3.50	3.12	3.5	3.25	3.40	3.39	3.33	4.20	4.41	4.42	4.3	4.19
6.	Total Hardness	235	240	242	229	260	270	278	255	275	285	291	278	289	290	298	281	260	278	279	255	266	288	281	272
7.	Ca-Hardness	162	166	166	16.0	182	116	186	161	176	189	204	171	186	266	195	175	172	174	169	124	163	176	172	165
8.	Mg-Hardness	73	74	76	69	78	94	92	88	99	96	87	1.01	1.03	84	103	106	88	104	11.0	131	103	112	109	1.07
9.	Chlorides	18.8	17.2	17.5	17.1	20.2	20.6	20.8	20.5	22.9	22.95	23.1	22.8	23.6	23.5	23.7	23.6	21.2	21.3	21.3	21.3	19.3	19.4	19.5	19.2
10.	TDS	1060	1068	1073	1058	1102	1112	1129	1128	1190	1198	1197	1192	1220	1223	1223	1223	1600	1664	1684	1631	11650	1658	1641	1638
11.	Sulphates	1.10	1.12	1.11	1.09	1.23	1.24	1.24	1.23	1.61	1.69	1.67	1.59	2.12	2.12	2.12	2.10	1.90	1.95	1.96	1.95	1.98	1.98	1.98	1.96
12.	Phosphates	1.95	1.96	1.96	1.95	2.01	2.12	2.12	2.03	2.35	2.37	2.33	2.39	2.80	2.89	2.92	2.78	2.65	2.68	2.68	2.66	2.99	2.99	3.01	2.99
13.	Nitrate	2.9	3.01	3.04	3.00	3.02	3.11	3.8	3.02	4.01	3.9	3.74	3.81	4.80	4.98	4.1	4.5	4.6	4.3	5.2	5.8	5.01	5.6	5.4	5.10

* All parameters are in mg/l except pH and Temp. (°C).

The phytoplankton communities were represented mainly by four groups. Chlorophyceae, Cyanophyceae, Bacillariophyceae and Charophyceae. Chlorophyceae was represented by *Spirogyra, Clostridium, Cosmarium etc.* showed its maximum value during the month of April and minimum during the month of Aug. Bacillariophyceae was represented by *Navicula, Cymbella, Diatoma vulgare etc.* Its maximum value was noted down during April and May and minimum was in Aug. Cyanophyceae was a significant group, this group includes Anabaena, Oscillatoria, Nostoc, Microcystis *etc.* It shows higher appearance during the month of Jan and minimum in month of June. Charophyceae was represented by Chara and Nitella during present investigation. They stand fourth in their dominance. Kumar (1990) estimated that density of phytoplankton is greater during summer, post monsoon and winter and is lowest in monsoon. In the present study also peak of the phytoplankton was observed during summer and lowest during monsoon. Verma and Mohanty (1995) recorded three peaks March, July and Jan for phytoplankton at Danmukundpur pond. In present study the phytoplankton shows their dominance as follows:

Bacillariophyceae > Chlorophyceae> Cyanophyceae> Charophyceae

The zooplankton communities were presented mainly by four groups Rotifers, Ostracoda, Copepoda and Cladocera. In the present observation Rotifer were found maximum in the month of April and May and minimum was in the month of Sept. They are represented by *Asplanchnopus, Brachionus, Licane etc.* The Cladocera represented by *Moinodaphnia, Bosmina, Moina etc.* They showed maximum value during the month of January and minimum in month of July. Copepoda was represented by Diaptomus, Cyclops *etc.* They showed their maximum value during month of December and minimum during month of August. Ostracoda represented by Cypris showed its maximum value during month of August and minimum during month of May.

In present investigation zooplankton showed their dominance as follows:

Rotifers > Cladocera > Copepoda > Ostracoda

The average value revealed that the physico-chemical and biological parameters were in permissible range showing no much variation, indicates better quality of water.

Acknowledgement

The authors are thankful to Dr. N.H. Khatri, Principal, A.C.S. College, Tukum, Chandrapur for providing laboratory facilities.

References

APHA, AWWA and WPCF, 1998. *Standard Method for Examination of Water and Wastewater*, 20th edn. American Public Health Association, Washington, DC.

Bhosle, L.J., Sabala, A.B. and Mulik, N.G., 1994. Survey and status report of project submitted to Shivaji University, Kolhapur, India, pp. 60.

Khatavkar, S.D., Kulkarni, A.V. and Goel, P.K., 1989. Observation on the dial cycle of phytoplankton and some nutrient during summer in the surface water of a shallow mesotrophic Lake. *Geobios*, 16: 210–214.

Kumar, S., 1990. Limnology of Kunjwan pond with reference to plankton and macrophytes. *M.Phil Disser.*, Jammu, India.

Verma, J.P. and Mohanty, R.C., 1995. Phytoplankton and its correlation with certain physico-chemical parameters of Damnukundpur pond. *Poll. Res.*, 14: 233–242.

Water and Wastewater Management in 2 Vols (2011) *Pages 513–516*
Editors: **D.R. Khanna, R. Bhutiani & Gagan Matta**
Published by: **DAYA PUBLISHING HOUSE, NEW DELHI**

Chapter 77

Status of Water Quality of Masala Lake Durgapur, District Chandrapur, M.S.

P.M. Telkhade[1], N.R. Dahegaonkar[1], S.B. Zade[2] and P.N. Charde[3]
[1]A.C.S. College Tukum, District Chandrapur, M.S.
[2]M.B. Patel Science College Sakoli, District Bhandara, M.S.
[3]Sevadal Mahila Mahavidyalaya, Nagpur, M.S.

ABSTRACT

Masala lake is situated 6 km away from Chandrapur. The present study has been made to investigate the physico-chemical parameters of the lake. The present study was carried out from the month of August 2006 to January 2007. Samples were collected on monthly basis from three sampling stations S_1, S_2 and S_3. The physico-chemical parameters like temp, pH, conductivity, DO, COD, BOD, free CO_2, chloride, total hardness, Ca hardness, Mg hardness, phosphate, sulphate, nitrate, nitrites, carbonates and bicarbonates were analyzed during the course of study. The study revealed that as compared to site S_1 and S_3, site S_2 showed higher concentration of sulphate, phosphate and bicarbonates during month of Nov and Dec 2006 and were recorded low during the month of August and September 2006 may be due to dilution, site S_2 shows more TDS value because of domestic activities. In all there is definite impact on the water quality due to domestic and agricultural activities.

Keywords: *Masala lake, Water quality, Physico-chemical.*

Introduction

The water quality of lakes and tanks are deteriorating, mainly due to rapid increase in human settlement nearer to these places. Deforestation, grazing and removal of vegetation cover in the water shed results in silting of reservoir. The agricultural practices in the catchment area not only increased the siltation problem but are also responsible for the addition of large quantities of nutrients, pesticides and organic matter. The knowledge of physico-chemical characteristics of water bodies helps in planning and successful management of water bodies. To fulfill these criteria present investigation has been carried out.

Materials and Method

Water samples were collected from three sampling stations *viz.* Station S_1, S_2 and S_3 once in a month during study period for estimation of various physico-chemical parameters. Samples were collected and analysed according to APHA (1998). The samples were analyzed within few hours of collection. The pH, Temp. and DO were measured on the spot. Determination of chloride, sulphate, phosphate, carbonates, bicarbonates *etc.* were carried out by following standard methods given in the Table 77.1.

Table 77.1: Methods Used to Evaluate Parameters

Sl.No.	Parameters	Methods/Equipment Utilised
1.	Temp.	Direct measurement on site by thermometer
2.	pH	pH meter (model EQ 610)
3.	Conductivity	Conductivity meter (model EQ 660)
4.	Alkalinity	Titrimmetric method
5.	DO	Modified Winkler's method
6.	CO_2	Titrimmetric method
7.	BOD	Modified Winkler's method
8.	COD	Modified Winkler's method
9.	TDS	Gravimetric method
10.	Hardness	Titrimmetric method
11.	Ca-Hardness	Titrimmetric method
12.	Mg-Hardness	Titrimmetric method
13.	Phosphate	Spectrophotometric (EQ. 820)
14.	Sulphate	Spectrophotometric (EQ. 820)

Results and Discussion

Variations of physico-chemical characteristic during August 2006 to January 2007 are given in Table 77.2. Temperature is basically important for its effects on a certain chemical and biological activities. The surface water temperature of the lake ranges from 24.8°C to 29.4°C, maximum temp was recorded in month of August at site S_3 and minimum in month of January at site S_1 and S_3. The temp showed an inverse relationship with dissolved oxygen, which is consistent with result reported by Das and Shrivastava (1956) and Khanna (1997). The pH of lake water is alkaline and varied from 7.1 to 8.3. During the period of study, maximum pH value recorded in month of September 2006 at site S_2 and minimum was recorded in month of August 2006 at site S_2. Conductivity was minimum 0.25 µmhos/cm in month of January 06 at site S_1 and maximum conductivity 0.411 µmhos/cm was observed in month of September 06 at site S_3. The dissolved oxygen is one of the most important factor in any aquatic system. Self purification of water system depends on presence of sufficient amount of oxygen

Table 77.2: Physico-chemical Parameter of Masala Lake during August 2006 to January 2007

Sl.No.	Months	August			September			October			November			December			January		
	Sites / Parameters	S_1	S_2	S_3	S_1	S_2	S_3	S_1	S_2	S_3	S_1	S_2	S_3	S_1	S_2	S_3	S_1	S_2	S_3
1.	Temp.	29.3	29.3	29.4	28.4	28.3	28.1	28.1	28.0	28.1	27.1	27.3	27.2	25.4	25.9	25.7	24.8	24.9	24.8
2.	pH	72	7.1	7.2	8.1	8.3	8.2	7.99	7.98	7.91	7.98	7.88	7.97	8.01	8.03	7.99	7.3	7.2	7.4
3.	Conductivity	0.390	0.398	0.401	0.405	0.410	0.411	0.404	0.409	0.404	0.380	0.382	0.378	0.262	0.279	0.281	0.250	0.260	0.262
4.	CO_2	3.52	3.56	3.41	2.19	2.27	2.38	2.22	2.42	2.38	2.58	2.67	2.59	3.01	3.04	3.03	2.97	2.81	2.86
5.	DO	5.97	6.02	5.88	6.00	7.02	6.81	6.03	7.08	6.18	6.58	7.68'	7.02	6.89	7.33	6.91	7.09	7.81	7.02
6.	Chloride	202	21.0	20.5	19.2	20.2	19.3	19.4	20.1	19.0	18.6	19.12	18.4	16.3	16.5	16.2	17.8	17.9	17.7
7.	Total Hardness	238	245	231	240	257	245	189	198	186	151	159	157	131	140	139	132	150	139
8.	Ca- Hardness	192	181	183	180	171	176	132	143	137	112	106	105	89	112	91	98	104	94
9.	Mg- Hardness	46	64	48	60	86	69	47	45	49	49	51	52	42	38	48	33	46	35
10.	TDS	980	1020	996	940	918	860	689	702	691	554	629	562	565	602	557	358	492	371
11.	Alkalinity	162	178	162	161	171	158	187	192	180	191	205	189	182	219	179	212	187	182
12.	Carbonate	98	102	78	69	90	62	102	56	70	68	64	62	81	78	62	90	86	79
13.	Bicarbonate	64	76	83	92	81	96	82	136	110	123	142	127	101	141	114	122	101	103
14.	BOD	4.12	4.09	4.08	3.69	3.42	3.54	4.12	4.00	4.02	4.97	4.02	5.02	4.86	3.86	4.76	4.86	3.82	4.09
15.	COD	12.8	12.9	13.1	10.5	10.6	11.0	12.9	12.6	12.7	15.0	13.1	15.1	15.2	13.2	15.1	14.1	12.1	12.3
16.	Sulphate	1.0	1.08	1.02	1.09	1.08	1.07	1.42	1.68	1.52	1.32	2.41	1.29	1.28	2.62	1.42	1.33	2.12	1.3
17.	Phosphate	1.8	1.92	1.81	1.2	1.3	1.25	2.01	2.07	2.01	1.85	2.89	1.84	1.78	2.84	1.61	2.03	2.07	2.05
18.	Nitrate	7.02	7.06	7.0	704	709	7.02	7.62	7.81	7.06	8.03	8.14	8.11	8.29	8.31	8.20	8.89	8.96	8.76
19.	Nitrites	0.01	0.03	0.01	0.02	0.03	0.02	0.2	0.4	0.1	1.02	1.04	1.03	1.01	1.31	1.2	1.32	1.48	1.31

* All values are in mg/l except pH, Temp. (ºC) and Conductivity (μmhos/cm).

dissolved in it. The DO content fluctuated between minimum 5.88 mg/l in August 06 at site S_1 and maximum 7.81 mg/l recorded at site S_2 in month of January 2007. Bahura (1998) reported an inverse relationship of DO, with temperature. While comparing dissolved oxygen data with that of above authors, it is observed that DO is inversely proportional to temp and free CO_2.

In present study BOD shows its maximum value 5.02 mg/l in month of November 2006 at site S_3 and minimum value 3.42 mg/l at site S_2 in month of September 06. The maximum COD value was recorded 15.20 mg/l during month of December 2006 and minimum was recorded 10.50 mg/l during the month of September 2006. TDS showed its high value during the whole study of site S_2 may be due to agricultural and domestic activities. TDS showed maximum value 1020 mg/l in month of August at site S_2 and minimum value was recorded 358 mg/l at site S_1 in month of January 2007. Chloride content was found fluctuating between 16.02 mg/l to 21.01 mg/l. The maximum value was recorded in month of August at site S_2 and minimum was at S_1 in month of December 2006. Total hardness range between 131 mg/l to 257 mg/l. The maximum value was observed in month of September 06 at site S_2 and minimum was in the month of December 06 at site S_1. It was may be due to presence of carbonates and bicarbonate.

Among the important nutrients Nitrates showed the range of 7.00 mg/l to 8.98 mg/l. The level was maximum in the month of January 2007 at site S_2 and minimum was in August 06 at site S_3. The level of phosphate ranges between 1.20 mg/l to 2.89 mg/l, values were maximum in November 2006 at site S_2 and minimum in September 06 at site S_1 may be due to dilution factor. Similar study was shown by Sanjeev (1991). Value of sulphate range between 1 mg/l to 2.62 mg/l. The maximum value was recorded in month of December 06 at site S_2 and minimum was recorded at site S_1 in month of August 06. In present study total alkalinity range between 158 mg/l to 219 mg/l. The maximum value was recorded at site S_2 in month of December 06 and minimum was at site S_1 in month of September 06. The values are within the permissible limit as suggested by WHO.

Acknowledgement

The authors are thankful to Dr. N.H. Khatri, Principal, A.C.S. College, Tukum Chandrapur for providing laboratory facilities.

References

APHA, AWWA and WPCF, 1998. *Standard Methods for Examination of Water and Wastewater*, 20th edn. American Public Health Association, Washington, DC.

Bahura, C.K., 1998. A study on physico-chemical characteristics of a highly eutrophic temple tank Bikaner, Rajasthan. *J. Aqua. Bio.*, 13(142): 47–51.

Das, S.M. and Shrivastava, V.K., 1956. Some new observation of freshwater plankton, Pt. I: Plankton of fish tanks of Lucknow. *Sci. and Cult.*, 21(8): 466–467.

Khanna, D.R., 1997. Physico-chemical parameters of Ganga canal at Haridwar. *Him. J. Env.*, 12(83): 193–197.

Sanjeev, M., 1991. Impact of habitation in hydrobiology of lake Picholar, Udaipur, India. *J. Env. Protection*, 11: 853–856.

Water and Wastewater Management in 2 Vols (2011) *Pages 517–521*
Editors: **D.R. Khanna, R. Bhutiani & Gagan Matta**
Published by: **DAYA PUBLISHING HOUSE, NEW DELHI**

Chapter 78

Assessment of Pollution Level in Hindon River in the Stretch of Meerut–Greater Noida, Due to Rapid Industrialization

Kinshuki Sharma[1], Garima Singhal[1],
Babita Sharma[1] and Ashish Ranjan Sharma[2]
[1]*Department of Biotechnology and Microbiology,*
Institute of Applied Medicines and Research, 9th Milestone, Delhi–Meerut road, Ghaziabad
[2]*Department of Biotechnology, Jaipuria Institute, Vasundhara, Ghaziabad*

ABSTRACT

This chapter deals with the study of pollution trend in Hindon river between Meerut and Greater Noida stretch along with change in their population density, types of industries and living standards. It was observed that heavy metal pollution increases when we move from Meerut to Mohan Nagar and decreases when we further move towards Greater Noida. Organic pollution and BOD level increased many folds than the permissible limits in Meerut and Greater Noida. Concentration of inorganic ions gradually increases from Meerut to Greater Noida. Where population density is high, organic pollution dominates and where industries are more, heavy metal and inorganic ions load is more. With increase in pollution level, risk of exposure to diseases also increased so, we should emphasize on treatment and control.

Keywords: *BOD, Heavy metal, Pollution, Meerut, Greater Noida.*

Introduction

Water has become the most commercial product of the 21st century. This may sound bizarre, but true. In fact, in the closing decades of the 20th century environmental pollution emerged as a major concern for the survival of mankind through out the world. Modern civilization, armed with rapid advancing technology and fast growing economic system is under threat due to its own activities. Water is the basic component of life and therefore it is of vital importance. Water is to the 21st century, oil was to the 20th century. The stress on the water resources are results of multitude of factors. On one hand, the rapid increasing population and changing lifestyles have increased the need for freshwater and on the other hand, intense competitions among agriculture practitioners, industrial sectors and domestic sectors are pushing the groundwater table deeper. This study throws light on the physico-chemical properties of Hindon river in the Meerut-Greater Noida, Uttar Pradesh, stretch. This area is highly populated, have major industries and their number increases as we move towards Greater Noida. A polluted river is severely restricted in its ability to meet any of the above functions. Hindon river is of much importance as it is a major tributary of the Yamuna River. Deteriorated water quality of rivers, reflect lack of respect not only for environmental and health factors, but also for the river's religious significance.

The severe contamination of the Hindon river is not an isolated incidence of river water quality degradation but heavy bacteriological and chemical contamination of rivers are occurring through out India. The following are just a few of these sites, while many other rivers remain just as polluted, as yet uninvestigated. The Hindon river, a tributary of the Yamuna, is heavily used for discharge of chemically contaminated effluents from large number of industries within the vicinity. Groundwater within the region is also subjected to high level of chemical contamination. Groundwater contamination is not only result of hydraulic continuity with the contaminated river, but also as a result of dumping industrial effluents directly to the groundwater through boreholes. Groundwater contamination in Hindon river has been identified to a depth of 185 meters in Meerut (Lewis, 2007). Increased industrialization and chemical based agriculture within the western U.P. is identified as a responsible factor for considerable pollution in the river and underlying groundwater aquifer.

Materials and Method

Hindon river originating in Saharanpur and terminating in Yamuna river (Delhi) is located at Lat. 28°40' North and Long. 77°25' East and 268 m above mean sea levels. Water was collected from different locations in this stretch in sterilized bottle. pH and temperature were measured at spot.

Sampling Sites

1. Baleni bridge on road from Meerut to Ghaziabad
2. Hindon Bridge at Mohan Nagar Ghaziabad
3. Greater Noida near Momnath village before confluence with River Yamuna.

Sample Collection Analysis

Water samples were collected once in a month for one year from above mentioned sampling sites and were tested for various physical and chemical parameters like total solids, turbidity, hardness, heavy metal concentration (Cd, Cr *etc.*), inorganic ions *etc.* by using standard method of APHA (1998) and Khanna and Bhutiani (2005).

Results and Discussion

Study carried on different stretches of Hindon river showed mixed results in its physico-chemical conditions. Table 78.1 represent results of various physico-chemical parameters at different sampling site in this river. The metal pollution in Hindon river water was assessed for Cd, Cr, Fe and Pb. The metal concentration in water showed wide temporal variation compared with bed sediment because of variability in water discharge and variations in suspended solid loadings. Metal ratios for the bed sediments of the River Hindon were determined and the general trend of relative mobility was observed as Cr > Pb >Fe> Cd.

Table 78.1: Comparison Between the Standards of BIS (1991) and the Value of Physico-chemical Parameter Observed in Hindon River Within Meerut–Greater Noida Stretch

Sl.No.	Parameters	Desirable Limits	Meerut	Mohan Nagar	Greater Noida
1.	Color	-	Brownish	Muddy	Muddy
2.	pH	6.5-8.5	7.44	7.56	7.42
3.	Turbidity (NTU)	5.0	0.83	5.06	5.77
4.	DO (mg/l)	-	6.60	6.80	7.22
5.	COD (mg/l)	250	175.72	177.24	176.22
6.	BOD (mg/l)	-	29	42	48
7.	Total Solid (mg/l)	-	200	235	182
8.	TDS (mg/l)	500	563	570	578
9.	Conductivity (μmhos/cm)	-	0.34	0.67	1.22
10.	Chloride (mg/l)	250	153.98	152.66	157.66
11.	Nitrate (mg/l)	45	10.45	15.01	15.78
12.	Fluoride (mg/l)	1.00	0.197	0.295	0.285
13.	Hardness (mg/l)	300	210.0	330.0	356.0
14.	Calcium (mg/l)	75	114.2	118.23	118.75
15.	Magnesium (mg/l)	50	60.71	78.9	78.4
16.	Phosphorus (mg/l)	-	0.878	BDL	BDL
17.	Iron (mg/l)	0.3	0.269	0.233	0.264
18.	Chromium VI (mg/l)	0.05	0.0068	0.065	0.062
19.	Lead (mg/l)	0.01	0.03	0.08	0.1
20.	Cadmium (mg/l)	0.003	0.006	0.017	0.001

The analytical results of various physico-chemical parameters have been compared with the Bureau of Indian Standards (BIS, 1991)– Drinking water desirable/permissible limits are given in the Table 78.1. The water quality monitoring results obtained indicate that the organic and heavy metal contamination are continued to be critical in water body and this is mainly due to discharge of domestic waste-water and industrial waste mostly in untreated form. As we go from Meerut to Greater Noida, the form of pollution shifts from organic to heavy metal due to increase in industries in the vicinity of the Hindon river. Ghaziabad is a growing industrial city, its population had increased from 5,81,886 (in 1901) to 27,03,933 (in 1991) and still continue to grow mainly on account of its rapid industrialization and its proximity to Delhi. A large number of people reside here but carry on their trades in Delhi or are employed there. The municipal corporations are unable to treat this increasing load of municipal sewage flowing into river water. Secondarily receiving river water also do not have adequate water for dilution. Therefore, the oxygen demand and bacterial population is increasing day by day. This is mainly responsible for water borne diseases. Since, discharge of untreated domestic waste-water is the predominant source of pollution of aquatic resources in India. CPCB is regularly monitoring the status of water supply, waste-water generation, collection, treatment and disposal in Class I cities (Population> 1,00,000). The urban areas are responsible for more than 25 per cent of the sewage generation in the region. The small towns and rural areas do not generate significant amount of sewage obviously due to low per capita water supply. The waste-water generated from such areas, percolate into the soil or evaporate, and thus does not contribute to the pollution of water resources. Hence the focus was laid on large urban and industrialized areas.

Conclusion

The pollution increasing trend indicates that as the population density and number of industries increases from Meerut to Greater Noida heavy metal load in river becomes more alarming. High level of pollution at Mohan Nagar site is probably due to presence of discharge from industries around it as well as occurence of heavy soil erosion due to less vegetation in this region which contributes lot in raising of organic contaminants in the river. Heavy metal accumulation in human beings are detrimental to his health as, heavy metals causes following defects in human body:

Cadmium

Toxic to humans, it can enter through ingestion, intraperitoneal, subcutaneous, intramuscular and intravenous routes. Highly toxic to freshwater and marine organisms. Increased exposure can increase risk of lung cancer.

Chromium

The metal exists in two forms, *i.e.* trivalent and hexavalent. Hexavalent chromium in high doses has been implicated as the cause of digestive tract cancers, cutaneous and nasal mucous membrane ulcers.

Lead

Affects human central nervous system, moderate irritation occured when ingested. Lead is a cumulative poison. Increased amount in the body eventually cause disability. Lead can cause irreversible behavioral disturbances, neurological damage and other developmental problems in young children and babies.

The contribution of different point sources to the River Hindon has also been assessed. The highest metal loads were related to the highest flow of the river and thereby increased both by surface runoff and sediment resuspension. Industries producing inorganic chemicals, fertilizers, dyes, paints, pharmaceuticals and battery were identified as hazardous as their waste is non-degradable and tedious to recycle (CPCB, 1982-83).

The Biochemical-Oxygen demand (BOD) is one of the most important indicators of pollution level, between Meerut and Greater Noida stretch was within 29-48 mg/l.

Hindon is a major source of water to the highly populated and predominantly rural population of western Uttar Pradesh. It drains a catchment area of about 5,000 km of farmland while also flowing through a number of towns and villages. About 60 functional industrial units are located along the vicinity of River Hindon and its two main tributaries, the Kali and Krishna rivers. These industries abstract large volumes of water from the river for their manufacturing process, and discharge their industrial effluents, often with nominal or without treatment, directly into the river.

Major industries in this area include paper and textiles, dairy units and slaughter houses. "Hindon river no longer serve for domestic purposes as it is too polluted," says the study. The river is now only used for the watering and washing of livestock. Use of the river for disposal of untreated human sewage is one of the primary cause of poor water quality within the Hindon river. The river receives large volumes of untreated sewage and municipal waste. The river receives a high load of degradable and non-degradable domestic litter. The river water is odorous and become the breading house for disease causing pathogen and vector. The Hindon as well as its tributaries are consistently and massively exceeding the permissible limits provided for surface water and for potable water. Water from the river and tributaries is unable to support a functioning aquatic ecosystem nor is safe for drinking due to the presence of toxic heavy metals, as the entire length of the river apparently has only one water treatment plant, but this does not have adequate capacity. Any person using this water for domestic purposes will exhibit symptoms of heavy metal poisoning. Because of the contaminated river and groundwater, villagers along the Hindon river suffer from serious illnesses such as cancer, neurological disorders, stomach and digestive disorders, skin lesions and respiratory disorders.

References

APHA, AWWA, WEF, 1998. *Standard Methods for the Examination of Water and Wastewater*, 18th edn. American Public Health Association, Washington, DC.

BIS-1991. *Indian Standard Drinking Water*–Specification (BIS 10500 : 1991).

CPCB, 1982–83. *Assessment and Development Study of River Basin*, Series: ADSORBS/7, Part 2, Basin, Sub-Basin Inventory of Water Pollution, The Ganga Basin.

Khanna, D.R. and Bhutani, R., 2005. *Water Analysis at a Glance*. Action for Sustainable Efficacious Development and Awareness (ASEA) Rishikesh.

Lewis, H., 2007. *Hindon River: Gasping for Breath*. Janhit Foundation Meerut.

Water and Wastewater Management in 2 Vols (2011) *Pages 522–524*
Editors: **D.R. Khanna, R. Bhutiani & Gagan Matta**
Published by: **DAYA PUBLISHING HOUSE, NEW DELHI**

Chapter 79

A New Report of *Characiosiphon* from West Nimar District of Madhya Pradesh

S.K. Mahajan
Ex-Professor of Botany, Govt. P.G.College, Khargone, M.P.

ABSTRACT

Characiosiphon belongs to the order Chlorococcales of the class Chlorophyceae. It was reported for the first time from India in 1936 by M.O.P. Iyengar who collected it from a stream situated near Tiruchirapalli (Tamil Nadu). Later this green alga was reported from various parts of the country *i.e.* Gwalior (Agarkar, 1953), Tirupati (Rao, 1954), Jodhpur (Bhandari, 1955) and Indore (Sharma, 1958). The thallus of this algae shows wide variations with respect to its length, width and the number and shape of protoplasts. The material collected from Banihar dam of West Nimar of Madhya Pradesh shows much resemblance with that of *C. rivularis* Iyengar accept that it possesses larger dimensions. The abnormalities described by Bhandari (1955) in *C. rivularis* has not been observed in the present material. The occurrence of *Characiosiphon* is a new report from west Nimar district of Madhya Pradesh.

Keywords: *Phycology, Freshwater algae, Chlorophyceae, Chlorococcales.*

Introduction

West Nimar is one of the tribal districts of Madhya Pradesh. Khargone is the district headquarter which is situated at 21°45′N lattitude, 75°37′ E Longitude and 250.38 m above mean sea level. It has number of water bodies both of temporary and permanent nature. Very little is known about the algal flora of West Nimar except few reports (Seerwani, 1963; Mahajan, 1986, 1987, 1988 a and b, 1990, 1991 a and b, 1994, 2004, 2005a and b and 2006). As early as in 1936, Iyengar had reported *Characiosiphon* from a stream situated near Tiruchirapalli. Later on it was reported from various parts of India *viz.*

from Gwalior (Agarkar, 1953), Tirupati (Rao, 1954), Jodhpur (Bhandari, 1955) and Indore (Sharma, 1958). This alga generally prefers standing water and temporary stream. Bhandari (1955) has described certain abnormalities in *C.rivularis*,collected from Umed bund near Jodhpur (Rajasthan).The present investigation deals with the occurrence of *Characiosiphon* for the first time from this area.

Materials and Method

During algal collection by Post-graduate students of Botany of Govt.P.G. College, Khargone in 2006-07, this new algae was collected from Banihar dam which is about 5 km away from Khargone bus stand. The material was identified after consulting the standard literature (Philipose,1967 and Fritsch, 1945).The collected material is deposited in the Botany Department, Govt.P.G. College, Khargone for record under Regd. No. PGDB 954.

Results and Discussion

The thallus of the present alga is 560 micron long and 38 micron broad with round apex and gradually tapering towards the base where it is attached to the substratum. Protoplasts in the majority of the thalli are separated from each other and appear round to ellipsoidal in surface view but in few thalli they are compactly arranged in the upper part due to mutual pressure and look angular (pentagonal to hexagonal) in surface view. Cell wall is thick and lamellated. Protoplasts are 14 to 18 micron in diameter. The thallus shows wide variations with respect to the length, width and number of protoplasts. The present material shows much resemblances with that of *C. rivularis* Iyengar accept that it possesses larger dimensions, abnormality in shape reported by Bhandari (1955) has not been observed in the material collected from Banihar dam.

Acknowledgement

The author is grateful to the Principal, Govt.P.G.College, Khargone for facilities and encouragements.

References

Agarkar, M.S., 1953. Note on *Characiosiphon rivularis* from Gwalior. *Curr. Sci.*, 22(8): 245–246.

Bhandari, M.M., 1955. Occurrence of *Characiosiphon rivularis* Iyeng. *Curr. Sci.*, 24(1): 16–17.

Fritsch, F.E., 1945. *The Structure and Reproduction of the Algae*, Vol. 1. Cambridge University Press.

Iyengar, M.O.P., 1936. *Characiosiphon*: A new member of Chlorococcales (preliminary note). *J. Indian Bot. Soc.*, 15(5): 313–318.

Mahajan, S.K., 1986. A preliminary survey of the algal flora of Khargone, Madhya Pradesh. In: *56th Annual Session, Nat. Acad. Sci., India* (Biol. Sect.), Abst. pp. 51.

Mahajan, S.K., 1987. Algal flora of Pawagiri Oon, Madhya Pradesh. *Phykos*, 26(1 and 2): 61–62.

Mahajan, S.K., 1988a. Phycological studies at Aurangpura, Madhya Pradesh. In: *Proc. 75th Ind. Sci. Cong.*, Part 3 (Bot. Sect.) Abst. pp. 16.

Mahajan, S.K., 1988b. Comparative studies of algal flora in lentic and lotic systems at Khargone, Madhya Pradesh. In: *58th Annual Session, Nat. Acad. Sci., India*, (Biol. Sect.), Abst. pp. 87.

Mahajan, S.K., 1990. Cyanobacteria of Khargone, Madhya Pradesh. In: *Proc. Natl. Symp. on Cyanobacterial Nit. Fixation*, NFBGAC, IARI, New Delhi, pp. 586–587.

Mahajan, S.K., 1991a. Occurrence of *Spirulina* at Khargone, Madhya Pradesh. In: *Natl. Symp. on Spirulina*, ETTA, AMM, MCRC, Tharamani, Madras.

Mahajan, S.K., 1991b. Algal flora of Choolgiri Hills, Madhya Pradesh. *Phykos*, 30(1–2): 87–89.

Mahahan, S.K., 1994. Chlorophyceae of Khargone, Madhya Pradesh. In: *Proc. 81st Ind. Sci. Cong.*, Part III (Bot. Sect.), Abst. pp. 5.

Mahajan, S.K., 2004. A new species of *Spirulina* (= *Arthrospira*) *Mahajani* Mahajan from Khargone, Madhya Pradesh. *J. Bombay Nat. Hist. Soc.*, 101(2): 294–295.

Mahajan, S.K., 2005a. Similarity index of phytoplankton species of River Kunda at Khargone, Madhya Pradesh. In: *Aquatic Biodiversity in India: The Present Scenario*, (Eds.) D.R. Khanna, A.K. Chopra and G. Prasad. Daya Publishing House, Delhi, pp. 350–353.

Mahajan, S.K., 2005b. Limnological assessment of phytoplankton in a reservoir at Khargone, Madhya Pradesh. In: *Aquatic Biodiversity in India: The Present Scenario*, (Eds.) D.R. Khanna, A.K. Chopra and G. Prasad. Daya Publishing House, Delhi, pp. 354–358.

Mahajan, S.K., 2006. Taxonomic studies of algae in Madhya Prdesh: A review. In: *Recent Advances in Algal Taxonomy, Vol. 2: Taxonomical and Cultural Studies*, (Eds.) Vidyavati and A.K. Mahato. Associated Publishing House, New Delhi, pp. 201–217.

Philipose, M.T., 1967. *Chlorococcales*. ICAR, Krishi Bhawan, New Delhi.

Rao, K.V.S., 1954. *Characiosiphon* from Tirupati. *Curr. Sci.*, 23(6): 91.

Seerwani, A.B., 1963. *Enteromorpha* at Khargone, Madhya Pradesh. *Curr. Sci.*, 32(10): 182.

Sharma, R., 1958. A sexual reproduction in *Characiosiphon rivularis* Iyengar. *Curr. Sci.*, 27(12): 500–501.

Water and Wastewater Management in 2 Vols (2011) *Pages* 525–529
Editors: D.R. Khanna, R. Bhutiani & Gagan Matta
Published by: DAYA PUBLISHING HOUSE, NEW DELHI

Chapter 80

A Study on Organic Pollution Based on Algal Distribution in Some Rural and Temple Ponds of Balrampur, U.P.

S.P.M. Tripathi, Madhavi Pathak and J.P. Tiwari
Department of Botany, M.L.K. P.G. College, Balrampur

ABSTRACT

The present chapter deals with the algal distribution and organic pollution in fifteen rural and temple ponds in Balrampur. Algae belonging to Chlorophyceae, Cyanophyceae, Euglenophyceae and Bacillariophyceae were studied. Occurrence of *Microcystis aeruginosa* was observed in many ponds. Among the ponds studied one pond showed confirm high organic pollution, five ponds moderate organic pollution, five probable high organic pollution and four ponds no organic pollution.

***Keywords**: Organic pollution, Algal distribution, Temple pond.*

Introduction

Freshwater resources of Balrampur are facing destruction due to urbanization and development. Land filling and no reliance on traditional water sources have led to the destruction or impairment of traditional ponds. Due to this, quality of water in many ponds have drastically deteriorated. The rural and temple ponds of Balrampur have traditionally served as source of water for drinking and recreational purposes. No reports are available on the quality of water in these ponds. Algae often serve as good indicator of the quality of the water, hence a study was undertaken to understand the floristic structure and to evaluate the water quality by employing the water quality index of Palmer (1969).

Materials and Method

The present study was carried out in fifteen ponds located in and around Balrampur. These ponds are–(1) Tulsidas pokhara (2) Mewalal talab (3) Pajawa tal (4) Jharkhandi talab (5) Motisagar tal (6) Purainiya tal (7) Jalwania pond (8) Burhwa tal (9) Suraj kund (10) Radha kund (11) Jhali tal (12) Seta dwar pond (13) Bihar pond (14) Chikani pond and (15) Mouri pond.

Collections were made during October to December 2006. The samples were preserved in 4 per cent formaldehyde and the phytoplankton were identified using relevant monographs. Palmer's pollution index was employed to find out the level of organic pollution in these ponds. For quantitative estimation, haemocytometer method was used and percentage composition of all the groups was calculated.

Results and Discussion

In the present study, the algal flora was dominated by diffrent groups *viz.* cyanophyceae, chlorophyceae, euglenophyceae and bacillariophyceae. The distribution pattern of phytoplankton in different ponds is summarized in Table 80.1.

Table 80.1: Distribution of Phytoplankton in Fifteen Pond

Algal Genera	*Pond Name*														
	1	*2*	*3*	*4*	*5*	*6*	*7*	*8*	*9*	*10*	*11*	*12*	*13*	*14*	*15*
Chlorococcus	+	+		+	+	+	+	+					+		+
Merismopedia	+				+	+		+						+	+
Microcystis		+	+	+		+	+				+			+	
Oscillatoria				+	+	+		+			+				+
Anthrospira	+														
Alabaenopsis															
Chlamydomonas														+	
Chlorella													+		
Ankistrodesmes	+	+	+		+	+		+	+	+	+		+	+	
Coelastrum		+			+	+		+						+	+
Crucigenia			+					+						+	
Dictyosphaerium	+	+	+		+			+						+	+
Scelenastrum		+													
Kirchneriella	+							+						+	
Micractinium											+				
Pediastrum	+	+	+		+	+		+	+		+			+	
Schoroederia	+						+								
Tetraedron		+							+	+	+			+	
Tetrastrum	+					+									
Scenedesmus	+		+	+	+	+	+	+	+	+	+			+	+
Cosmarium	+														

Contd...

Table 80.1–Contd...

Algal genera	*Pond Name*														
	1	*2*	*3*	*4*	*5*	*6*	*7*	*8*	*9*	*10*	*11*	*12*	*13*	*14*	*15*
Staurastrum					+										
Euglena	+			+	+				+				+		+
Phacus					+						+		+		
Lepocinclis													+		
Trachelomons	+	+	+						+	+	+		+		+
Cyclotella															
Melosira				+		+				+	+	+	+		
Synedra															
Cymbella	+	+													
Navicula					+										
Nitzschia				+		+	+	+			+	+	+	+	+

About thirty two algal genera dominated the flora. Among them six belonged to *Cyanophyceae* sixteen to *chlorophyceae,* four to *euglenophyceae,* and six to *bacillariophyceae.* The Jhali tal showed dominance of *bacillariophyceae.* The *chlorococcales* were widely distributed a most of the ponds studied. Among them *Scenedesmus* represented in thirteen and *Ankistrodesmus* in twelve ponds. The ponds offer a suitable habitat for these algal forms. The percentage composition also showed variation. The algal percentage composition of various algal groups in the 15 ponds is given in Table 80.2.

Table 80.2: Percentage Composition of Algae in Fifteen Ponds

Algal Class	*Pond Name*														
	1	*2*	*3*	*4*	*5*	*6*	*7*	*8*	*9*	*10*	*11*	*12*	*13*	*14*	*15*
Cyanophyceae	12	49	40	59	14	47	75	18	0	0	55	7	7	32	67
Chlorophyceae	84	13	40	24	30	33	15	72	38	25	19	26	26	59	20
Euglenophyceae	0	33	20	7	0	0	0	0	62	13	12	62	62	0	3
Bacillariophyceae	4	5	0	10	56	20	10	10	0	62	14	5	5	9	10

The Palmer's algal pollution index is employed in all the ponds and the results are summarized in Table 80.3. The maximum index was observed in Motisagar Tal with an index value of 22. Five ponds showed moderate organic pollution, five showed probable high organic pollution and four ponds showed no organic pollution.

Aquatic habitat assessment with relation to the resident algal community helps in assessing the quality of the water resource. Algal community dominated by *Chlorococcus, Microcystis, Ankistrodesmus, Dictyosphaerium, Pediastrum, Scenedesmus, Trachelomons, Melosira and Nitzschia* were found in many ponds. Their presence is an indication of the organic pollution of the water as established (Hosmani and Bharati, 1980). The genus *Scenedesmus* is represented in thirteen ponds, its occurrence in polluted water especially in eutrophicated water is established by Palmer (1969) also gave a high ranking for this genus. The absence of desmids is often considered as an indication of pollution in pond. Absence of desmids was observed in thirteen ponds. All these ponds except one showed moderate levels of organic pollution. Singh (1973) used *Microcystis* (*Anacystis*) as the best single indicator of pollution. About fifty per cent of the ponds studied showed the presence of *Microcystis* suggesting the deteriorated quality of the water.

Table 80.3: Palmer's Algal Generic Pollution Index Values in Fifteen Ponds

Algal Genera	*Index*	*Pond Name*														
		1	*2*	*3*	*4*	*5*	*6*	*7*	*8*	*9*	*10*	*11*	*12*	*13*	*14*	*15*
Anacystis	1		1	1	1		1	1				1			1	
Oscillatoria	4				4	4	4		4			4				4
Phormidium	1															
Chlamydomonas	4														4	
Pandorina	1															
Scenedesmus	4	4	4	4	4	4	4	4	4	4	4	4			4	4
Micratimium	1											1				
Ankistrodesmus	2	2	2	2			2		2	2	2	2		2	2	
Chlorella	3													3		
Closterium	1															
Stigeoclonium	2															
Cyclotella	1															
Melosira	1		1		1	1	1				1	1	1	1		
Gomphonema	1															
Navicula	3	3														
Nitzschia	3				3		3	3	3			3	3	3	3	3
Synedra	2					2										
Euglena	5	5		5	5	5				5				5		5
Phacus	2			2								2		2		
Nepocinelis	1													1		
Total		14	8	14	18	22	15	8	13	11	7	18	4	17	14	16

Biological indices offer a cheap, fast and effective means of finding the pollution status of the water. By employing Palmer's algal pollution index it was found at about seven per cent of the ponds showed confirmed organic pollution, thirty three per cent showed probable high organic pollution, thirty three per cent showed moderate organic pollution and twenty seven per cent showed no organic pollution. It is found that quality of water in majority of the ponds are in a poor state.

Acknowledgement

Authors sincerely acknowledge Professor G.L. Tiwari, Ex-Head Department of Botany, Allahabad University, Allahabad for their constant encouragement during course of investigation.

References

Hosmani, S.P. and Bharati, S.G., 1980. Algae as indicator of organic pollution. *Phykos.*, 1: 23–26.

Palmer, C.M., 1969. A composite rating of algae tolerating organic pollution. *J. Phycol.*, 5: 78–82.

Singh, R.N., 1973. Liminological relations of Indian inland waters with special reference to water blooms Verh. Int. Verein. *Theor Anen Limnol.*, 12: 831–836.

Water and Wastewater Management in 2 Vols (2011) *Pages 530–533*
Editors: **D.R. Khanna, R. Bhutiani & Gagan Matta**
Published by: **DAYA PUBLISHING HOUSE, NEW DELHI**

Chapter 81

Rotifer Diversity in Ramala Lake, Dist. Chandrapur, Maharashtra

P.J. Khinchi[1], P. M. Telkhade[2], N.R. Dahegaonkar[2] and S.B. Zade[3]
[1]*Janta Mahavidyalaya, Chandrapur, M.S.*
[2]*Arts, Commerce and Science College, Tukum, Chandrapur, M.S.*
[3]*M.B. Patel Science College, Sakoli, Dist. Bhandara, M.S.*

ABSTRACT

An extensive study of the ecology and systematics of Rotifers in the Ramala Lake of Chandrapur was made from May-2006 to April-2007. The Rotifers constituted 34.6 per cent of total zooplankton and their density ranged from 300 to 2000 ind/ltr. The higher density of Rotifers was recorded during winter season whereas, minimum observed in summer. Dominance of Family Brachionidae was observed during the whole period of investigation indicates the eutrophication in the ecosystem.

Keywords: *Rotifers, Ecology, Diversity, Eutrophication, Ramala lake.*

Introduction

Globally, around 2000 species of rotifers are known. Present study of rotifers in India have 325 species belonging to 25 families and 63 genera. Thus rotifer species from this country comprise about 13 per cent of the global Rotifera (Sharma, 1998). Rotifers play the role as link between the nanoplankton and carnivorous zooplankton. Rotifer exhibits ability to colonise diversified freshwater biotopes and they are apparently the most sensitive indicators of water properties.

Rotifers can populate vacant niches with extreme rapidity and convert primary production into a form, usable for secondary consumer, producing upto 50 per cent of the total plankton biomass.

Rotifers are amongst some of the most abundant and important members of the freshwater fauna. According to Sladeck (1983) rotifer population are very useful in indicating water quality, particularly in pollution studies.

The Ramala Lake is oldest of its kind and is situated in the central part of the City, the lake is influenced by anthropogenic activities like washing, bathing, boating, idol immersion and commercial fishery.

Materials and Method

The study was concluded from May-06 to April-07, the zooplankton samples were collected monthly by towing 41 mm mesh net, brought to laboratory and preserved in 4 per cent formalin for identification and counting. The rotifers were observed with the help of binocular microscope and counted with Sedgwick Rafter Cell and identified by Edmondson (1959), Tonapi (1980) and Dhanapathi (2000).

Results and Discussion

Detailed microscopic examination of rotifera under binocular microscope has carried out and revealed that, there were total 18 species observed throughout the study period. In rotifers, Brachionus genera showed their dominance during winter season and least occurrence during monsoon and moderate during summer.

Rotifers are aquatic microscopic (100-1000 mm) organisms with the anterior end modified into a ciliary organ, the corona. The coronal disc with synchronous beating of its cilia look like a rotating wheel hence rotifers are commonly called wheel animalcules.

Sharma (2001) in a review article focus the role of freshwater rotifera as a biological indicators and bio-monitor in water quality assessment. A rational approach to maintain desirable quality of our freshwater resources is to recognize the bio-indicator species dwelling in them, this offers option and opportunity.

For environmental engineers and biologist to designate and recruit appropriate rotifer species and operate the bioreactors at peak efficiency in water treatment and recycling.

In the present investigation, 18 species of rotifer population were observed belonging to 10 genera. The rotifera population in Ramala Lake has higher magnitude during winter season and lower magnitude during monsoon. Their diversity and density is mainly controlled by availability of food as a favourable water quality.

The similar study was reported by Abdus Saboor and Altaff (1995) in a tropical pond during summer and rainy season. Seshagiri Rao *et al.* (1984) observed in Manjira reservoir, Sangareddy (A.P.).

The higher density as well as biomass of brachionoids in the zooplankton community clearly indicate that most of the ponds are polluted by the organic pollution. Most of which is contributed by the domestic sewage, Sladecke (1983).

In the present investigation, brachionoids showed high density in the rotifers. *Brachionus forticula* showed their dominance as far as *B. calyciflorus* observed minimum. The *B. calyciflorus* species considered as freshwater indicator.

Seasonal changes in Rotifera density depends particularly on physical condition *i.e.* water quality, partially on available food supply and partially on predatory pressure exerted by the carnivorous zooplankton. In present investigation, the less number of genera observed during monsoon may be due to high water level and less nutrients in the pond. The reduction in the number of genera may also be due to predation.

Table 81.1: Monthly Variation in Different Species of Rotifers

Sl.No.	Rotifers	May	Jun	Jul	Aug	Sep	Oct	Nov	Dec	Jan	Feb	Mar	Apr
1.	*Filina longiseta*	03	01	–	–	06	06	09	06	11	05	04	01
2.	*Filina opoliensis*	06	02	–	–	00	02	–	–	–	03	–	03
3.	*Keratella tropica*	25	39	05	02	17	19	19	25	32	12	13	19
4.	*Asplanchna* sp.	13	12	02	–	–	07	07	11	11	03	–	09
5.	*Trichocerea longiseta*	12	04	03	01	02	11	11	17	21	07	05	07
6.	*Trichocerea cylendrica*	04	02	–	–	–	02	03	06	06	05	02	04
7.	*Trichocerea porcellus*	05	04	–	02	02	02	–	–	04	04	04	04
8.	*Brachionus calyciflorus*	04	–	–	–	03	02	02	06	06	–	–	07
9.	*Brachionus quadricornis*	10	02	02	–	–	03	05	05	03	03	16	09
10.	*Brachionus forticula*	27	17	07	07	13	18	18	25	27	14	12	14
11.	*Brachionus bidenta.*	09	03	–	–	12	12	12	02	–	–	12	06
12.	*Brachionus plicatilis*	05	02	–	–		02	02	08	07	–	–	05
13.	*Lepadella* sp.	02	02	–	–	–	02	02	04	06	06	–	–
14.	*Lecane bidentata*	08	–	–	02	02	06	06	06	–	–	05	08
15.	*Lecane monostyla*	06	05	–	–	–	–	02	07	07	–	–	02
16.	*Monostyla bulla*	04	02	–	01	–	–	02	04	06	–	–	02
17.	*Synchaeta* sp.	07	–	02	01	05	08	–	–	08	08	06	06
18.	*Epiphanes* sp.	02	01	–	–	01	–	01	01	02	01	02	03

In the monsoon, fall of rotifers density can be attributed in the distributed effects (Bhati and Rana, 1987). Brachionus species are the characteristic feature of eutrophic status, as far as Ramala Lake shows brachionus species dominance, indicating higher eutrophic nature of lake.

Acknowledgement

The authors are thankful to Dr. M.M. Wankhede, Principal, A.C.S. College, Tukum, and Dr. Subhash, Principal, Janta Mahavidyalaya, Chandrapur. (M.S.) for their kind co-operation during the present work.

References

Abdus, Saboor and Altaf, K., 1995. Qualitative and quantitative analysis of zooplankton population of a tropical pond during summer and rainy season. *J. Ecol. Biol.*, 7(4): 269–275.

Bhati, D.P.S. and Rana, K.S., 1987. Zooplankton in relation to biotic component in the fort moat of Bharatpur. *Proc. Nat. Acad. Sci. India*, 57(B): 237–242.

Dhanapati, M.V.S.S., 2000. Further studies on rotifers from Andhra Pradesh India. *Indian J. Aqu. Biol.*, 15(1 and 2): 6–15.

Edmondson, W.P., 1959. *Freshwater Biology*. Seattle, Washington, pp. 420–495.

Sharma, P.D., 1998. *Ecology and Environment*. Rastogi Publication, Meerut, pp. 1–139.

Sharma, B.K., 2001. Biological monitoring of freshwater with reference to role of freshwater rotifer as biomonitor. In: *Water Quality Assessment, Biomonitoring and Zooplankton Diversity, Proc. Young Scientists National Workshop,* (Ed.) B.K. Sharma. Deptt. of Zoology, Northern-Eastern Hill Univ., Shillong (Megalaya), India, pp. 83–97.

Sladecke, V., 1983. Rotifers an indicator of water quality. *Hydrobiologia*, 101: 111–114.

Seshagiri, Rao and Khan, M.A., 1984. Ecology of the rotifers in the plankton of the Manjira reservoir Sangareddy, Andhra Pradesh India. *J. Aqua. Biol.*, 2(1): 23–31.

Tonapi, G.T., 1980. *Freshwater Animal of India and Ecological Approach.* Oxford and IBH Publishing Company, New Delhi, pp. 341.

Water and Wastewater Management in 2 Vols (2011) *Pages 534–539*
Editors: D.R. Khanna, R. Bhutiani & Gagan Matta
Published by: DAYA PUBLISHING HOUSE, NEW DELHI

Chapter 82

Toxic Effect of Malathion on Acetylcholinesterase Activity of Liver, Brain and Gills of Freshwater Catfish *Heteropneustes fossilis*

Sudhish Chandra
P.G. Department of Zoology, B.S.N.V. College, Lucknow

ABSTRACT

The toxic effects of malathion were evident in the inhibition of acetylcholinesterase activity of liver, brain and gills of freshwater catfish *Heteropneustes fossilis*. Maximum inhibition of 77.12 per cent and 72.83 per cent were recorded in brain and gills respectively after 72 hour of exposure to 4.80 mg/l pesticide. However, in liver highest inhibition of 67.81 per cent in enzyme level was noticed at 6.50 mg/l pesticide concentration after 24 hours of exposure, beyond which fish could not survive. Pesticide repressed the enzyme activity so intensely it showed no sign of return to normalcy. The fish also elicited dissociated behaviour with increasing concentrations of pesticide toxicity.

Keywords: AchE, Malathion, Toxicity, Concentration, Inhibition.

Introduction

Fish inhabiting a diverse environment are sensitive to toxic environment, being exposed *via* extensive delicate respiratory surface. Wide variety of pesticides and due to their persistance and nonbiodegradable nature, they got accumulated in different tissues of the fish and affect their growth and survival (Kaur and Toor, 1977; Kulshrestha and Arora, 1985). Acetylcholinesterase is the enzyme responsible for terminating the action of acetylcholine at cholinergic synopsis (Ronald *et al.*, 1999). It acts as a key transducer at cell membrane level, an alteration in the level of its functional activity may result in microdeformation and configurational change. Acetylcholinesterase is the target enzyme for organophosphate pesticides which is inhibited by these pesticides and its inhibition leads to blockage of neurotransmission (O'Brien, 1976) leading to physiological alterations in non targeted organisms of the aquatic system.

Present observation deals with the effect of organophosphate pesticide malathion on the Acetylcholinesterase (AchE) activities of liver, brain and gill tissues of freshwater stinging catfish *Heteropneustes fossilis,* highly popular and expensive table fish, due to its higher nutritional quality and medicinal value.

Materials and Method

Freshwater stinging catfish *Heteropneustes fossilis* (length 20.5 – 24.0 cm, weight 160 – 200 grams) were collected from local resources, transported to laboratory and treated with $KMnO_4$ (2 mg/l). Apparently healthy looking fishes were acclimatized for seven days in glass aquaria under standard laboratory conditions and fed them with small pieces of goat liver, but starved for a period of 24 hours prior to the experiment. Static bioassay tests were followed as given by Doudoroff *et al.*, 1951; APHA, 1992. The pesticide was commercially formulated product (Cynamid India Ltd.), extensively used in India. Pesticide concentrations were selected on the basis of 80 – 100 per cent survival of fishes for that period. Control fish in experiment were also maintained in identical conditions.

The controlled and exposed fishes were taken out from each preparation after every 24 hours interval and washed with distilled water. Liver, brain and gills tissues were taken out and 2.5 per cent homogenate was prepared in 0.7 per cent saline solution, at nearly freezing temperature. Biochemical analysis was performed following the method of Hestrin (1949), using Bausch and Lomb spectronic–20 spectrophotometer at 540 μm against blank. The physico-chemical properties of the tap water used in the experiment were as follows: temperature (22±1.4 °C), pH (7.4±0.02), Dissolved oxygen (6.6±0.8 mg/l), alkalinity (106±10.0 mg/l) and hardness (118±12.4 mg/l).

Results and Discussion

Acetylcholinesterase activity was significantly inhibited by malathion toxicity in *H. fossilis*. The results obtained on enzymal activity of liver, brain and gill tissue of fishes, exposed to different concentrations of pesticide have been summarized in Table 82.1. At 3.55 mg/l pesticide concentration AchE levels were inhibited 58.62 per cent, 53.16 per cent, 44.82 per cent and 39.08 per cent in liver, 58.51 per cent, 56.38 per cent, 60.63 per cent and 60.10 per cent in brain and 62.34 per cent, 58.02 per cent, 49.38 per cent and 41.97 per cent in gills, after exposure for 24, 48, 72 and 96 hours respectively, from control levels. The maximal inhibition was found in liver after 24 hours, in brain after 72 hours and in gills after 24 hours. At 4.80 mg/l concentration the enzyme level continuously fell by 48.27 per cent, 57.47 per cent and 66.66 per cent in liver, 51.06 per cent, 55.58 per cent and 77.12 per cent in brain and 33.33 per cent, 49.38 per cent and 72.83 per cent in gills after exposure to 24, 48, 72 hours respectively, versus control levels. At 5.80 mg/l concentration AchE activity decreased significantly by 56.89 per cent and 62.06 per cent in liver, 41.48 per cent and 79.78 per cent in brain and 52.77 per cent and 60.49 per cent in gills after exposure for 24 and 48 hours respectively, as compared to control. At the highest concentration of 6.50 mg/l, all the fishes died after 24 hours of exposure, when decline of 67.81 per cent in liver, 75.00 per cent in brain and 56.79 per cent in gills had occurred from initial control levels.

Table 82.1: Effect of Malathion Toxicity on Acetylcholinesterase levels of *H. fossilis*

Pesticide Conc. (mg/l)	Acetylcholinesterase (μ moles AchE hydrolysed/100 mg f.wt./hr.) Mean ± Standard Deviation Time of Exposure in Hours			
	24	48	72	96
LIVER – Control – 69.60±8.15				
3.55	28.80±7.05	32.60±6.40	23.20±3.72	41.36±5.05
4.80	36.00±7.75	29.60±2.82		
5.80	30.00±3.72	26.40±5.05		
6.20	22.40±5.40	38.40±3.82		
BRAIN – Control – 75.20±9.21				
3.55	31.20±6.56	32.80±6.10	29.60±5.05	30.00±6.32
4.80	36.80±7.83	33.40±4.25	17.20±4.07	
5.80	44.00±8.84	15.20±5.05		
6.20	18.80±6.09			
GILLS – Control – 64.80±6.64				
3.55	24.40±5.04	27.20±2.70	32.80±1.91	37.60±3.20
4.80	43.20±7.21	32.80±3.72	17.60±4.36	
5.80	30.60±3.82	25.60±4.10		
6.20	28.00±4.60			

No. of observations – 8 in each experiment.

Pesticide repressed the enzymal activity so intensely that it showed no sign of return to normalcy and finally proved fatal to fish after 96 hours, even at the lowest concentration of pesticide used. Exposure at higher concentration of pesticide, fish appeared restless, hyperactive showing erratic movements, rapid opercular movements, gulping air at the surface, followed by loss of balance, became calm and finally died. Esterases and transaminases are the target enzyme involving every tissue. Interaction of these enzymes with toxicant is a complex phenomenon involving interplay between several metabolic pathways. Repressed enzyme levels caused by toxicant and other associated compounds have been attributed to decreased neurosecretory and hepatosecretory activities (Praveen *et al.*, 2004). The toxicant on entering into the fish body affect their metabolism, leading to physiological, pathological and biochemical disorders (Bais and Arasta, 1995; Arasta *et al.*, 1999; Karuppasamy, 2002; David *et al.*, 2003). Organophosphate pesticide are known to metabolise into their corresponding oxygen analogue which are potent inhibitors of acetylcholinesterase in invertebrates and vertebrates (Natarajan, 1984). Varied LC_{50} have been reported with different organophosphate pesticides in different fishes under varying situations (Kumar *et al.*, 1995; Thannipon *et al.*, 1995; Rawat and Bhargava, 1997; Nath *et al.*, 2000; Jeyarthi Shanti and Jebanesan, 2001; Sharma *et al.*, 2001 and Singh, 2003) noticed marked decline in acetylcholinesterase activity. Present investigation clearly indicated significant inhibition in AchE enzyme levels in liver, brain and gill tissues of *H. fossilis*, due to the toxicity of malathion, even at lowest concentration. Sudheer *et al.* (2006) reported maximal inhibition in AchE in brain followed by muscle, gill and liver in *Tilapia mossambicus*, exposed to chlorophyriphos.

Progressive decrease in AchE activity in Nile *Tilapia* by Thannipon *et al.* (1995) exposed to monocrotophos. Fish *H. fossilis* exposed to different concentrations of malathion for different periods revealed comparatively higher abatement in brain showing a maximal inhibition of 77.12 per cent, followed by gills (72.83 per cent) and liver (67.81 per cent). Gills and body surface being the primary sites of absorption (Murphy, 1971; Kumar *et al.*, 2000) were the first to be affected by toxicant. In order to get rid off the toxic environment fish exhibited, increased opercular movements copious mucus secretion followed by restlessness, hyperactivity and erratic movements showing fish in acute stress. Intensity in such evident behavioral changes are dose dependent (Matsmura, 1980) and their median survival time seems to be directly related to body size, weight and age. Profuse secretion of mucus by gills with oedema and fusion of hyperplastic secondary gill lamellae were observed by Mishra *et al.* (2005). It appears that fish *H. fossilis* exposed to lowest concentration of pesticide did not reach up to the exhaustion, rather they were able to accommodate and acclimatize with the developed stress.

Biochemical changes due to the utilization of organic resources of fish owing to the shift in respiratory metabolism in toxic environment, possibly yielded excess energy to compensate the stress. A shift in cellular respiratory metabolism towards anaerobiosis as a prelude towards adaptability to cope with the enhanced energy demand (Obula and Neerja, 2001). Chandra (1988) opined that malathion may lead to inhibition of cholinesterase at neuro effector sites in adrenal medulla, leading to hypersecretion of adrenalin which resulted in hypermetabolic conditions to meet stress condition. Tandon and Dubey (1983) reported increased aldolase activity, an important gluconeogenic enzyme to withstand the stress condition caused by pesticide malathion and dimicron in *Clarias batrachus*. Marked inhibition in AchE in different tissues of fish following exposure to higher concentration of sumithion has been reported by Koundiya and Ramamurthy (1978), Dubey (1980), Natrajan (1984), Thannipon *et al.* (1995) and Sudheer *et al.* (2006). It has been observed that the cause of death in organophosphate pesticide poisoning is always asphyxiation. The pesticide penetrates brain which results in the failure of respiratory centre of brain after that of respiratory organs (O'Brien, 1976). In these observations AchE levels were prominently hindered in brain and gills at lower but maximal inhibition occurred in liver at highest malathion concentration. Since the fishes have ability to detoxify xenobiotics in their liver like mammals, probably that is why liver might have the last target of toxicant in accumulating more acetylcholine. The dissociated behaviour of fish *H. fossilis* may be related due to the accumulation of acetylcholine which resulted in neuromuscular paralysis of the bronchial pump and oxygen diffusion across the gill and lead to excitatory effects like tremors and convulsions.

References

APHA, 1992. *Standard Methods for the Examination of Water and Wastewater*, 17th edn. American Public Health Association.

Arasta, T., Bais, V.S. and Thakur, P.B., 1999. Changes in selected biochemical parameters in the liver and muscles of fish, *Mystus vitattus* exposed to aldrin. In: *Environmental Pollution Management*, (Ed.) V.S. Bais. Creative Pub., Sagar, pp. 109–112.

Bais, V.S. and Arasta, T., 1995. Effect of sublethal concentration of Aldex on protein, lipid and glycogen levels in the liver and muscles of a catfish *Mystus vitattus*. *J. Freshwat. Biol.*, 7: 151–154.

Chandra, S., 1988. Effect of malathion on the cholesterol levels of different tissues of freshwater catfish *Clarias batrachus*. *J. Rec. Adv. Appl. Sci.*, 3: 500–503.

David, M., Mushiger, S.B., Prashant, M.S. and Mathad, S.G., 2003. Hepatotoxicity of malathion on protein metabolism in *Catla catla*. *Adv. Biol.*, 22: 115–120.

Doudoroff, P.A., Burdick, B.G., Galtsoff, P.S., Hart, W.B., Patrick, R., Strong, E.R. and VonHorn, W.M., 1951. Bioassay methods for the evaluation of acute toxicity of industrial wastes to fish. *Sewage Ind. Wastes*, 23: 1380–1397.

Dubey, A., 1980. Effect of environmental pollution on the physiology of fishes. *Ph.D. Thesis*, Luck. Univ. Lucknow.

Hestrin, S., 1949. The reaction of acetylcholine and other carboxylic acid derivatives with hydroxylamine and its analytical applications. *J. Biol. Chem.*, 180: 249–261.

Jeyarthi Shanthi, T. and Jebanesan, A., 2001. Effect of sublethal concentration of Chlorpyriphos on acetylcholinesterase activity in freshwater fish *Cyprinus carpio*. *Convergence*, 3: 31–35.

Karuppasamy, R., 2002. Short and long term effect of phenyl mercuric acetate on protein metabolism in *Channa punctatus*. *J. Natcon.*, 12: 83–93.

Kaur, K. and Toor, H.S., 1977. Toxicity of pesticides to embryonic stages of *Cyprinus carpio*. Part 2. Early Juveniles. *Int. J. Acad. Ichthyol.*, 6: 1–4.

Koundiya, P.R. and Ramamurthi, S.R., 1978. Effect of Sumithion on some selected enzyme system in the fish *Tilapia mossambica*. *Ind. J. Exp. Biol.*, 16: 809–811.

Kulshrestha, S.K. and Arora, N., 1985. Effect of Carbofuran Rogor and DDT on the development of *Cyprinus carpio*. *Ind. J. Exp. Biol.*, 23: 193–196.

Kumar, S., Lata, S. and Gopal, K., 1999. Acute toxicity of Deltamethrin to the freshwater teleost *Heteropneustes fossilis* and *Channa punctatus*. *Proc. Acad. Environ. Biol.*, 8: 83–85.

Kumar, S., Lata, S. and Gopal, K., 2000. Histopathological changes in gills of *Heteropneustes fossilis*, induced by Deltamethrin. *Proc. Acad. Environ. Biol.*, 9: 171–173.

Matsmura, F., 1980. *Toxicology of Insecticides*. Plenum Press. N. York.

Mishra, D.K., Bohidhar, K. and Pandey, A.K., 2005. Histopathological alterations in gills of freshwater teleost *Channa punctatus* exposed to sublethal concentration of Cartap. *J. Ecophysiol. Occup. Health.*, 5: 163–165.

Murphy, P.G., 1971. The effects of the size on the uptake of DDT from water by fish. *Bull. Environ. Contam. Toxicol.*, 6: 20–23.

Natarajan, G.M., 1984. Effect off lethal (Lc_{50}/48 Hr) concentration of Metasystox and some selected enzyme system in air breathing fish *Channa punctatus*. *Comp. Physiol. Ecol.*, 9: 29–32.

Nath, A., Chand, G.B., Sinha, A. and Verma, P., 2000. Thyroxine hormonal imbalance and related ultrastructural changes in the testis of Rogor treated *Clarias batrachus*. *J. Inland Fish. Soc. India*, 32: 87–93.

O'Brien, R.D., 1976. Acetylcholinesterase and its inhibition. In: *Insecticides Biochemistry and Physiology*, (Ed.) C.F. Wilkinson. Plenum Press, New York.

Obula Reddy, K.P. and Neerja, P., 2001. Certain metabolic modulations in carbohydrate metabolism of fry *Cyprinus carpio* on ammonia stress. *J. Nacton.*, 13: 191–194.

Praveen, M., Sharma, R. and Kumar, S., 2004. Effect of Neemta 2100 toxicity on Acetylcholinesterase and serum glutamate oxaloacetate transaminase enzyme in serum of fish *Oreochromis mossambica*. *J. Ind. Fish. Assoc.*, 31: 97–100.

Rawat, M. and Bhargava, S., 1997. Synergic action of pesticides on fish *Hetropneustes fossilis*. *J. Ecotoxicol. Env. Monit.*, 7: 177–180.

Ronald, B.P., Javier, S.V. and David, H.S., 1999. Inhibition of Neurite outgrowth from chick sympathetic neurons by Cholinesterase inhibitors not mediated by binding of cholinesterase. *Neurosci. Letters.*, 266: 77–80.

Sharma, R.R., Pande, A.K. and Shukla, G.R., 2001. Histopathological alterations in fish tissues induced by pesticide toxicity. *Aquacult*, 2: 31–43.

Singh, A., 2003. Toxicity of few chemicals to fish *Heteropneustes fossilis* using static bioassay procedure. *J. Nacton.*, 15: 255–260.

Sudheer Kumar, D.J., John Sushma, N., Sivaiah, V., Madhava Rao, S. and Jayantha Rao, K., 2006. Effect of chloropyrifos and azadirachtin on Ach and AchE of fish *Tilapia mossambica*. *Aquacult*, 7: 87–91.

Tandon, R.S. and Dubey, A., 1983. Toxic effect of two organophosphate pesticides on Fructose 1,6, Diphosphatase Aldolase activity of liver, brain and gills of freshwater fish *Clarias batrachus*. *Environ. Pollut.*, 31: 1–7.

Thannipon, W., Thannipon, W., Luappaiboon, P. and Chinabut, S., 1995. Effect of organophosphate insecticide monocrotophos on acetylcholinesterase activity in the nile *Tilapia* fish brain. *Neurochem. Res.*, 20: 587–591.

Water and Wastewater Management in 2 Vols (2011) *Pages* **540–553**
Editors: **D.R. Khanna, R. Bhutiani & Gagan Matta**
Published by: **DAYA PUBLISHING HOUSE, NEW DELHI**

Chapter 83

Physico-chemical Status of Soil Between Lowang to Gatti, Sewa Catchment, Kathua District, J&K

Amita Fotedar and Anil Kumar Raina
Department of Environmental Science, Jammu University, Jammu

ABSTRACT

Four stations were selected in Sewa river catchment, district Kathua, J&K for soil studies for 2 years on monthly basis. Geologically and from pH studies have revealed acidic character of soils. The soils are mostly clay-loamy in texture. Sewa catchment for most of its parts is barren and soil moisture is accounted for persistent snow cover on high peaks and frequent rain for 8 months during the year. Moisture ranges between 8.73 per cent to 12.90 per cent in all profiles and results in high ingress of water. The loss of nutrition among SiO_2, Al_2O_3, TiO_2, Fe_2O_3, CaO, MgO, Na_2O and K_2O and MnO by chemical analysis, revealed that TiO_2, Fe_2O_3, Na and K get depleted from composite samples and as such need compensation. Al_2O_3 and MnO get added to the composite soils, because of excessive erosion usually common most of the Himalayan catchments. SiO_2, CaO and MgO do not show any impoverishment in composite samples and remain almost same in both *in-situ* and composite samples. Organic carbon too, in all the four profiles indicates uniform distribution. The present studies shows soil of catchment area of River Sewa moderately to highly degraded and prone to soil erosion, landslides and rock falls– leading ultimately to loss of nutrients essential for maintaining the fertility of the soil. The catchment areas should properly be forested, planted with long rooted grass and vetiver grass to control erosion. Various land management practices, such as mixed cropping, intercropping, strip cropping, rotational cropping, mulching, application of organic manures will go a long way to reduce soil erosion and conserve sub-soil moisture effectively.

***Keywords**: Soil, Sewa river, Catchment, Kathua District, Erosion.*

Introduction

River Sewa, a snowfed river that originates from the Siwalik range, is one of the major tributaries of River Ravi. It traverses through the Bani and Basohli blocks of District Kathua, which is situated in the south-east of Jammu and Kashmir State. The northern part of District Kathua is situated in the foothills of Himalaya which includes the Siwalik range. The southern part consists of alluvial plain. Ravi river, which possesses rich power and irrigation potential, flows in the eastern part of the District. Sewa river originates from Sarthal hills, at an altitude of 4200 m above mean sea level on the south-western slopes of the Chattar watershed and after traversing through Bani and Basohli covers a distance of 53 km, then merges with River Ravi at Mashka at an altitude of 578 m.

Sewa river catchment is 481 sq km and the topography is diverse on account of altitudinal gradient from 578 to 4200 m above mean sea level. This catchment has a large number of small and big glaciers. There is a steep gradient at some places along the smaller stretches of River Sewa.

The soil studies in a catchment area of any river is very important dynamic component of ecosystem and is one of the natural resources required for the growth of land plants, the essential supply of water and the nutrients for the plants are available from it. The physical and chemical characteristics of the soil determine the type of vegetation which can be supported by it. The top layer of the soil is a vital component, since all the nutrients required by the plants are present in this layer. In fact, it is the source of 13 elements out of 16 elements essential for the plant growth.The state of Jammu and Kashmir has in all seven types of soils, namely Udalfs, Orthents, Ochrepts, Ochrepts-Orthents, Ochepts-Orthents-Udalfs, Udolls and soils under glaciers and snow caps. The Sewa river catchment has the ochrepts type of soil (Fotedar, 2006).

These soils are found in lower plains of Kathua District and are mostly alluvial and loamy consisting of little clay contents. The knowledge of physico-chemical composition of the soil and its variation during the different parts of the year is pre-requisite for the proper utilization and management of soil. Since no study has been carried out in respect of soils of Sewa catchment so far, it is thus important to collect the primary data about soils in different profiles. The data needs to be in respect of texture, chemical composition, soil moisture, pH, high and low altitude variation because of erosion and loss of nutrition *etc.*, all these have been studied in detail in this chapter. The four stations in the Sewa catchment have been included in the present study, namely Lowang, Bani, Sarthali and Gatti, all situated in Kathua District, J&K state.

Materials and Method

Soil samples were collected from four different sampling stations of Sewa watershed representing the higher elevations (*in-situ i.e.* higher elevation in the watershed area) and the stations below the drainage lines near the river banks at correspondingly lower elevations on monthly seasonal basis (Figure 83.1). Samples collected from higher elevations were named as higher elevation (*in-situ*) soil samples, and the one which were collected from the stations near the River banks at correspondingly lower elevations were named as composite soil samples. For collecting the soils at the higher elevation *i.e.* the *in-situ* soil samples, first the overburdened accumulated soils consisting of humus, litter, lots and pots of forest residues, were removed by scaling and digging upto 0-15 cm from the surface and then around 100 gm of soil samples were collected in the polythene bags. Similarly, composite soil samples were collected at the termination of gullies, near the banks of Sewa river. The sampling of soil was done in this way in order to assess the loss of nutrients.

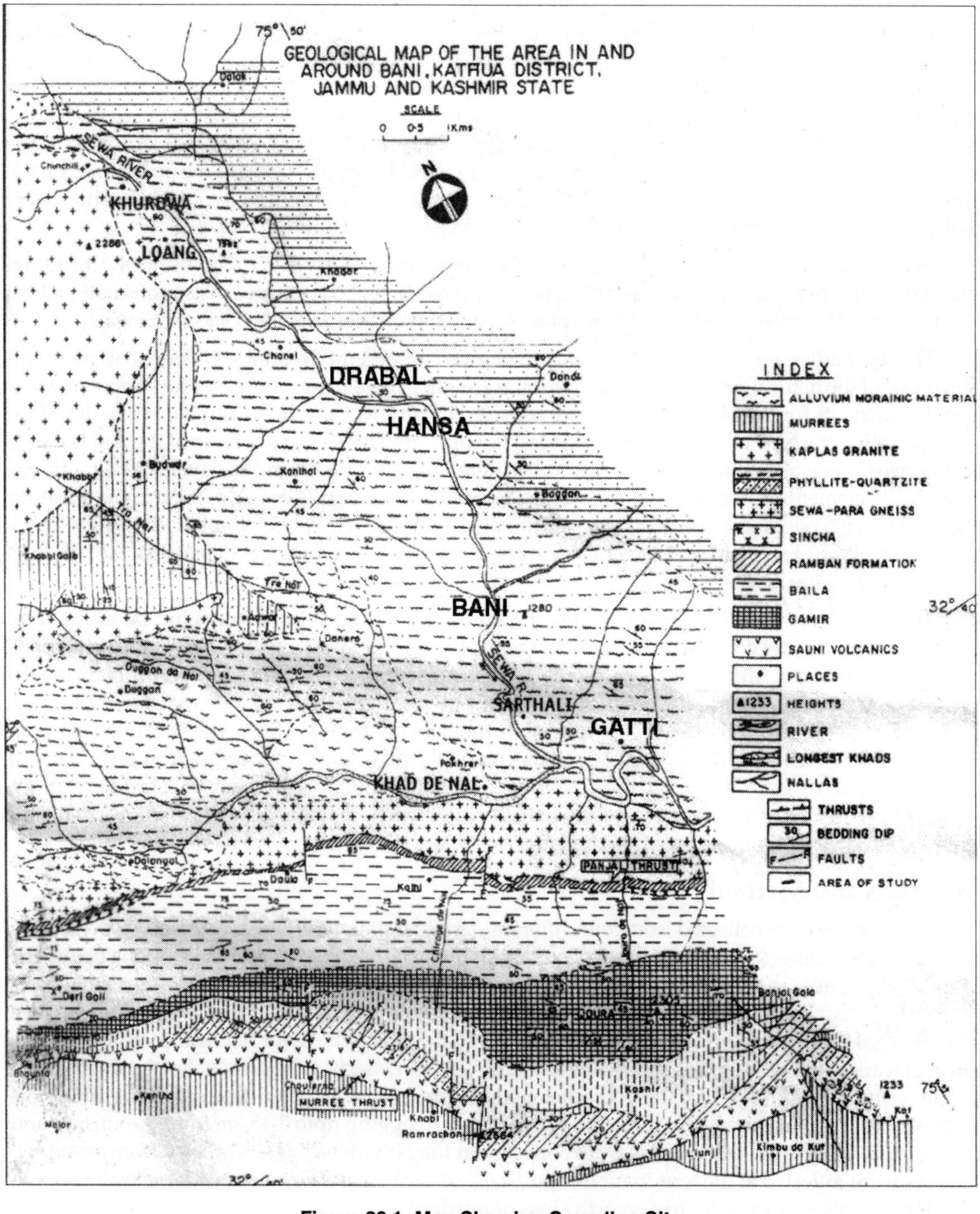

Figure 83.1: Map Showing Sampling Site

Soil texture of the samples was determined according to structural triangle method of "US Department of Agricultural Hand Book No. 18" after asserting the percentage of various soil components *viz.*, coarse sand, fine sand, silt, clay and loam by sieving the sample through electrically operated sieve set having sieve mesh numbers 25, 53, 72, 120, 200 and bottom pan. Soil moisture was determined by gravimetric method. Soil sample of known weight was dried in oven at 105 °C for 24 hours. The percentage moisture was calculated by the loss of weight of the sample. Soil pH meter "HANNA" make was used for recording the pH of the soil samples. For this, 1:2 soil water suspensions were prepared and then the pH was recorded. pH was further confirmed in the laboratory by Phillip pH meter.

Seven elemental oxides were determined by the following methods from 'A' solution and 'B' solution according to Shapiro and Brunnock Method (1956). Al_2O_3 was analyed volumetrically using EDTA and Zinc Sulphate (Shapiro and Brunnock, 1956). TiO_2 was determined calorimetrically using Ferric Ammonium Sulphate, $KMnO_4$ and Ammonium Thiocyanate (Shapiro and Brunnock, 1956). Total iron was determined calorimetrically using Tartaric Acid, *P*-Nitrophenol, 1-10 Phenanthroline and Hydroxylamine Hydrochloride (Shapiro and Brunnock, 1956).

CaO and MgO was determined volumetrically by EDTA Method (Vogel, 1962 and APHA, 1998). Na_2O and K_2O was determined on Flame Photometer (EEL-Model). MnO was determined by using Potassium Periodate (Vogel, 1962). Organic carbon was estimated according to Walkley and Black's rapid literation method (Piper, 1944).

Geology in Brief

The area from north to south falls in pre-cambrian terrain. In the extreme south, tertiary rocks bind the area. The area is highly folded, faulted and metamorphosed due to two granitic intrusions, one in north-west as Kaplas granite of Bhaderwah and second in the south-west as Dalhousie granite. Besides, two important thrusts, namely Murree Thrust in the south and Panjal Thrust in the north are responsible for formation of anticlines and synclines causing plications in the rocks. Bhaderwah-Chamba is an example of such a syncline and the rocks have been referred to as Bhaderwah Slates by Sharma (1973-75). Panjal Thrust passes east-west above Ramban and Sincha formations to the south of Gatti area. Sewa para gneiss is overlain conformably by phyllite-slate sequence of late-pre-cambrian Salkhala group (Figure 83.1). The present area of study falls strictly in late-pre-cambrian Bhaderwah Formation of Salkhala group of rocks. These rocks are acidic in character and constitute mostly granitic and reworked metamorphosed sediments slightly containing more ferromagnesian minerals soils (Dhar *et al.*, 1996).

Physico-chemical Characteristics of Higher Elevation and Composite Soil Samples

Physical Parameters

At station 1 (Lowang), coarse sand varied from a minimum value of 9.7 per cent to 16.0 per cent and the average being 12.6 per cent. Fine sand fluctuated between 8.2 per cent to 21.8 per cent with an average value of 14.0 per cent. Silt concentration varied from 21.7 per cent to 39.4 per cent with an average of 33.5 per cent. Clay concentration ranged between 1.4 per cent to 23.7 per cent with an average value of 18.4 per cent. Loam concentration fluctuated from 11.7 per cent to 30.7 per cent with an average value of 21.5 per cent. The texture class is clay-loam.

At station 2 (Bani), coarse sand varied from a minimum of 9.3 per cent to 16.0 per cent with an average value of 12.35 per cent. Fine sand fluctuated between 9.8 per cent to 17.0 per cent with an average value of 13.17 per cent. Silt varied between 3.8 per cent to 43.9 per cent with an average value of 36.06 per cent. Clay ranged between 11.3 per cent to 27.0 per cent with an average value of 18.20 per cent. Loam ranged between 11.7 per cent to 33.5 per cent with an average value of 20.2 per cent. The texture class is clay-loam.

At station 3 (Sarthali), coarse sand varied between 9.3 per cent o 16.7 per cent with an average value of 13.5 per cent. Fine sand fluctuated between 10.4 per cent to 23.2 per cent with an average value of 14.2 per cent. Silt concentration varied between 19.8 per cent to 45.8 per cent with an average value of 34.0 per cent. Clay fluctuates between 13.6 per cent to 24.1 per cent with an average value of 19.1 per cent. Loam ranged between 12.9 per cent to 34.6 per cent with an average value of 19.2 per cent. The texture class is clay-loam.

At station 4 (Gatti), coarse sand ranged between 8.31 per cent to 14.2 per cent with an average value of 11.9 per cent. Fine sand ranged between 9.4 per cent to 18.4 per cent with an average value of 13.8 per cent. Silt fluctuates between 25.8 per cent to 45.9 per cent with an average value of 35.7 per cent. Clay fluctuates from 15.6 per cent to 22.1 per cent with average value of 18.4 per cent. Lastly, loam ranged between 14.0 per cent to 33.5 per cent with an average value of 20.2 per cent. Texture class is clay-loam.

Soil Moisture

Station 1 (Lowang)–Soil moisture varied from 5.4 per cent to 19.2 per cent in case of higher elevation (*in-situ*) soils, whereas in composite soils, moisture varies from 5.4 per cent to 15.0 per cent. The average concentration of moisture does not show much variations.

Station 2 (Bani)–Soil moisture varied from 5.2 per cent to 20.6 per cent in case of *in-situ* soils, whereas in composite soils, it ranges between 5.2 per cent to 17.3 per cent. In higher and lower elevation soils, the moisture does not show much variation.

Station 3 (Sarthali)–Soil moisture varied between 4.5 per cent to 23.8 per cent in higher elevation soils, whereas in lower elevated soils (composite samples), it ranged between 4.5 per cent to 21.8 per cent. Concentration of moisture does not show much variation in this profile.

Station 4 (Gatti)–In higher elevation soils, the moisture fluctuates between 4.7 per cent to 18.2 per cent, whereas in composite soils, it ranges between 4.7 per cent to 17.8 per cent. The moisture condition in this profile for two sites does not show much fluctuation. The moisture condition in higher and lower elevation soils is same because of nearness of Panjal Thrust. The area is highly unstable and due to thrusting, uniform drainage flows occur from top to bottom of the rock sequences.

Soil pH

Station 1 (Lowang)–At this station, soil pH fluctuated between 5.6 to 7.1 for higher elevation and 5.9 to 7.0 in composite soils.

Station 2 (Bani)–Values ranged between 5.0 to 7.1 for higher elevation and 6.2 to 7.1 for composite soils.

Station 3 (Sarthali)–At this station, soil pH ranged between 5.6 to 7.0 in higher elevation (*in-situ*) samples, while as for composite samples, it ranged between 5.0 to 7.4.

Station 4 (Gatti)–At this station, the soil pH ranged between 5.6 to 7.0 for *in-situ* soil samples while as for composite soils it ranged between 5.7 to 7.0.

Results and Discussion

Physical Parameters

As far as textural classification of soils of Sewa catchment is concerned, the soils belong to clayey-loam in texture for all the four stations. From pH measurement data for all the four profiles, acidic nature of the soils is revealed. This is consistent with the acidic character of the rocks (Dhar *et al.*, 1996). The soil pH reflects both, the chemical properties of the mineral soils as well as of the organic residue deposited as litter (Wazir, 1984). However, along the longitudinal profiles for Station 1.0 to 4.0, pH has shown fluctuation from slightly acidic and slightly alkaline. This range of pH favours the breaking of silicate chain aiding the denudation (Wedepohl, 1978). Also, burial metamorphism in Salkhala group of rocks, occurring in Sewa catchment is an additional factor causing erosion.

Soil moisture appears to be uniformly distributed in forest soils and soils in the lower elevations. The ability of forest soils to retain water depends upon the amount of silt and clay present. The amount of silt and clay exists almost uniformly distributed (50 per cent or nearly so) and as such moisture does not show much variation. The frequent rains in the upper reaches and presence of glaciers keeps the moisture conditions almost uniform.

Chemical Parameters

The concentration of silica in composite soil samples in all the four stations is less at almost all stations as compared to higher elevation soil samples, which shows that the area of study has suffers from extensive erosion. Silica is rich in acidic soils as has been reported by Nesbitt and Young (1982), so under favourable pH and metamorphism, the acidic rocks correspond to high values of silica in the soils of Sewa catchment. Along the longitudinal profiles from Station 1 (Lowang) to Station 4 (Gatti), silica recorded high values at Lowang as compared to rest of the stations. The reason for this is that Lowang is surrounded by granitic rocks which are acidic in nature and silica is rich in acidic rocks.

Aluminium oxide in the present study has been found higher in case of composite soils than *in-situ* soils. Aluminium contents get added up in the composite soils, where intensity of erosion is more. The soils have got weathered to a large extent and this has resulted in more clayey type of composite soil rich in aluminium content. This observation is supported by studies of Wedepohl (1978) in the present case. More aluminium in soil samples give indirectly an idea of severe erosion having undergone by the rocks of Sewa catchment. According to longitudinal profile, TiO_2 in soils of Sewa catchment is lower in composite soils as compared to *in-situ* higher elevation soils. This indicates more of erosion suffered by the rocks of Sewa catchment. This influence is supported by the study of Nesbitt and Young (1982) as TiO_2 is considered one of the best indicators to assess the rate of weathering. In the present study, ilmenite has been best carrier of TiO_2 which has impoverished in composite soils with prolonged course of weathering.

CaO percentage in Lowang (Station 1), is the highest points in the area of study. In higher elevation (*in-situ*) soils, the average came to 1.93 per cent, while as in composite soils it was found as 1.95. In Bani (Station 2), CaO percentage average was found as 1.55 in higher elevation soils, while as in composite soils it came to 1.52 per cent, loss of nutrition being merely 3.0 per cent. In Sarthali profile (Station 3), CaO percentage was 1.03 in *in-situ* soils and 0.93 per cent in composite soils, loss of nutrition being merely 0.1 per cent. In Gatti (Station 4), CaO percentage registered a loss from *in-situ* to composite soils as 0.28 per cent. Albrecht and Smith (1952) considered that calcium-deficiency is prominent feature of the adverse effects of soil acidity upon plant growth. An insufficiency of calcium permits the accumulation of undesirable ions in plants. In all profiles of Sewa catchment, the loss of nutrition that has occurred is not much and cannot be considered undesirable.

Combined alkalies ($Na_2O + K_2O$) in Lowang were found to be 5.65 per cent in higher elevation soils, while as it was found as 3.74 per cent in case of composite soils. At Bani, the percentage fluctuated from 2.94 per cent to 2.66 per cent. At Sarthali, the percentage dropped from 7.05 per cent to 3.50 per cent and at Gatti, the percentage dropped from 6.79 per cent to 5.53 per cent. The nutrition loss in case of total alkalies, thus, came to be 0.91 per cent, 0.28 per cent, 3.55 per cent, 1.16 per cent, respectively in all the four profiles. Sarthali registers maximum loss of alkalies, because the whole catchment in this profile is barren for the most part.

Mg is very important component of chlorophyll pigment and its deficiency in general can cause chlorosis in plants. In the present study, MgO has been found to lie between 1.74 per cent in *in-situ* soils of Lowang to 2.21 per cent to composite soils; at Bani station, MgO has been found to lie between 1.10 per cent to 0.97 per cent for *in-situ* and composite soils, respectively. At Sarthali, MgO percentage for higher elevation soils was found to be 0.51 per cent and for composite soils as 0.44 per cent. At Gatti, MgO was found to be 0.40 per cent in *in-situ* soils and 0.38 per cent in composite soils.

The present values in all the profiles are in accordance with the values got by Wazir (1984) in the Bhaderwah soils existing in the north of the present area of study. Overall, MgO percentage in all the profiles of Sewa catchment does not show much impoverishment when we examine values passing from *in-situ* to composite soil samples.

Mg in forest soils is usually from 1/5th to 1/3rd of Ca (Wilde, 1958). In the present case, Mg has been recorded higher than this ratio. Mg in the composite samples on an average is lesser than *in-situ* samples by a margin of 0.15 per cent. The Mg ions in excess produce harmful effects in isolation, but in forest soils, these effects are neutralized by the amount of Ca present in the soils (Puri and Gupta, 1951; Ray Chaudhury and Datta, 1964; Champion and Seth, 1968; Wazir, 1984).

MnO percentage in all the profiles showed an increase from *in-situ* soils to composite soils. Close examination (Tables 83.2–83.5) revealed that in the longitudinal profile from Lowang to Gatti, Mn gets considerably added up. This is consistent with extensive studies done by Horowitz (1974) in down water soil profiles, which he ascribes to most of Mn getting concentrated by the mechanism of adsorption on fine grained sediments because of large surface area. Similar adsorption and concentration increasing in low profiles has been noticed in Jajjar Nalla, Udhampur, J&K by Fotedar and Loan (2004) and Tikoo (2004) in a tributary of Chenab in J&K.

Fe_2O_3 found in higher elevation soils of Lowang was 2.89 per cent and in composite soils as 1.79 per cent. At Bani, Fe_2O_3 was found as 4.6 per cent in *in-situ* samples and as 3.50 per cent in composite samples. At Sarthali, in high elevation soils (*in-situ*), Fe_2O_3 was found as 3.87 per cent going down to 2.10 per cent in composite soils. At Gatti, Fe_2O_3 in high elevation soils (*in-situ*) came to be as 2.61 per cent and 1.84 per cent in composite soils. The decrease in concentration observed in four profiles came to be 1.10 per cent, 1.06 per cent, 1.77 per cent, 0.77 per cent in Lowang, Bani, Sarthali and Gatti, respectively.

Organic carbon in Lowang for *in-situ* soils and composite soils is 0.29 per cent and 0.24 per cent, respectively. For Bani *in-situ* samples, it is 0.29 per cent and 0.25 per cent, respectively. In Sarthali higher elevation, organic carbon is 0.32 per cent for *in-situ* soils and for composite soils only 0.25 per cent. For Gatti, organic carbon is 0.27 per cent for *in-situ* samples and 0.24 per cent for composite samples. In all the four profiles from Lowang to Gatti, there is not much variation as far as organic carbon is concerned.

Table 83.1: Mean Monthly Percentage Values of Soil Texture from January 2002 to December 2004

Texture Separates	*Jan*	*Feb*	*Mar*	*Apr*	*May*	*Jun*	*Jul*	*Aug*	*Sep*	*Oct*	*Nov*	*Dec*	*Average*
Coarse sand	11.4	14.9	11.8	13.3	16.0	15.0	12.0	13.3	9.7	10.1	12.5	11.4	12.6
Fine sand	16.3	10.8	11.0	8.2	9.8	1.6	11.8	12.9	15.2	21.8	16.2	1.5	14.0
Silt	33.8	36.8	38.4	39.4	43.9	39.2	33.2	24.2	27.0	33.2	21.7	31.4	33.5
Clay	19.5	11.3	18.7	17.6	16.3	16.5	18.1	21.4	1.4	18.9	20.9	23.7	18.4
Loam	19.0	26.2	20.1	21.5	14.0	11.7	24.9	28.2	30.7	16.0	28.7	17.0	21.5
Texture Class	Clay-loam	Clay-loam	Clay-loam	Clay-loam	Clay-loam	Clay-loam	Clay	Clay	Clay	Clay-loam	Clay	Clay	Clay-loam
Coarse sand	9.3	16.0	10.0	13.4	15.0	12.0	14.9	11.8	12.5	11.2	12.0	10.1	12.35
Fine sand	15.8	9.8	11.3	15.9	17.6	11.8	10.8	11.0	15.2	13.9	12.5	12.5	13.17
Silt	34.8	43.9	36.8	31.0	39.2	33.2	3.8	38.4	27.2	25.8	41.1	42.6	36.06
Clay	19.9	16.3	27.0	19.1	16.5	18.1	11.3	18.7	17.1	15.6	19.8	19.1	18.20
Loam	20.2	14.0	12.9	20.6	11.7	24.9	26.2	20.1	28.0	33.5	14.6	15.7	20.2
Texture class	Clay-loam	Clay-loam	Clay-loam	Clay	Clay-loam	Clay	Clay-loam	Clay-loam	Clay	Clay	Clay-loam	Clay-loam	Clay-loam
Coarse sand	13.7	9.3	16.7	16.2	13.0	10.6	13.4	15.8	12.8	11.2	15.2	15.1	13.5
Fine sand	20.6	15.6	11.9	10.4	12.0	11.3	15.9	16.1	19.0	20.0	23.2	17.9	14.2
Silt	29.9	34.8	35.2	35.4	45.8	38.8	31.0	26.5	19.8	29.8	21.9	29.4	34.0
Clay	17.8	19.4	18.5	20.5	15.2	2.0	19.1	17.1	13.6	17.9	18.6	24.1	19.1
Loam	18.0	21.2	17.7	17.5	14.0	12.9	20.6	24.5	34.6	21.1	21.1	13.5	19.2
Texture class	Clay-loam	Clay-loam	Clay-loam	Clay-loam	Clay-loam	Clay-loam	Clay	Clay-loam	Clay	Clay-loam	Clay-loam	Clay-loam	Clay-loam
Coarse sand	12.8	12.9	12.2	13.5	12.5	9.6	13.5	12.5	11.2	8.3	9.0	14.2	11.9
Fine sand	17.5	9.9	11.4	9.4	12.7	10.9	16.2	15.2	13.9	12.8	18.4	16.8	13.8
Silt	35.0	36.6	39.3	41.5	41.2	45.9	35.8	27.2	25.8	32.0	37.9	30.7	35.7
Clay	15.8	19.4	22.1	18.0	18.1	16.0	20.5	17.1	15.6	21.4	19.0	18.3	18.4
Loam	18.9	21.2	15.0	17.5	15.5	17.6	14.0	28.0	33.5	25.5	15.7	20.0	20.2
Texture class	Clay-loam	Clay-loam	Clay-loam	Clay-loam	Clay-loam	Clay-loam	Clay-loam	Clay	Clay	Clay	Clay-loam	Clay-loam	Clay-loam

Note: All the readings represent the average of 2 years.

Table 83.2: Mean Monthly Values of Physico-chemical Parameters of Soil Collected from Higher Elevated (*in-situ*) and Composite Samples of Station 1 (Lowang) from January, 2002 to December, 2004

Month	Soil Moisture (Per cent)	Soil pH	SiO_2 (Per cent)	Al_2O_3 (Per cent)	TiO_3 (Per cent)	Fe_2O_3 (Per cent)	CaO (Per cent)	MgO (Per cent)	Na_2O (Per cent)	K_2O (Per cent)	MnO (Per cent)	Organic Carbon (Per cent)
Higher elevated (*in-situ*) soil samples												
Jan	14.8	7.1	75.47	12.10	0.19	4.60	2.12	1.82	2.90	3.34	0.03	0.19
Feb	15.0	7.0	75.48	12.12	0.21	1.62	2.14	1.84	2.91	3.35	0.03	0.22
Mar	10.0	7.0	74.43	13.13	0.20	2.61	2.10	1.82	2.91	2.45	0.09	0.59
Apr	6.5	6.6	74.44	13.14	0.22	3.62	2.12	1.84	2.92	2.46	0.09	0.20
May	5.9	5.9	74.45	13.15	0.21	1.63	2.14	1.86	2.93	2.47	0.07	0.30
Jun	5.4	5.7	75.49	12.99	0.20	2.40	2.38	1.82	2.30	2.22	0.03	0.30
Jul	5.6	6.7	75.50	13.00	0.22	2.42	2.40	1.83	2.32	2.27	0.03	0.24
Aug	7.8	6.6	75.51	13.01	0.24	3.44	2.42	1.84	2.34	2.32	0.03	0.22
Sep	13.6	7.0	75.40	11.96	0.14	1.57	1.08	1.48	2.87	3.50	0.03	0.45
Oct	11.9	6.3	75.45	11.97	0.19	1.59	1.10	1.50	2.89	3.53	0.04	0.35
Nov	10.1	6.5	75.50	11.98	0.24	4.61	1.12	1.52	2.91	3.56	0.05	0.14
Dec	19.2	5.8	75.46	12.80	0.20	4.58	2.10	1.80	2.89	3.33	0.03	0.28
Average	10.31	6.5	75.21	12.61	0.20	2.89	1.93	1.74	2.75	2.90	0.04	0.29
Composite soil sample												
Jan	10.8	6.0	62.10	23.10	0.10	3.00	2.10	2.20	1.28	3.00	0.06	0.19
Feb	15.0	6.4	61.90	23.11	0.10	2.10	2.11	2.21	1.29	3.10	0.07	0.18
Mar	9.0	6.0	62.00	23.30	0.11	2.01	1.85	2.20	1.27	2.00	0.06	0.38
Apr	6.5	6.9	62.10	23.10	0.12	2.02	1.90	2.21	1.30	2.00	0.06	0.17
May	5.9	7.0	62.20	22.90	0.13	1.03	1.70	2.10	1.22	2.40	0.06	0.30
Jun	5.4	7.0	62.11	22.09	0.09	1.11	1.79	2.24	1.10	2.29	0.07	0.23
Jul	5.6	5.9	62.12	22.08	0.10	1.12	1.80	2.23	1.08	2.24	0.08	0.24
Aug	7.8	6.6	62.13	22.07	0.11	2.13	1.81	2.22	1.22	2.00	0.08	0.22
Sep	13.6	6.8	60.92	24.15	0.10	2.03	1.90	2.16	1.20	2.16	0.07	0.38
Oct	7.9	6.9	60.93	24.10	0.10	2.04	2.00	2.10	1.24	3.10	0.06	0.35
Nov	8.1	6.6	60.94	23.95	0.10	1.05	2.10	2.24	1.10	3.00	0.06	0.14
Dec	11.2	5.9	62.00	23.09	0.10	1.90	2.09	1.27	1.00	2.00	0.07	0.19
Average	8.73	6.5	61.78	23.08	0.10	1.79	1.95	2.21	1.19	2.55	0.06	0.24

Note: All the readings represent the average of 2 years.

Table 83.3: Mean Monthly Values of Physico-chemical Parameters of Soil Collected from Higher Elevated (*in-situ*) and Composite Samples of Station 2 (Bani) from January, 2002 to December, 2004

Month	Soil Moisture (Per cent)	Soil pH	SiO_2 (Per cent)	Al_2O_3 (Per cent)	TiO_3 (Per cent)	Fe_2O_3 (Per cent)	CaO (Per cent)	MgO (Per cent)	Na_2O (Per cent)	K_2O (Per cent)	MnO (Per cent)	Organic Carbon (Per cent)
Higher elevated (*in-situ*) soil samples												
Jan	18.0	7.0	66.27	18.40	0.33	5.00	1.50	1.28	2.40	3.80	0.03	0.24
Feb	17.3	5.8	66.29	18.41	0.36	5.02	1.52	1.30	2.44	3.82	0.02	0.28
Mar	20.6	5.6	66.23	20.02	0.22	4.10	1.56	0.98	1.19	3.90	0.01	0.37
Apr	8.5	7.0	66.24	20.04	0.21	4.12	1.58	0.99	1.20	3.88	0.01	0.24
May	8.2	7.0	66.25	20.06	0.20	4.08	1.57	1.00	1.21	3.92	0.01	0.27
Jun	6.7	6.7	64.13	22.00	0.20	3.96	1.48	0.94	2.08	3.48	0.01	0.16
Jul	9.9	5.9	64.14	22.10	0.21	3.95	1.50	0.96	2.09	3.50	0.01	0.30
Aug	15.2	7.0	64.15	22.20	0.19	3.97	1.52	0.98	2.10	3.52	0.01	0.25
Sep	9.8	6.8	65.42	19.10	0.40	5.10	1.62	1.16	1.65	3.63	0.01	0.48
Oct	10.5	5.7	65.40	19.11	0.38	5.18	1.64	1.18	1.67	3.65	0.01	0.35
Nov	18.5	7.0	65.38	19.09	0.42	5.26	1.66	1.20	1.69	3.67	0.01	0.24
Dec	14.6	7.1	66.25	18.39	0.30	4.98	1.48	1.26	2.36	3.78	0.04	0.37
Average	12.9	6.5	65.51	19.91	0.28	4.56	1.55	1.10	1.84	3.71	0.02	0.29
Composite soil sample												
Jan	12.0	6.6	64.10	20.22	0.20	4.98	1.48	0.90	2.00	3.70	0.03	0.24
Feb	17.3	7.0	64.11	20.23	0.21	4.99	1.49	0.91	2.01	3.71	0.04	0.28
Mar	10.6	7.1	62.90	22.31	0.28	3.98	1.50	0.96	1.09	3.85	0.02	0.37
Apr	8.5	7.0	63.00	22.32	0.32	4.00	1.52	0.98	1.10	3.86	0.02	0.15
May	5.2	6.8	63.10	22.33	0.36	4.02	1.54	1.00	1.11	3.87	0.04	0.27
Jun	6.7	6.3	64.08	21.09	0.19	3.97	1.46	0.95	1.98	3.47	0.02	0.14
Jul	9.9	7.0	64.10	22.00	0.20	3.98	1.48	0.96	1.99	3.00	0.02	0.19
Aug	12.2	7.1	64.12	22.01	0.21	3.99	1.50	0.97	2.00	2.30	0.05	0.20
Sep	9.8	6.8	64.51	23.09	0.36	3.40	1.60	1.05	1.64	2.10	0.06	0.28
Oct	10.5	6.4	64.52	23.11	0.37	3.00	1.62	1.07	1.68	2.00	0.07	0.35
Nov	13.5	7.0	64.53	23.13	0.38	3.10	1.64	1.09	1.72	2.09	0.08	0.24
Dec	10.6	6.2	64.09	20.21	0.19	2.90	1.47	0.89	1.99	2.05	0.09	0.37
Average	10.31	6.7	63.93	21.83	0.27	3.50	1.52	0.97	1.69	2.97	0.05	0.25

Note: All the readings represent the average of 2 years.

Table 83.4: Mean Monthly Values of Physico-chemical Parameters of Soil Collected from Higher Elevated (*in-situ*) and Composite Samples of Station 3 (Sarthali) from January, 2002 to December, 2004

Month	Soil Moisture (Per cent)	Soil pH	SiO_2 (Per cent)	Al_2O_3 (Per cent)	TiO_3 (Per cent)	Fe_2O_3 (Per cent)	CaO (Per cent)	MgO (Per cent)	Na_2O (Per cent)	K_2O (Per cent)	MnO (Per cent)	Organic Carbon (Per cent)
Higher elevated (*in-situ*) soil samples												
Jan	23.8	6.5	65.08	22.35	0.21	3.08	1.08	0.50	2.70	4.30	0.04	0.21
Feb	15.0	5.6	64.90	22.37	0.20	3.04	1.01	0.51	2.78	4.31	0.03	0.32
Mar	9.0	5.6	64.61	23.00	0.18	3.95	1.28	0.52	2.83	3.97	0.02	0.43
Apr	6.5	7.0	60.62	23.01	0.17	3.96	1.30	0.51	2.84	3.98	0.03	0.28
May	6.9	7.0	60.63	23.02	0.19	3.97	1.32	0.53	2.85	3.99	0.01	0.30
Jun	5.4	6.7	61.00	22.20	0.22	4.97	1.11	0.53	2.87	4.00	0.02	0.38
Jul	6.6	5.9	64.10	22.00	0.23	4.98	1.10	0.54	2.86	4.11	0.03	0.24
Aug	7.8	7.0	63.20	21.80	0.21	3.99	1.09	0.55	2.85	4.22	0.01	0.26
Sep	9.6	6.4	66.17	22.20	0.22	3.50	0.72	0.52	2.81	4.56	0.03	0.47
Oct	11.9	5.7	63.18	22.21	0.21	3.52	0.70	0.50	2.82	4.55	0.03	0.39
Nov	10.1	6.8	60.19	22.22	0.20	3.54	0.68	0.48	2.80	4.57	0.03	0.24
Dec	16.2	7.0	64.00	22.36	0.19	3.96	1.03	0.49	2.80	4.29	0.03	0.28
Average	10.73	6.4	63.13	22.39	0.20	3.87	1.03	0.51	2.82	4.23	0.03	63.13
Composite soil sample												
Jan	21.8	6.2	63.68	23.70	0.19	1.98	0.98	0.47	2.78	4.12	0.03	0.21
Feb	16.0	5.0	63.69	23.71	0.20	1.99	0.99	0.48	2.79	4.13	0.05	0.21
Mar	8.0	5.6	64.64	23.65	0.16	1.94	1.00	0.39	2.80	3.98	0.03	0.24
Apr	4.5	7.0	64.60	23.66	0.17	1.96	1.02	0.40	2.83	3.99	0.05	0.28
May	4.9	7.4	64.56	23.67	0.18	1.98	1.04	0.41	2.86	4.00	0.07	0.30
Jun	5.4	6.7	64.90	23.64	0.16	1.90	1.00	0.40	2.80	4.08	0.08	0.29
Jul	6.6	5.9	65.00	23.67	0.18	1.97	1.04	0.41	2.00	2.04	0.08	0.20
Aug	7.8	7.0	65.10	23.70	0.20	2.04	1.08	0.42	2.00	2.01	0.05	0.23
Sep	9.6	6.4	62.27	24.66	0.15	2.48	0.66	0.47	1.80	2.00	0.05	0.29
Oct	10.9	5.7	62.28	24.68	0.20	2.50	0.68	0.49	1.80	1.90	0.04	0.34
Nov	8.1	6.3	62.29	24.70	0.25	2.52	0.70	0.51	1.70	1.70	0.08	0.24
Dec	7.2	7.0	63.67	23.69	0.18	1.97	0.97	0.46	1.00	1.03	0.07	0.28
Average	9.23	6.3	63.89	23.92	0.19	2.10	0.93	0.44	1.42	2.08	0.05	0.25

Note: All the readings represent the average of 2 years.

Table 83.5: Mean Monthly Values of Physico-chemical Parameters of Soil Collected from Higher Elevated (*in-situ*) and Composite Samples of Station 4 (Gatti) from January, 2002 to December, 2004

Month	Soil Moisture (Per cent)	Soil pH	SiO_2 (Per cent)	Al_2O_3 (Per cent)	TiO_3 (Per cent)	Fe_2O_3 (Per cent)	CaO (Per cent)	MgO (Per cent)	Na_2O (Per cent)	K_2O (Per cent)	MnO (Per cent)	Organic Carbon (Per cent)
Higher elevated (*in-situ*) soil samples												
Jan	12.0	6.6	67.13	18.20	0.20	2.85	1.50	0.44	2.84	4.15	0.02	0.14
Feb	10.3	5.8	67.15	18.18	0.22	2.86	1.51	0.43	2.85	4.16	0.02	0.24
Mar	8.6	5.6	67.29	19.20	0.12	1.84	1.54	0.36	2.80	4.15	0.03	0.28
Apr	5.4	7.0	67.28	19.21	0.16	1.86	1.56	0.38	2.78	4.16	0.04	0.18
May	5.2	7.0	67.30	19.22	0.20	1.88	1.58	0.40	2.76	4.17	0.05	0.30
Jun	4.7	6.7	69.11	16.76	0.13	2.83	1.56	0.40	2.12	4.10	0.05	0.19
Jul	8.9	5.9	69.12	16.78	0.12	2.85	1.57	0.41	2.14	4.09	0.06	0.18
Aug	18.2	6.4	69.13	16.80	0.14	2.87	1.58	0.39	2.16	4.08	0.07	0.25
Sep	17.8	6.8	66.67	18.36	0.26	2.88	1.48	0.38	2.76	4.28	0.03	0.40
Oct	10.5	5.7	66.68	18.37	0.24	2.90	1.52	0.41	2.78	4.24	0.04	0.35
Nov	17.5	6.9	66.66	18.35	0.22	2.92	1.56	0.44	2.80	4.20	0.05	0.44
Dec	9.6	5.6	67.14	18.22	0.18	2.84	1.49	0.45	2.88	4.17	0.02	0.38
Average	10.72	6.3	67.55	18.13	0.18	2.61	1.53	0.40	2.63	4.16	0.04	0.27
Composite soil sample												
Jan	9.00	6.7	63.38	23.78	0.19	1.84	1.52	0.39	2.80	4.14	0.02	0.11
Feb	10.30	5.7	63.39	23.80	0.20	1.86	1.54	0.40	2.82	4.16	0.04	0.24
Mar	8.60	6.0	66.00	23.00	0.14	1.80	1.50	0.37	2.75	4.12	0.06	0.20
Apr	5.40	6.9	66.01	23.03	0.15	1.85	1.51	0.38	2.77	4.08	0.07	0.18
May	5.20	7.0	66.02	23.06	0.16	1.90	1.52	0.39	2.79	4.10	0.08	0.24
Jun	4.70	7.0	65.10	24.25	0.09	1.80	0.46	0.34	2.02	3.99	0.08	0.19
Jul	8.90	6.9	65.12	24.30	0.10	1.83	0.48	0.36	2.04	4.00	0.09	0.18
Aug	9.20	6.6	65.14	24.35	0.11	1.86	0.50	0.38	2.06	4.01	0.08	0.25
Sep	17.80	6.8	63.20	24.67	0.18	1.88	1.48	0.36	2.05	3.00	0.07	0.32
Oct	10.50	6.9	63.28	24.68	0.20	1.86	1.50	0.40	2.00	1.80	0.08	0.35
Nov	8.50	6.6	63.36	24.90	0.22	1.84	1.52	0.44	2.00	1.78	0.10	0.24
Dec	9.60	5.9	63.37	23.76	0.18	1.82	1.50	0.38	2.00	1.78	0.05	0.38
Average	8.97	6.5	64.44	23.94	0.16	1.84	1.25	0.38	2.40	3.13	0.06	0.24

Note: All the readings represent the average of 2 years.

In conclusion, loss of nutrition of four element oxides has taken place in Sewa catchment, namely TiO_2, Fe_2O_3, Na_2O and K_2O, which can be mitigated by improving the soil by modern silvicultural methods using adequate organic manures.

Suggestions for Control

1. The enthusiastic endeavours by the Forest Department, J&K Government alongwith the local communities, will surely pave a way in checking enormous erosion occurring in the whole area.
2. Diversion of land suitable for sustainable farming to non-farming uses should be checked by legislation.
3. The stoniness in all the transects persists due to high intensity of erosion in the area of study. The stones can be removed manually as there exists manpower and this can make the soil worth fruit cultivation. There exists a huge promise for planting fruit trees in the whole area of study. The climate and soil are both congenial for starting fruit industry.
4. Modernized farming and use of manures is necessary to make soils completely fertile for three crops (rice, wheat and maize) that are mainly cultivated on repose slopes, on both sides of river Sewa.
5. Long rooted grass and vetiver grass should be used as a soil builder in the watershed areas. In many Himalayan terrains, vetiver grass is being used successfully for checking erosion (Rao, 2002; Lavania, 2004).
6. Afforestation in the catchment area is needed on a largescale so as to combat erosion. Forests in barren areas should be started on regeneration principle. It will check erosion and besides this, it will screen all the contaminants from entering into the solution of the Sewa river, thus protecting riverine ecology.
7. It is necessary to construct a series of staggered contour trenches on slopes, series of stone check dams in gullies and plantation of ecologically suited soil builder species (Grewal *et al.*, 1983).
8. Besides the above, various land management practices such as mixed cropping, intercropping, strip cropping, rotational cropping, mulching, application of organic manures, appropriate residue management, go a long way to reduce soil erosion and to conserve sub-soil moisture effectively.

Acknowledgement

The authors are highly thankful to the Head of the Department of Environmental Sciences, Jammu University, Jammu for giving the laboratory facilities for the analytical work used in this chapter.

References

Albrecht, W.A. and Smith, G.E., 1952. *Trans. Ind. Soc. Soil Sci.Comm.* II and IV; 1: 19.

APHA, 1998. *Standard Methods for the Examination of Water and Wastewater*, 20th edn. American Public Health Assoc., Washington.

Champion, H.G. and Seth, H.K., 1968. *Forests of India*. Govt. of India Publ. Division, New Delhi, pp. 104.

Dhar, B.L., Raina, A.K., Fotedar, B.K. and Singh, R., 1996. Petrology of Doda-Granites, Jammu, Lesser Himalaya. *Ind. J. Nepal Geol. Soc.*, 13: 49–56.

Fotedar, A., 2006. Ecological studies of river Sewa and its catchment (J&K). *Ph.D. Thesis,* Jammu University, Jammu.

Fotedar, B.K. and Loan, B.A., 2004. Physico-chemical studies of Jajjar nallah waters, District Udhampur, Jammu Himalaya. In: *Sem. Geohazards of North-West Himalaya*, Jammu University, Jammu, 9–11 October (Abstract).

Grewal, 1983. In: Ecological studies of river Sewa and its catchment (J&K) (Fotedar, A. 2006). *Ph.D. Thesis*, Jammu University, Jammu.

Horowitz, A., 1974. The geochemistry of sediments from the Northern ReykJanes Ridge and the Iceland–Faeroes Ridge. *Marime Geology.*, 17: 103–122.

Lavania, U.C., 2004. Vetiver system ecotechnology for environmental protection and sustainable development. *Curr. Sci.*, 86(1): 11–14.

Nesbitt, H.W. and Young, G.H., 1982. Early proterozoic climates and plate motions inferred from major element chemistry of lutites. *Nature*, 299: 715–717.

Puri, G.S. and Gupta, A.C., 1951. Himalayan conifers. II. The ecology of humus in conifer forest in Kulu Himalaya. *Ind. For.*, 77: 124–129.

Rao, K., 2002. Soil erosion and conservation. *Environment and People*, pp. 23–26.

Ray Choudhury, S.P. and Dutta, N.P., 1964. Phosphorus and potassium status of Indian soil. *ICAR Review Series Bull.*, 36: 43.

Shapiro, L. and Brunnock, W.W., 1956. Rapid analysis of silicate rocks (revised) from Circular 165. *USGS Bull.* 1936–C, 19–56.

Tikoo, V.J., 2004. Geological aspects of environmental degradation along the national highway–IA between Jammu and Banihal, J&K State. *Ph.D. Thesis*, Jammu University.

Vogel, I. Arthur., 1962. *A Textbook of Quantitative Inorganic Analysis*, 3rd edn.

Wazir, P.K., 1984. Physico-chemical studies of soils from Bhaderwah forests, Jammu. *M.Phil. Dissertation*, Jammu University

Wedepohl, K.H., 1978. *Handbook of Geochemistry*. Springer Verlag, Hedelberg.

Wilde, S.A., 1958. *Forest Soils: Their Properties in Relation to Silviculture*. The Ronald Press Co., New York.

Water and Wastewater Management in 2 Vols (2011) *Pages 554–561*
Editors: **D.R. Khanna, R. Bhutiani & Gagan Matta**
Published by: **DAYA PUBLISHING HOUSE, NEW DELHI**

Chapter 84

Water Quality Status of Dharampuri Ward Lake, Aheri, District Gadchiroli, M.S.

L.H. Rohankar[1], N.R. Dahegaonkar[2], P.M. Telkhade[2], P.J. Khinchi[3], A.N. Lonkar[4] and S.B. Zade[5]

[1]*R.D. Science College, Aheri, Dist. Gadchiroli, M.S.*
[2]*Arts, Commerce and Science College, Tukum, Dist. Chandrapur, M.S.*
[3]*Janata Mahavidyalaya, Chandrapur, M.S.*
[4]*Nutan A.A.C and Smt. M.H.Wegad Science College, Umrer, Dist. Nagpur, M.S.*
[5]*M.B. Patel, Science College, Sakoli, Dist. Bhandara, M.S.*

ABSTRACT

The Dharampuri Ward lake is the oldest lake of Aheri, build up by the Gond Maharaja for domestic purposes. The lake is now used for domestic, irrigation and pisciculture on lease basis. Monthly variations of different physico-chemical characteristics were studied for period of one year from July-06 to June-07 on three sites to enhance the water quality status. Various physico-chemical parameters revealed that the fluctuation in water temperature, pH, DO and sulphate are within desirable limits, slight increase was found in the values of Hardness, Phosphate and Nitrate indicating the mesoeutrophic nature of the lake at site S_2.

Keywords: *Physico-chemical characteristics, Dharampuri Lake, Water quality.*

Introduction

Water resources are of critical importance to both natural ecosystem and human development. The study of inland water bodies has gained immense importance in recent years because of their multiple uses for the human consumption, agriculture and industry. The men's influence on lotic and lentic ecosystems caused increase in silt and nutrient load, pushing lakes towards eutrophied situation at a very early stage and lead to changes in their trophic status and render them unsuitable for aquaculture and domestic purpose.

Dharampuri Ward lake is situated in eastern region of Aheri and is extensively used for domestic purpose. The total area of lake is 7.12 hectare. In India and abroad large number of studies on lentic water bodies have been carried out among those are Welch (1952), Munnawar (1970) and Kadam (2007). Study of different water parameters is very important for understanding metabolic events in aquatic ecosystem. The parameters influence each other and also sediment parameters, as well as they govern the abundance and distribution of flora and fauna, hence the regular monitoring of physico-chemical parameters is essential to determine the status of water body.

Materials and Method

Dharampuri Ward lake is located in Aheri city of Gadchiroli district, Maharashtra State having depth of 3.4 meters. Physico-chemical characteristics of water were studied monthly during the period of July-2006 to June-2007. For the sampling, three sites were selected *viz.* S_1, S_2, and S_3. Samples were collected in 5 litres dried plastic cans during 9.00 am. to 10.30 am., brought to laboratory and analyzed for alkalinity, hardness, free CO_2, TS, TDS, chloride, sulphate, phosphate and nitrate. For determination of dissolved oxygen, sample was fixed in field and brought to laboratory in an ice box for further process, the parameters like temperature, pH, conductivity, and transparency were analyzed on the spot. The parameters were analyzed with the help of standard methods APHA (1986) and Khanna (1993).

Results and Discussion

Temperature is one of the most important physical parameter for the water body since the rise in temperature speed up the bio-chemical reactions, reduces the solubility of gases and affects population fluctuation. Air temperature of the area ranged from 25 °C to 44 °C. High atmospheric temperature was recorded during summer, moderate in monsoon and slightly lower in winter season. Similar pattern of temperature fluctuation was noted by Swarnalatha and Rao (1991) from Saroornagar lake, Hyderabad, Sawane (2002) in river Erai, Chandrapur.

Water Temperature

Water temperature of Dharampuri lake ranged from 23.5°C to 35.0°C. The water temperature varied between 23.8 °C to 34.8 °C, 23.6 °C to 34.9 °C and 23.89 to 34.9 °C at site S_1, S_2 and S_3 respectively. It was highest in the summer months (May) and relatively low in monsoon and winter months. The minimum water temperature was recorded at S_2 and Maximum at S_3. Water temperature was consistently lower than atmospheric temperature of the lake, to summaries, air and water temperature followed a common pattern. It was higher in summer and relatively lower in monsoon and winter (Salve and Hiware, 2007). A similar pattern with regard to temperature variation is in conformity with Singh *et al.* (1979), Vyas *et al.* (1979) and Swarnalatha (1942) in Saroornagar lake, Hyderabad.

pH

pH is the value expressed as the negative logarithm of hydrogen ion concentration. The pH varied between 7.14 to 8.10, 7.56 to 7.99 and 7.56 to 8.02 at site S_1, S_2 and S_3 respectively. Water which has a pH value more than 9.6 or less than 2.5 becomes unsuitable for most life forms. In present investigation, pH was observed to be alkaline, ranged from (7.54 to 8.01) throughout the period of investigation with no definite seasonal variation. Generally, slightly alkaline conditions are favourable for growth of algal species in lotic systems (Welch, 1952; Blum, 1956) while acidic conditions are

Table 84.1: Average Value of Monthly Variation in Physico-chemical Parameters of Dharampuri Ward Lake, Aheri

Parameter	July	August	September	October	November	December	January	February	March	April	May	June
Temp. (°C)	30.6±0.09	29.3±0.08	28.7±0.09	28.6±0.04	26.6±0.14	24.4±0.14	23.7±0.09	44.0±23.4	29.9±0.09	32.6±0.09	34.8±0.04	33.6±0.12
pH	7.9±0.00	7.7±0.00	7.6±0.01	7.5±0.00	7.5±0.00	7.8±0.00	7.9±0.03	7.9±0.01	7.8±0.00	7.8±0.01	7.9±0.03	7.8±0.24
Cond. (μmhos/cm)	0.152±0.0	0.127±0.0	0.121±0.0	0.122±0.0	0.446±0.46	0.093±0.3	0.096±0.4	0.110±0.0	0.145±0.0	0.168±0.0	0.173±0.2	0.167±0.0
Trans. (cm)	55.5±0.81	34.6±0.84	38.9±0.29	45.5±0.40	46.8±0.47	54.1±2.01	66.0±1.08	73.8±2.32	75.3±1.64	76.3±0.23	66.3±1.24	62.8±1.02
Alka. (mg/l)	137±1.63	105±0.84	84.3±1.88	79.3±1.24	55.3±0.94	88.6±0.47	92.0±0.84	111±1.24	153±49.8	139±9.89	158±0.81	183±2.44
Hard. (mg/l)	155±2.62	152±2.62	90.6±1.69	78.3±2.62	94.0±1.63	103±1.69	71.8±49.7	147±1.69	189±2.16	220±1.63	227±6.64	227±6.64
T.S. (mg/l)	646±39.4	619±46.67	452±72.6	415±47.7	415±47.7	324±16.9	188±114.3	259±7.31	279±3.85	298±4.02	386±17.72	450±1.69
T.D.S. (mg/l)	129±2.16	118±1.24	74.3±51.8	71.3±20.7	76.8±0.6	61.8±0.62	58.0±0.81	75.3±0.94	105±0.94	112±1.24	130±2.49	123±2.05
D.O. (mg/l)	6.03±0.17	7.13±0.17	7.37±0.20	7.37±0.17	7.8±0.04	8.30±0.16	9.07±0.26	10.3±0.66	9.27±0.68	6.67±1.15	5.57±0.89	5.17±0.41
CO_2 (mg/l)	5.60±0.21	5.30±0.21	4.37±0.30	4.33±0.12	3.97±0.17	3.23±0.41	2.87±0.24	5.33±0.30	5.53±0.09	5.97±0.17	6.27±0.47	6.70±0.16
B.O.D. (mg/l)	3.50±0.28	2.67±0.25	2.30±0.21	2.03±0.09	1.80±0.81	1.60±0.28	1.47±0.18	1.63±0.09	2.27±0.12	2.7±0.21	3.67±0.20	3.93±0.20
C.O.D. (mg/l)	14.4±0.04	14.6±0.04	12.8±0.45	18.53±0.3	18.8±0.04	19.6±0.08	19.7±0.14	16.6±0.14	17.5±0.16	14.1±0.49	11.2±0.04	11.3±0.04
Sulp. (mg/l)	21.7±0.35	20.0±0.83	15.4±1.32	13.3±0.1	11.5±0.26	12.3±0.08	11.1±0.56	15.7±1.39	27.5±0.47	30.3±0.89	36.6±1.48	38.3±0.63
Phos. (mg/l)	0.36±0.03	0.28±0.02	0.23±0.01	0.19±0.00	0.17±0.00	0.18±0.00	0.29±0.00	0.36±0.02	0.40±0.00	0.43±0.00	0.52±0.92	0.60±0.02
Nitrate (mg/l)	0.32±0.00	0.39±0.06	0.20±0.00	0.19±0.00	0.40±0.41	0.36±0.37	0.56±0.35	0.36±0.36	0.16±0.01	0.22±0.00	0.60±0.46	0.24±0.04

detrimental. The similar studies were observed by, Vyás and Kumar (1968), Sreenivasan (1974) and Bohra (1976).

Conductivity

Conductivity ranged between 0.89 to 0.199 µmhos/cm. The conductivity varied between 0.89 to 0.171 µmhos/cm., 0.101 to 0.175 µmhos/cm. and 0.91 to 0.177 µmhos/cm. at site S_1, S_2 and S_3 respectively. It was high in summer months (May) and relatively low in winter months (December). The minimum conductivity was recorded at site S_1 and maximum at site S_3. In present study, maximum conductivity was observed in summer may be due to evaporation of water and minimum during winter season due to dilution from large volume of rainwater.

Transparency

The extent to which light can penetrate into the water column depends on the transparency of standing water column. Further, transparency of water is inversely proportional to turbidity created by suspended inorganic and organic matter (Saxena and Baskaran, 1981).

Transparency ranged between 34.6 cm to 76.5 cm. It varied between 35.3 cm to 76.5 cm, 33.5 to 76.00 cm. and 35.00 to 77.00 cm at site S_1, S_2 and S_3 respectively. The maximum transparency observed during summer season and minimum during monsoon and winter. Kaushik and Saxena (1999) have reported low transparency in summer and rainy season in different water bodies in the country. The low transparency during late summer is attributed to decline in water level due to evaporation and plankton growth while its lower values in monsoon is due to addition of silt and clay particles along with runoff water from the catchment area (Pailwan, 2005).

Alkalinity

Alkalinity is the capacity of water or sediments to neutralize strong acid and is characterized by presence of hydroxyl ions capable of combining with hydrogen ions. In natural water, carbonates and bicarbonates mainly contribute to alkalinity. In general, alkaline water is more productive and supports the diversity of aquatic life.

Alkalinity ranged between 54.00 to 186 mg/l. It varied between 56 to 183 mg/l, 54 to 186 mg/l, and 56 to 180 mg/l, at site S_1, S_2 and S_3 respectively. The maximum alkalinity was during summer months and the minimum during winter season. The acceptable limit for alkalinity is 200 mg/l, observed values are not more than acceptable limit. Similar findings observed by Pailwan (2005).

Hardness

The ecological significance of major cations *i.e.* calcium and magnesium in the biotic dynamics of aquatic fauna and flora is well-established fact. Hardly a group of freshwater animals exists in which the distribution of some species has not been related to calcium concentration in the environment (Macan, 1961). Magnesium is essential for flora, for chlorophyll biosynthesis and enzymatic transformations, particularly the phosphorylation, in algae, fungi and bacteria (Wetzel, 1975). Alkaline earth metals mainly Ca and Mg present in it impart hardness of water. The total hardness of water recorded between 76 to 227 mg/l. The hardness varied between 76.00 mg/l, to 227, 81.00 to 239 mg/l and 77.00 to 232 mg/l, at site S_1, S_2 and S_3 respectively. Total hardness may be correlated with alkalinity was found to be maximum during summer months while minimum during winter. These observations are similar to that of Vyas *et al.* (1979) in Gordhan Vilas lake near Udaipur, India.

Total Solids

All matter in water except water is called as solids. The TS is the matter remains after the temperature of water at 103°C to 105°C. During the present study, total solids ranged between 254 to 615 mg/l, 270 to 702 mg/l and 255 to 622 mg/l, at site S_1, S_2 and S_3 respectively. The maximum values were recorded during monsoon and minimum during winter months. These findings were in accordance with Tripati and Pandey (1990), Ugale (2002) in Jagatunga Samudra reservoir. The total solids recorded low during winter may be due to sedimentation and high during monsoon due to high surface run-off, silt-sewage from catchment area.

Total Dissolved Solids

Total dissolved solids include salts and variety of organic substances. TDS level plays an important role in community structure due to limiting impact on primary production of trophodynamics. In present study TDS values ranged between 57 to 134 mg/l, 58 to 130 mg/l. and 59 to 130 mg/l at site S_1, S_2 and S_3 respectively. The lower values were recorded in winter months and high during summer. This may be due to low water level and similar in accordance with Chandrashekhar (1996) and Dhanapati (2000).

Dissolved Oxygen

Dissolved oxygen is one of the most important parameter of water quality affecting survival and distribution of flora and fauna and reflects the physical and biological processing prevailing in water (Trivedy and Goel, 1984). In present investigation DO values ranged between 5.1 to 10.2 mg/l, 5.7 to 9.7 mg/l and 4.7 to 11.2 mg/l at S_1, S_2 and S_3 respectively. The minimum values recorded during summer months. Pahawa and Mehrotra (1966) and Salodia (1996) have reported minimum DO in summer and maximum in winter. Normally high DO is encountered in unpolluted while lower levels of the same in polluted areas of an aquatic ecosystem. It is a well known fact that the oxygen balance of lake is tagged with the photosynthetic and respiratory activities of biota and chemical oxidization on one hand and prevailing physico-chemical condition on the other, High winter values of DO could be attributed to relatively stable abiotic conditions and higher algal-biomass stimulating rate of photosynthesis. Intermediate values in monsoon could be attributed to dilution factor that adversely affects algal biomass and photosynthetic replenishment of oxygen . Kulshrestha *et al.* (1989) recorded DO from 6.9 to 12.9 mg/l in lower lake and from 4.8 to 9.9 mg/l in Chunnabhatti lake.

Free CO_2

Free CO_2 is directly proportional to the bi-carbonates and indirectly to carbonates and is highly soluble in water. The respiratory activity of aquatic organisms and decomposition of organic matter are important sources of carbon dioxide in freshwater bodies. The free CO_2 is released during the decomposition of certain substances and metabolic activities of the living organism since; higher temperature accelerates the decomposition of organic substances as well as the respiratory activity of the biota. A direct relationship was established between water temperature and free CO_2.

In present study, CO_2 ranged between 2.8 to 6.7 mg/l, 3.02 to 6.4 mg/l and 3.25 to 6.9 mg/l. at site S_1, S_2 and S_3 respectively. Minimum values recorded during winter and maximum during summer. The high utilization by more algal blooms in winter causes depletion of CO_2 in winter.

Biochemical Oxygen Demand

Biochemical oxygen demand (BOD) can be defined as the quality of dissolved oxygen in mg/l required under the condition for complete oxidation of the organic matter in sample. In unpolluted water, BOD is lower while it is higher in the polluted water (Hynes, 1960; Willaughly, 1976). In present study, the BOD values recorded 1.7 to 3.9 mg/l., 1.5 to 4.2 mg/l. and 1.6 to 3.7 mg/l. at site S_1, S_2 and S_3 respectively. The high BOD values were recorded during summer months and low during winter. Biochemical Oxygen Demand (BOD) is an important parameter that indicates the magnitude of water pollution by oxidizable organic matter. The main sources of organic pollution include untreated domestic sewage; agricultural run off containing residual fertilizers and certain industrial effluents, similar observation was made by Bobdey (2002).

Chemical Oxygen Demand

The chemical oxygen demand is a parameter that indicates the extent of chemical pollution mainly from industrial effluents, low level of D.O. is usually built up of COD and BOD. The chemically polluted sample is the sample in which microbiological degradation is not possible due to presence of toxicant and chemicals. The measurement of oxygen required in oxidizing organic compounds by involving a chemical oxidant such as potassium chromate or potassium permanganate. In present study, COD ranged between 11.2 to 19.5 mg/l. 11.3 to 19.9 mg/l and 11.2 to 19.6 mg/l at site S_1, S_2 and S_3 respectively. The maximum seasonal value was recorded during winter and minimum during summer, a similar observation was reported by Bobdey (2000).

Sulphate

Sulphates are produced by biological oxidation of sulphur containing organic matter and weathering of rocks. Sulphate exists in number of oxidation state from the most oxidized sulphate to the most reduced sulphide. The biological reduction of sulphur can takes place in both aerobic and anaerobic condition. Under the aerobic condition the reduction of sulphate is an assimilatory process (Tuttle, 1980).

The sulphate values at three sampling stations S_1, S_2 and S_3 in the present investigation were recorded between 10.30 to 37.9 mg/l, 11.2 to 39.2 mg/l and 10.7 to 37.8 mg/l. The maximum value was recorded in the summer months and minimum during winter. Angadi (2005) reported minimum concentration in month of November and maximum in March. Higher concentration of sulphate in summer may be due to activity of biodegradation (Munnawar, 1970).

Phosphate

In natural water, phosphates are present in small quantities. Generally aquatic ecosystems receive excess of this nutrient through untreated domestic sewage and agriculture run off. Normally phosphate acts as a limiting nutrient in the process of eutrophication and lakes can be aesthetically classified into good, fair, bad, very bad, and lawful on the basis of per cent phosphates loading (Edmondson, 1991). Like nitrate, phosphate plays an important role in determining trophic status of water body. In the present study the phosphate ranged between 0.169 to 0.632 mg/l. The phosphate values varied between 0.169 to 0.589 mg/l, 0.170 to 0.632 mg/l and 0.169 to 0.579 mg/l at site S_1, S_2 and S_3 respectively. The maximum value of phosphate was noted down during summer months due to less volume of water and minimum during winter.

Nitrate

The determination of nitrate is very important as it helps in measuring the pollution status and gives relative picture of availability of decomposable organic matter. Aquatic ecosystems in urban environment receives excess of nitrates through untreated domestic sewage and along with phosphates, are responsible for the process of lake degradation called eutrophication (Kodarkar and

Chandrasekhar, 1995). The determination of nitrate is very important as it helps in measuring the pollution status. In the present investigation, nitrate ranged between 0.077 to 0.342 mg/l, 0.92 to 0.348 mg/l and 0.69 to 0.339 mg/l. at site S_1, S_2 and S_3 respectively. The minimum value of nitrate was recorded in summer and maximum in monsoon months. During monsoon maximum nitrate value is linked to heavy run off of organic matter from the catchment area (Golterman, 1975). Nitrate peaks followed by rain in the catchment area have also been reported by various workers (Venkateshwarlu, 1969; Adwant, 1981). In the present investigation, the Dharampuri Ward lake show physico-chemical parameters within a permissible limit at site S_1 and S_3 however slight increase in values of hardness, phosphate and nitrate indicating that the lake is towards mesoeutrophic situation at site S_2.

Acknowledgment

The authors are thankful to Dr. M.M. Wankhade, Principal, A.C.S. College, Tukum, Chandrapur for providing laboratory facilities.

References

Adwant, M.F., 1981. Limnological studies on Godavari river basin at Nanded, Maharashtra, India. *Ph.D. Thesis,* Marathwada University, Aurangabad.

APHA, 1986. *Standard Methods for the Examination of Water and Wastewater*, 16th edn. American Public Health Association, New York.

Bohra, O.P., 1976. Some aspects of limnology of Padamsagar and Ranisagar, Jodhpur (Raj). *Ph.D. Thesis,* Univ. of Jodhpur, Jodhpur.

Chandrasekhar, S.V.A., 1996. Ecological studies on Saroornagar Lake, Hyderabad. *Ph.D. Thesis,* Osmania University, Hyderabad.

Khanna, D.R., 1993. *Ecology and Pollution of Ganga River*. Ashish Publication House, Delhi, pp. 1–241.

Kulshreshtra, S.K., Adholia, U.N. Khan, Altaf A., Bhatnagar, Anita and Bangali, Manju, 1989. A pollution study on the river Chambal near Nagaland with reference to phytoplankton community. In: *Proc. Natl. Symp. Envi., Pollution and Pesticide Toxicology*, p. 119–126.

Munnawar, M., 1970. Limnological studies on freshwater ponds of Hyderabad. I. The biotope. *Hydrobiologia*, 1: 127–162.

Pahwa, D.V. and Mehrotra, S.N., 1966. Observation of fluctuations in the abundance of plankton in relation to certain hydrobiological conditions of river Ganga. In: *Proc. Nat. Acad. Sci.*, 36: 157–189.

Saxena, M.M. and Baskaran, 1981. Diurnal rhythm in physical, chemical and zooplanktonic components in a polluted water tank, Gulbsagar (Jodhpur). In: *Proc. Nat. Symp. Eval. Environ.*, Sp. Vol. *Geobios*, p. 86–90.

Singh, S. and Swarup, K., 1979. Limnological studies of Suraha lake (Ballia). II. The periodicity of phytoplankton. *J. Indian Bot. Soc.*, 58: 310–329.

Sreenivasan, A., 1974. Limnological features of a tropical impoundment, Bhavanisagar Reservoir (Tamil Nadu), India. *Int. Revue. Ques. Hyderabad*, 59: 327–342.

Swarnalatha, N., 1994. Seasonal rhythm of various physico-chemical characteristics of Saroornagar Lake. *J. Mendel.*, 11(1 and 2): 65–66.

Swarnalatha, N. and Rao, A. Narsing, 1991. Investigations of Saroornagar lake with reference to water pollution. *J. Phytol. Res.*, 4(2): 121–129.

Toronta, Wuhrmann, K. 1938, Lake eutrophication and its control. In: *Proc. Shiga Conference on Conservation Management of World Environment*, p. 26–37.

Tripati, A.K. and Pandey, S.N., 1990. *Water Pollution*. Ashish Publishing House, New Delhi.

Trivedy, R.K. and Goel, P.K., 1984. *Chemical and Biological Methods for Water Pollution Studies*. Environ. Media Pub. Karad (India), p. 215.

Vankateshwarlu, V., 1969. An ecological study of algae of Musi, Hyderabad with special reference to water pollution II. Factors including distribution of algae. *Hydrobiologia*, 33(3 and 4): 352–363.

Vyas, L.N. and Jain, S.C., 1979. Observation on the hydrobiology of Gordhan Vilas lake near Udaipur, India. In: *Paper Presented at 5th Int. Symp. of Tropical Ecology*, Kulalampur, Malaysia during 16–21 April, Abst., pp. 215–216.

Welch, P.S., 1952. *Limnological Methods*. The Blankston Co., Philadelphia.

Water and Wastewater Management in 2 Vols (2011) *Pages 562–565*
Editors: **D.R. Khanna, R. Bhutiani & Gagan Matta**
Published by: **DAYA PUBLISHING HOUSE, NEW DELHI**

Chapter 85

Occurrence of Zooplankton in a Perennial Freshwater Reservoir of Wadgaon Dam during Monsoon Season at Nagpur District

Suresh B. Zade[1], Shashikant R. Sitre[2] and Rajendra V. Tijare[2]
[1]Department of Zoology, M.B. Patel College, Sakoli Distt. Bhandara, M.S.
[2]Govt. Institute of Science, R.T. Marg, Nagpur, M.S.

ABSTRACT

The zooplanktons occupy a central position between the autotrophs and heterotrophs and form an important link in the food chain of a freshwater ecosystem. The occurrence and abundance of zooplankton is always influenced by physico-chemical factors and the level of nutrients present in the water body. In this context a perennial logographic freshwater reservoir Wadgaon dam is studied with respect to zooplankton population in monsoon months *i.e.* July and August to assess the types of zooplankton present. Qualitative studies on the zooplankton in the Wadgaon dam reservoir situated in the Nagpur district were undertaken in monsoon months to assess the extent of forms present. The reservoir water is uncontaminated and clean with no influence of man made activities in the vicinity. The zooplankton population of the reservoir water is found to be represented by five different groups *viz.* protozoa, rotifera, cladocera, copepoda and ostracoda represented by about 21 different forms. The present study indicate the uncontaminated nature of reservoir water due to absence of pollution indicator species in the reservoir water.

Keywords: Zooplankton, Autotroph, Heterotroph, Wadgaon dam, Physico-chemical.

Introduction

The inland water bodies are closed ecosystem in which zooplankton hold a key position in the trophic level, food chain and energy flow of the ecosystem. As producers and consumers, plankton play an important role in the transformation of energy from one trophic level to the next higher trophic level ultimately leading to fish production which is the final product of aquatic environment.

The occurrence and abundance of zooplankton in freshwater ecosystems depends on its productivity, which in turn is governed by the physico-chemical parameters and level of nutrients available in the ecosystem. A large amount of work has been done on plankton world-wide by various researches on various water bodies like Ranga Reddy (2001), Kodarkar (1994), Malin (1984), Pai and Berde (2005), Pawar and Pulle (2005), Schindler and Noven (1971). But still there are many reservoirs and water bodies worldwide where there is no work reported still. So keeping this in view and after literature search the present work on zooplankton were undertaken on a water body situated in Nagpur district of Maharashtra state.

Materials and Method

The Wadgaon dam is a big reservoir situated near Butibori on state highway connecting Nagpur to Chandrapur. This beautiful water body has abundant water available through out all the seasons and is a oligo-trophic type of water body with clear water. The zooplankton samples were collected from the littoral zone at two sites by filtering 50 liters of water through plankton hand net of bolting nylon cloth (mesh size 45 μm) in early morning hours between 9 to 11 a.m. twice a month during July and august month. The procedure for collection storage and analysis of samples were followed as per Standard Methods (APHA, 1989). The zooplankton samples were preserved in 4 per cent formalin. The samples were identified using Standard literature (Battish, 1992; Edmondson, 1992; Dhanpathi, 2000; Michael and Sharma, 1988).

Table 85.1: List of Zooplankton Species in Wadgaon Reservoir during Monsoon Months

Types of Species Present in the Reservoir Water	*Month*	
	July	*August*
Protozoa		
Difflugia sp.	+	+
Rotifera		
Asplanchna sp.	-	+
Anuropsis fissa	-	+
Brachionus diversicornis	-	+
Brachionus calyciflorus	+	+
Brachionus falcatus	+	+
Epiphanes senta	+	+
Filinia sp.	-	+
Keratella tropica	-	+
Testudinella sp.	+	+

Contd..

Table 85.1–Contd...

Types of Species Present in the Reservoir Water	Month	
	July	August
Cladocera		
Bosmina	+	+
Chydorus	+	+
Macrothrix sp.	+	+
Simocephalus	+	+
Sida crystalina	-	+
Alona costata	+	+
Copepoda		
Cyclops sp.	+	+
Diaptomus	+	+
Copepod nauplius	+	+
Total forms present in the reservoir water	13	1

+: Present; –: Absent.

Results and Discussion

The zooplankton in Wadgaon dam reservoir is composed of five distinct groups *viz.* protozoa, rotifera, cladocera, copeopda and ostracoda represented by about 21 different species. In July month five different groups of zooplankton are present represented by 15 different forms while in August month 5 different groups of zooplankton are represented by about 21 forms. The protozoa is represented by only one form, rotifera by 9 different forms, cladocera by 6 different forms, copepoda by 3 forms and ostracoda by 2 forms. The ostracoda thrives on fine detritus and as detritus is available in monsoon in abundance the ostracods occur in large number. Tonapi (1980) has reported higher population of ostracods during monsoon months due to their abundance of fine detritus during this period of the year.

The rotifers play an important role as grazers and suspension feeders within the zooplankton community. The difference in periodicity and population density of different rotifer species can be analyzed by considering the nutritional ecology and biotic interactions. The rotifer species exhibit marked differences in their tolerance and adaptability to changes in physico-chemical and biological parameters. Chandrasekhar (1996) observed that in summer and monsoon the factors like water temp, turbidity, transparency and dissolved oxygen play an important role in controlling the diversity and density of rotifers. In the present study 9 different kinds of rotifers are present in the water body.

The water of the lake was clear with very less turbidity observed in monsoon months *i.e.* July and August from the sampling points. The present study confirms that the lake is uncontaminated and free from human interference as no indicator species is found in the water body.

References

APHA, 1989. *Standard Methods for the Examination of Water and Wastewater*, 17th edn. American Public Health Association, Washington D.C., pp. 2–1193.

Battish, S.K., 1992. *Freshwater Zooplankton of India*. Oxford and IBH Publishing Company, pp. 223.

Chandrasekhar, S.V.A., 1996. Ecological studies on Saroornagar Lake, Hyderabad with special reference to zooplankton communities. *Ph.D. Thesis*, Osmania University, Hyderabad.

Dhanpati, M.V.S.S.S., 2000. *Taxonomic Notes on the Rotifers from India (from 1889–2000)*. Indian Association of Aquatic Biologists (IAAB), Hyderabad.

Edmondson, W.T., 1992. *Freshwater Biology*, 2nd edn. International Books and Periodicals Supply Service, New Delhi.

Kodarkar, M.S., 1994. Biodiversity of zooplankton in Saroornagar Lake, Hyderabad. *J. Aqua. Biol.*, 9(1 and 2): 30–33.

Malin, M.A., 1984. The plankton community of an acid backwater South Carolina power plant impoundment. *Hydrobiologia*, 112: 167–177.

Michael, R.G. and Sharma, B.K., 1988. *Fauna of India and Adjacent Countries: Cladocera*. Zoological Survey of India, Calcutta, pp. 262.

Pai, I.K. and Berde, Vikrant, 2005. Comparative studies on limnology of freshwater bodies located in coastal and high altitude of Goa and Maharashtra. *J. Aqua. Biol.*, 20(2): 95–100.

Pawar, S.K. and Pulle, J.S., 2005. Qualitative and quantitative study of zooplankton in Pethwadaj dam Nanded district (Maharashtra) India. *J. Aqua. Biol.*, 20(2): 53–57.

Ranga, Reddy Y., 2001. Zooplankton diversity: Freshwater planktonic copepoda with key to common calanoid and cyclopoid genera in India. In: *Water Quality Assessment: Biomonitoring and Zooplankton Diversity*, (Ed.) B.K. Sharma, pp. 174–189.

Schindler, D.W. and Noven, B., 1971. Vertical distribution and seasonal abundance of zooplankton in two shallow Ontario lakes. *J. Fish. Res. Bd. Can.*, 28: 245–256.

Water and Wastewater Management in 2 Vols (2011) *Pages 566–570*
Editors: D.R. Khanna, R. Bhutiani & Gagan Matta
Published by: DAYA PUBLISHING HOUSE, NEW DELHI

Chapter 86

Zooplankton Diversity of Tadoba Lake, Tadoba Andhari Tiger Reserve (TATR) Dist. Chandrapur, M.S.

P.M. Telkhade[1], N.R. Dahegaonkar[1], P.J. Khinchi[2] and S.B. Zade[3]
[1]*Arts, Commerce and Science College, Tukum, Chandrapur, M.S.*
[2]*Janta Mahavidyalaya, Chandrapur, M.S.*
[3]*M.B. Patel Science College, Sakoli, Dist. Bhandara, M.S.*

ABSTRACT

Tadoba Andhari Tiger Reserve exhibits rich floral and faunal diversity. However, no consolidated account is available on its rich biodiversity. In considering the view of great Tadoba Lake, the present study was investigated; the study reveals that the biodiversity of zooplanktonic organisms has been fairly good which is evident from the identification of four groups with 33 species. The rich faunal density indicates highly productive nature of pond.

Keywords: *Tiger reserve, Biodiversity, Zooplanktonic.*

Introduction

Tadoba Andhari Tiger Reserve (TATR) in situated in the Chandrapur District in Eastern Maharashtra. It represents unique habitat for wildlife in central India and is oldest National Park of the state. Tadoba has been named after the local Gond Chieftan "Taru", locals offer prayer to *Tadoba Deo* located on the edge of the Tadoba Lake. There is a belief that sprinkling water of Tadoba Lake on their farmland deeps pests and disease under control. Tadoba Lake is a big lake about 120 ha. and it is a very prominent tourist spot of Maharashtra State. Zooplankton are the most fascinating group of micro-organisms found in aquatic body. The zooplankton in water belongs to four main taxonomic groups such as Rotifera, Copepoda, Cladocerans and Ostracoda. They are abundant in the shallow area of the lake; they play a vital role as primary consumers. The occurrence and abundance of zooplankton in a lake depends on its productivity. Zooplankton occupies an intermediate position of food webs. Zooplankton has been used as an indicator for monitoring the water quality. The zooplankton that plays a role of converting phytoplankton into food, it is suitable for fish and aquatic animals. Zooplankton has been a subject of study in India and several worked by some workers, Dhanpati (2000), Tonapi (1980), Pennak and Water (1955), and Kodarkar (1992). The chapter deals with studies on zooplankton diversity in Tadoba Lake. Seasonal qualitative and quantitative analysis of zooplankton diversity were carried out.

Materials and Method

During the period of investigation, samples were collected by plankton net made by silk bolting cloth no. 25 (mesh size 56), water sample was filtered through the net from littoral and open water zone, carefully transferred to 50 ml. bottle and preserved in 4 per cent formaline. Preserved samples were examined under a binocular microscope by using Sedgwick Rafter Counting Cell. Qualitative analysis was done by Tonapi (1980), Pennak (1989), Dhanpati (2000) and Kodarkar (1992).

Results and Discussion

The seasonal abundance of zooplankton (average) were noted in Tables 86.1 and 86.2. Seasonal zooplankton study was done and 33 species were identified, belonging to different taxonomic groups, 15 were Rotifera, 9 Cladocerans, 8 Copepoda and one from Ostracoda. In present investigation, Rotifera showed dominance and Ostracoda showed their least appearance, Cladocerans and Copepoda were moderate.

Table 86.1: Average Seasonal Values of Zooplankton in Tadoba Lake During 2005–06

Sl.No.	*Zooplankton*	*Monsoon*	*Winter*	*Summer*
	Rotifers			
1.	*T. cylendrica*	1.25±1.118	2.50±1.12	1.50±1.12
2.	*L. longiseta*	6.25±1.920	6.75±3.200	1.50±1.12
3.	*B. calyciflorus*	2.75±1.479	7.50±2.693	2.50±1.500
4.	*B. fulcatus*	4.75±3.112	10.2±6.300	3.75±2.487
5.	*B. quadricornis*	5.00±2.345	10.5±2.291	2.25±1.479
6.	*B. forticula*	2.25±2.487	5.50±3.041	3.25±1.920
7.	*B. rubens*	2.25±2.487	7.75±2.278	1.50±1.12
8.	*B. plicatus*	2.50±2.598	7.75±2.947	5.25±3.112
9.	*B. diversicornis*	3.50±1803	4.25±2.681	2.75±2.165
10.	*F. longiseta*	0.75±0.829	1.15±1.785	0.00±0.00
11.	*Monostyta*	5.25±1.920	9.00±1.450	3.75±2.487
12.	*Keratella*	3.00±2.236	12.2±5.914	1.50±1.658
13.	*Asplanchna*	2.00±0.707	1.80±1.12	2.00±1.871
14.	*Lecane*	3.00±1.871	7.00±1.358	3.50±1.118
15.	*Epiphanes*	0.00 = 0.00	0.25±0.50	0.00±0.00

Contd...

Table 86.1–Contd...

Sl.No.	Zooplankton	Monsoon	Winter	Summer
	Ostracoda			
1.	*Cypris*	9.50±7.500	0.00±0.00	10.5±10.595
	Cladocera			
1.	*Moinodaphnia*	7.50±7.826	13.0±4.301	4.25±1.090
2.	*M. branchiata*	2.75±1.479	8.50±3.571	4.25±3.832
3.	*B. longirostris*	2.25±2.487	0.00±0.00	0.00±0.00
4.	*Simocephalus*	4.25±2.278	4.50±2.50	5.50±1.803
5.	*Alona*	4.00±2.915	7.50±3.20	3.50±1.80
6.	*Chydorus*	4.75±1.92	6.00±1.41	1.75±1.48
7.	*P. denticulatus*	4.75±2.385	7.25±2.947	3.75±1.299
8.	*C. reticulata*	3.25±4.548	7.75±5.309	4.25±3.031
9.	*Macrothrix rosea*	0.00±0.00	0.00±0.00	0.00±0.00
	Copepoda			
1.	Cyclops male	3.25±2.947	1.75±1.785	0.00±0.00
2.	Cyclops female	6.75±1.479	9.00±4.64	3.25±1.785
3.	Mesocyclops	1.00±1.25	0.00±0.00	0.00±0.00
4.	Microcyclops	1.50±0.866	9.25±2.17	2.25±2.278
5.	Eucyclops	4.00±3.317	11.0±4.12	3.00±1.225
6.	Diaptomus male	3.25±2.861	2.50±1.50	1.75±1.920
7.	Diaptomus female	6.75±4.815	10.2±3.63	2.25±1.920
8.	Nauplius	1.00±1.225	1.50±1.12	0.00±0.00

Table 86.2: Average Seasonal Values of Zooplankton in Tadoba Lake during 2005–06

Season	Monsoon	Winter	Summer	Average
Rotifera	43.50±11.41	129.00+27.74	35.00±11.81	69.17±42.45
Ostracoda	9.50±7.50	0.00±0.00	10.50±50.59	6.67±4.73
Cladocera	33.50±11.19	54.50±8.08	27.25±4.3.9	38.42±11.65
Cooepoda	33.50±11.19	45.25±7.79	12.50±4.56	30.42±13.54
Total Zooplankton	120.0±18.17	238.00±34.59	100.0±12.44	152.6±60.88

In seasonal variation, the maximum average density of total zooplankton (238.0 ± 34.59) ind/lit. was noted during the winter season of 2005-06 and minimum seasonal average density (100 ± 12.44) ind/lit. was noted during summer season of 2005-06.

The study of zooplankton fauna especially zooplankton, even of a particular area is an extensive and complicated phenomenon due to environment, physical, geographical, and chemical variations involving ecological, extrinsic and intrinsic factors. The members of zooplankton community are important for their role in trophic dynamics and in energy transfer in the aquatic ecosystem. The seasonal fluctuations of the zooplankton population are a well-known phenomenon. Welch (1952) mentioned that, the fluctuation in zooplankton population is greatly influenced by the variation in temperature along with many other factors. Among the several factors, temperature seems to exhibit the greatest influence on the periodicity of zooplankton (Battish and Kumari, 1996). In the present investigation, seasonal zooplankton study shows that qualitatively 33 species were identified, belonging to different taxonomic groups, 15 were Rotifera, 9 Cladocerans, 8 Copepoda and one from Ostracoda. In present investigation, Rotifera shows dominance and Ostracoda shows their least appearance, Cladocerans and Copepoda were moderate. Similar findings were observed in Jagtunga samudra reservoir, Kandhar, Dist. Nanded (Ugale, 2002). The zooplankton population in Tadoba Lake was of higher magnitude during winter season and the lower magnitude during summer. Their diversity and density is mainly controlled by availability of food as favourable water quality (Chandrashekher, 1997).

In the present study, the results indicated that the maximum number of species occurred during winter season than summer and monsoon, which was also reported by Abdus Saboor and Altaf (1995), Kumar (2001). The less number of genera might be attributed to the less nutrient in the pond which consequently result in less productivity or might be due to the depletion of important factors such as dissolved oxygen and pH. According to Welch (1952) in most of the tropical freshwater bodies, pH is always correlated with photosynthetic activity and phytoplanktonic biomass.

Rotifera group showed their dominance, maximum population densities were observed during winter season and minimum population density was during summer season. Dhanapathi (1997) had suggested that, temperature plays a conspicuous role in the occurrence of variations in rotifers in tropical region.

Cladocerans species showed maximum population density during winter season and minimum population density during monsoon. The success of Cladocerans mainly depends on ability to feed efficiently on wide range of particular organic matter. Number of Copepods species recorded in present study were eight. Maximum population density of copepods occurs in winter season. Among the copepods *Cyclops* sp. and *Diaptomus* sp. were dominant. Ostracoda species presence was least in number; only Cypris species was recorded during investigation. Maximum population density of Ostracoda was observed during monsoon and total absence during winter season. Environmental factors like temperature, salinity, DO, and sediment composition seem to influence cumulatively the distribution of Ostracods. The higher population of Ostracoda during monsoon could be due to abundance of fine detritus to which omnivorous organism switch over during this period from their natural benthic habitat and bacteria, mould and algae as food (Tonapi,1980).

References

Abdus, Saboor and Altaf, K., 1995. Qualitative and quantitative analysis of zooplankton of a tropical pond during summer and rainy season. *J. Eco. Biol.*, 7(4): 268–276.

Battish, S.K. and Kumari, P., 1996. Effect of physico-chemical factors on seasonal abundance of Cladocera in tropical pond at village Reqba, Ludhiana. *Ecol.*, 13(1): 146–151.

Chandrasekhar, S.V.A., 1997. Macro-benthic fauna of lake, Hussainsagar, Hyderabad with reference to molluscs and chironomid larvae. *Rec. Zool. Surv. India*, 96(1–4): 1–5.

Dhanapathi, M.V.S.S., 1997. Variations in some rotifers of the family Brachionidae. *J. Aqua. Biol.*, 12(1–2): 8–15.

Dhanpati, M.V.S.S.S., 2000. *Rotifers from India.* IAAB Publ., Hyderabad, pp. 1–200.

Kodarkar, M.S., 1992. *Methodology for Water Analysis: Physico-chemical, Biological and Microbiological Methods.* Indian Association of Aquatic Biologists, Hyderabad, Publ. 2., pp. 50.

Kumar, K. Siva, 2001. Studies on the freshwater copepod and cladocera of Dharmapuri, Dist. Tamil Nadu. *J. Aqua.Biol.*, 1 and 2: 5–10.

Pennak, R. and Water, 1955. Comparative limnology of eight Colorado mountain lakes. *J. Fish Res. Bd. Can.*, 29: 1451–1462.

Pennak Robert, W., 1989. *Freshwater Invertebrates of the United States*, 3rd edn. John Wiley and Sons Inc., New York, pp. 204–430.

Tonapi, G.T., 1980. *Freshwater Animals of India: An Ecological Approach.* Oxford and IBH Publ. Co., New Delhi, pp. 314–315.

Ugale, B.I., 2002. Hydrobiological studies of Jagantunga Sumudra Reservoir from Kandhar, Dist. Nanded. *Ph.D. Thesis*, Dr. Babasaheb Ambedkar Marathwada University Aurangabad, (M.S.).

Welch, P.S., 1952. *Limnology*, 2nd edn. McGraw Hill Book Company, New York, pp. 1–238.

Water and Wastewater Management in 2 Vols (2011) Pages 571–576
Editors: D.R. Khanna, R. Bhutiani & Gagan Matta
Published by: DAYA PUBLISHING HOUSE, NEW DELHI

Chapter 87

Dispersion of Heavy Metals in Textile Effluent and Pond Environment in Panipat Industrial Area

D.S. Malik, P.K. Bharti, Rashmi Yadav, Pawan Kumar and Praveen Chauhan
Department of Zoology and Environmental Science, Gurukula Kangri University, Haridwar

ABSTRACT

The Panipat region is well known textile industrial zone of Northern India and have a large number of dye houses and textile processing units. The dyeing house industries uses a large volume of water for wet dip coloring process and drain most of the wastewater as effluent. The effluent containing composite mixture of different pollutant in terms of heavy metals. Metal accumulation in sediments provides a record of the spatial and temporal history of pollution from surface water to groundwater. The dispersion of heavy metals from textile effluents to adjoining pond system ultimately affect quality of water. The present study revealed that pond sediment found highly polluted regarding heavy metals load. Transfer factor for Fe, Mn, Ni and Pb was found very high in pond through open effluent drainage system.

Keywords: Textile effluent, Heavy metals, Pond sediment, Aquatic environment.

Introduction

Textile industries consume large volume of water of high purity and discharge large quantities of effluent that normally exhibit polluting characteristics. Approximately 40,000 different dyes and pigments are used in the industry and over 7×10^5 tons of these dyes are produced annually worldwide

(Zollinger, 1987). 10–15 per cent of the dyes used in textile processing were lost in the effluent during the dyeing processes (Vaidge and Datye, 1982). The effluents from dye industries contain high amount of cation, anion, organic pollutants in the form of colour and heavy metals. Both natural processes and anthropogenic activities are responsible for introducing metals into the aquatic system. Many contaminants discharged into surface water rapidly become associated with the particulate matter and incorporated in sediments. Metal contaminated sediments may release heavy metals back to the overlying water column and, thus, pose risk to aquatic life and ecosystems (Forstner and Wittman, 1981).

Composite effluents from textile industries and dyeing houses of Panipat industrial area normally discharged openly into a common-effluent drain without any adequate treatment and commences in a large pond near Binjhole village near to industrial area, which may change the quality of bottom sediments of drain and pond sediment and the physico-chemical characteristics. Sediments reflect the current quality of the water system and can be used in detecting the presence of contaminants that do not remain soluble after discharge into surface water (Forstner, 1985). As a result of complex physical, chemical and biological processes a major fraction of heavy metals are found to be associated with bottom sediments (Baruah *et al.*. 1996). Metal accumulation in sediments provides a record of the spatial and temporal history of pollution. Hence, sediment monitoring can provide important information on various pollution events.

Materials and Method

Geographically, Panipat city is situated between 29°09′50″ and 29°50′ North latitude and 76°31′15″ and 77°12′45″ East longitude with altitude of 255 msl. Panipat town is located on the National Highway No. 1, about 90 km toward north of Delhi having a population of about 0.27 million. Study point (Textile Dye houses) in Panipat city is situated on Jatal road in industrial area near GT Road. The effluent samples were collected from common effluent drain and surface water samples were collected from a pond of village Binjhole situated near industrial area, where the effluent drain ends up and delay for a long time. Sediment samples were collected from the bottom of pond. All the samples were collected and preserved for further analytical work. All samples were analyzed by the standard methods (APHA, 1998) and Trivedi and Goel (1984). Transfer factor between textile effluents and surface water was calculated for each metal according to Lokeshwari and Chandrappa (2006).

Results and Discussion

The concentrations of heavy metals in water samples from effluent drain, pond water and pond sediment are presented in Tables 87.1–87.3 respectively.

Transfer factor values of all metals are depicted in Table 87.4. Cadmium concentrations in textile effluent (0.012–0.052 ppm) and pond water (0.001–0.009 ppm) turn down in rainy season due to the dilution factor. Cadmium slightly fluctuated and remains mostly constant in pond sediment (4.0–5.7 ppm). Cadmium transfer is due to some other non-point sources of cadmium pollution. During the last decade, the industrial use of Cd has increased and can create both acute and chronic cases of clinically identifiable toxicity in humans (Mani *et al.*, 2005). When the animal kept on contaminated feed along with the industrial polluted environment for long time, they suffer from its toxicity symptoms which may be sub clinical or clinical. The retention time of Cd is quite high (half life being 40 years). Bordas and Bourg (1998), Vazquez *et al.* (2007) and Korfali and Davies (2004) observed the similar Cd concentrations in sediments.

Table 87.1: Seasonal Variations of Heavy Metals in Textile Effluents (in ppm)

Heavy Metals	*Pre-monsoon*		*Monsoon*		*Post-monsoon*	
	Mean	*SD*	*Mean*	*SD*	*Mean*	*SD*
Cadmium (Cd)	0.052	0.042	0.025	0.037	0.012	0.008
Copper (Cu)	0.440	0.085	0.343	0.058	0.165	0.066
Iron (Fe)	0.527	0.093	0.253	0.130	0.266	0.079
Manganese (Mn)	0.153	0.050	0.127	0.047	0.224	0.072
Nickel (Ni)	0.037	0.016	0.016	0.015	0.031	0.013
Lead (Pb)	0.500	0.207	0.250	0.085	0.422	0.133
Zinc (Zn)	0.173	0.059	0.110	0.012	0.238	0.132

Table 87.2: Seasonal Variations of Heavy Metals in Pond Water (in ppm)

Heavy Metals	*Pre-monsoon*		*Monsoon*		*Post-monsoon*	
	Mean	*SD*	*Mean*	*SD*	*Mean*	*SD*
Cadmium (Cd)	0.009	0.001	0.001	0.002	0.004	0.002
Copper (Cu)	0.427	0.045	0.397	0.136	0.252	0.084
Iron (Fe)	9.700	3.035	4.233	1.450	4.820	2.961
Manganese (Mn)	0.530	0.061	0.463	0.070	0.172	0.062
Nickel (Ni)	0.127	0.025	0.073	0.033	0.054	0.045
Lead (Pb)	1.800	0.557	0.610	0.412	0.504	0.364
Zinc (Zn)	0.156	0.031	0.110	0.020	0.158	0.066

Table 87.3: Seasonal Variations of Heavy Metals in Pond Sediments (in ppm)

Heavy Metals	*Pre-monsoon*		*Monsoon*		*Post-monsoon*	
	Mean	*SD*	*Mean*	*SD*	*Mean*	*SD*
Cadmium (Cd)	5.71 7	0.4 5 I	4.780	0.455	4.054	0.595
Copper (Cu)	51.520	5.098	43.880	6.877	43.834	7.332
Iron (Fe)	5640.333	2179.050	4312.667	463.092	6013.400	856:579
Manganese (Mn)	942.000	48 135	724.000	56.471	836.400	38.708
Nickel (Ni)	63.667	8.083	52.333	9.074	52.800	3.768
Lead (Pb)	69.000	6.245	55.333	9.292	44.000	5.385
Zinc (Zn)	240.000	92.601	142.333	59.341	123.200	42.032

Concentrations of copper in textile effluent are due to the residues of some dyes especially used for blue or green colours. Copper concentrations in textile effluent (0.16–0.44 ppm) were found parallel or sometimes higher than iron. Most of the part of copper content reaches in pond water (0.25–0.43 ppm) and rest settle down in drain sediment. Copper in pond sediment (43.80–51.50 ppm) also settled down after some time. Copper may accumulate in living organisms and their various body parts

(Kudesia, 1992). High amount of heavy metals like Cu and Zn may harm the living organism of existing ecosystem (Aslam *et al.*, 2004). The Cu concentration level in study area has been compared with those observed by Korfali and Davies (2004) in sediments. Iron is one of the most abundant heavy metal in rocks and soil, ranking fourth by weight. However, iron was not found in high concentration in textile effluents (0.24–0.52 ppm). Iron was present in high quantity in pond sediment (4312.00–6013.00 ppm). Iron's transfer factors were found very high from textile effluent to pond water up to 16.00. Manganese was found as a second abundant element in the pond sediment after iron. However, manganese was comparatively low in textile industrial effluent (0.13–0.22 ppm). Transfer factor of Mn from textile effluent to pond water was noticed from 1 to 4 almost in all seasons, while below 1 in post monsoon season. Nickel is one of the second trace elements of the present study after cadmium. As it is very toxic to plants and animals, even in low concentrations and has a tendency to get accumulated in different body parts. However Ni was presented in very low concentration in textile effluent (0.016–0.037 ppm) but in pond water (0.054–0.127 ppm), it was found comparatively high due to the regular input. Ni concentration in pond water was noticed higher during the pre-monsoon season due to the loss of water by vaporization from pond system. Ni was found very high like other metals due to the settling of metals on bottom in pond sediment (52.30–63.70 ppm). Highest transfer factor of Ni from textile effluent to pond water was found in monsoon seasons.

Table 87.4: Transfer Factor for Heavy Metals from Textile Effluent to Pond Water

Heavy Metals	*Pre-monsoon*	*Monsoon*	*Post-monsoon*
Cadmium (Cd)	0.17	0.05	0.32
Copper (Cu)	0.97	1.16	1.53
Iron (Fe)	18.42	16.71	18.12
Manganese (Mn)	3.46	3.66	0.77
Nickel (Ni)	3.39	4.52	1.74
Lead (Pb)	3.60	2.44	1.19
Zinc (Zn)	0.90	1.00	0.66

Nickel may accumulate in aquatic life but its presence is not magnified along food chains. This trend of Ni concentration in present study has been supported on the basis of recommendation of Nasr *et al.* (2006) in sediment. It may be pointed out that the builds up of heavy metal like Cr, Pb, Ni warrant toward continuous monitoring and suitable measures are needed before these become toxic.

Lead was found highest in textile effluent. Lead mostly remained in effluent (0.25–0.50 ppm) and fluctuated in pond (0.5–1.8 ppm). Rest quantity of lead settled down on bed sediment of pond. Lead was the second most abundant metal species after iron. Pb was found slightly higher in pond water (0.50–1.8 ppm). It should not be more than 0.01 mg/l for health views, while it should be less than 0.1 mg/l generally in drinking water (WHO, 2006).

Zinc was not found so high in textile effluent (0.11–0.24 ppm) in comparison to lead. Similarly, it was found in less quantity in pond water (0.11–0.16 ppm) and accumulated in very high quantity in bed sediment (123.20–240.00 ppm) of pond. Zinc concentration was fluctuated in pond sediment (116.2–240.0 ppm) season to season. High amount of Zn may harm to living organism of that ecosystem (Aslam *et al.*, 2004). The presence of heavy metals in the aquatic environment has been of great concern to scientists and biologist because of their toxic nature (Bharti, 2007). Heavy metals travel from one level to another in the ecological system by being accumulated in abiotic and biotic components. In the studied area, basically anthropogenic source of heavy metals are the dye houses in textile industrial area. Heavy metals reach in effluent drain and further into pond and bed sediment of pond near village Binjhole adjoining to textile industrial area of Panipat. Heavy metals percolate slowly from the bed sediment of Binjhole pond towards the groundwater table and there may be a threat to deep aquifers also by leaching process of metals. Heavy metals have a tendency to accumulate and store in different trophic levels. The storage of heavy metals has created harmful effects for biotic components. Heavy metals in pond water were observed slightly lower than textile industrial effluents.

The present study reveals that there was heavy metals pollution in pond water and sediment. Due to the regular discharging of metals residues in the textile industrial area, a threat have been appear on plant vegetation, livestock and human beings of the region (Malik *et al.*, 2006). It is therefore obvious that scientific treatment is necessary to minimize the pollution effects before textile industrial effluent is discharge on the land. There is a need to take effective steps for improving wastewater use, its efficiency through renovation and modernization of recycling treatments of wastewater.

References

APHA, 1998. *Standard Methods for Examination of Water and Wastewater*, 19th edn. American Public Health Association, 1015, Fifteenth Street, NW, pp. (1–1)–(10–150).

Aslam, M.M., Baig, M.A., Hassan, I., Qazi, LA., Malik, M. and Saeed, H., 2004. Textile wastewater characterization and reduction of its COD and BOD by oxidation. *Electron. J. Environ. Agri. Food Chem.*, 3(6): 804–811.

Baruah, N.K., Kotoki, P., Bhattacharya, K.G. and Borah, G.C., 1996. Metal speciation in Jhanjhi river sediments. *Sci. Total Environ.*, 193: 1–12.

Bharti, P.K., 2007. Effect of textile industrial effluents on groundwater and soil quality in Panipat region (Haryana). *Ph.D. Thesis*, Gurukula Kangri University, Haridwar.

Bordas, S. and Bourg, C.M., 1998. A critical evaluation of sample pretreatment of contaminated sediments to be investigated for the potential mobility of their heavy metal load. *Water, Air and Soil Pollution*, 103: 137–149.

Forstner, U., 1985. Chemical forms and reactivities of metals in sediments. In: *Chemical Methods for Assessing Bioavailable Metals in Sludge and Soils*, (Eds.) R. Leschber, R.D. Davies and L.P. Hermite. Elsevier, London, pp. 1–30.

Forstner, U. and Wittman, G.T.W., 1981. *Metal Pollution in the Aquatic Environment*, 2nd edn. Springer, Berlin, pp. 486.

Korfali, S.L and Davies, R.D.E., 2004. The relationship of metals in river sediments (Nahr–Ibrahim, Lebanon) and adjacent floodplain soil. *The CIGR J. of Scientific Research and Development*, manuscript LW04010, 6: 1–22.

Kudesia, Y.P., 1992. *Water Pollution*. Pragati Prakashan, Meerut, pp. 407.

Lokeshwari, H. and Chandrappa, G.T., 2006. Impact of heavy metal contamination of Bellandur Lake on soil and cultivated vegetation. *Current Science*, 91(9): 622–627.

Malik, D.S., Bharti, P.K. and Sumit Grover, 2006. Alteration in surface water quality near textile industries at Panipat (Haryana). *Environment Conservation J.*, 7(2): 65–68.

Mani, Y., Kaur, H. and Mohini, M., 2005. Toxic metals and environmental pollution. *J. Ind. Poll. Cont.*, 21(1): 101–107.

Nasr, S.M., Okbah, M.A. and Kasem, S.M., 2006. Environmental assessment of heavy metal pollution in bottom sediments of Aden port, Yemen. *International Journal of Oceans and Oceanography*, 1(1): 99–109.

Trivedi, R.K. and Goel, P.K., 1984. *Chemical and Biological Methods for Water Pollution Studies.* Environmental Publication, Karad, pp. 1–251.

Vaidge, A.A. and Datye, K.Y., 1982. Environment pollution during chemical processing of synthetic fibers. *Colourage*, 14: 3–10.

Vazquez, N.N., Gil, J.L., Esteves, and Narvarte, M.A., 2007. Monitoring heavy metal pollution in San Antonio Bay, Rio Negro, Argentina. *Bull. Environ. Contam. Toxicol.*, DOI–10.1007/s001 28–007–9084–z.

WHO, 2006. *Standards for Drinking Water*, available at http//:www.lenntech.com.

Zollinger, H., 1987. *Colour Chemistry Synthesis Properties and Application of Organic Dyes and Pigment.* VCH, New York, pp. 275.

Water and Wastewater Management in 2 Vols (2011) *Pages 577–587*
Editors: **D.R. Khanna, R. Bhutiani & Gagan Matta**
Published by: **DAYA PUBLISHING HOUSE, NEW DELHI**

Chapter 88

Effect of Tourism on the Physico-chemical Parameters of Stream Nalhota at Guchu Pani, District Dehradun

D.R. Khanna[1], R. Bhutiani[1], S.K. Pathak[2], Prashant Tyagi[1] and Bharti Tyagi[1]
[1]Department of Zoology and Environmental Science, G.K.V., Haridwar
[2]Govt. Girls P.G. College, Mhow

ABSTRACT

The present chapter deals with the study of the effects of tourism on the physico-chemical parameters of Stream Nalhota, a spring fed stream originates from the Robber's cave (Guchu Pani) Dehradun (Uttarakhand). This site is one of the famous picnic spots in Dehradun. The parameters studied were temperature, velocity, TS, TDS, TSS, pH, free CO_2, DO, hardness, calcium, magnesium, BOD, COD, chloride and alkalinity. The average value of all the parameters were found higher at sampling site C in comparison to sampling site A and B which may be due to higher activities of tourists and local people of near by villages. Maximum values of the studied parameters were observed during summer except DO and minimum values of these parameters obtained during winter due to less activities of localities and tourists in water, while higher values of DO were observed during winter which may be due to super saturation of oxygen at lower ambient temperature. Nature of water of stream Nalhota with respect to pH is alkaline. The study indicates that the water quality of stream Nalhota is not suitable for drinking purpose with respect to BOD and TS.

Keywords: *TDS, Hardness, Chloride, DO, BOD*

Introduction

Stream Nalhota is a spring fed stream originated from the Robber's cave (Guchu Pani) which is located at a distance of 8 km from the clock tower of Dehradun (Uttarakhand). The study site is in the Cantt. area, near a village named Guchu Pani under the panchayat of Vijaypur Gopiwala (Anarwala). Geographically the stream Nalhota and the Robber's cave are situated in the globe on longitude–30° 22'30" and latitude–78°2' 30". The total length of this stream is around 1 km including 600 m cave area with several natural springs and 400 m as an open area till it merges into the river Tons which in turn is the tributary of River Yamuna. This is an ultimate source of water for drinking and irrigation purpose for the people of near by villages named Guchu Pani, Vijaypur Gopiwala (Anarwala), Kotlawali, Purkal and Chandroroti. Thousands of people come here daily from April to June to enjoy this place and to have a good adventurous trekking beneath this cave. Many workers have studied the water quality of Ganga, Song and Suswa rivers which flow through the district Dehradun (Khanna and Bhutiani, 2003 and Khanna *et al.*, 2006; 2007). But no study has been conducted to assess the water quality of stream Nalhota till now. Viewing the importance of stream Nalhota for the villagers and tourists this study has been carried out.

Materials and Method

For the study of physico-chemical parameters of stream Nalhota in Uttarakhand, the water samples were collected from three sampling sites- site "A" at the starting point of Robber's cave to know the primary physico-chemical condition of the stream, site "B" at 600 m distance from site "A" where stream Nalhota comes out from the Robber's cave and site "C" at the point where the Nalhota stream merges into the river Tons. Some of the physico-chemical parameters like TS (total solid), TDS (total dissolved solid), TSS (total suspended solid), DO (dissolved oxygen), BOD (biochemical oxygen demand), COD (chemical oxygen demand), chloride, hardness, calcium, magnesium, alkalinity and phytoplankton were observed in the laboratory. Temperature, velocity, pH, and free CO_2 were observed at the spot. Grab sampling collection method was adopted through out the study. Sampling preservation and analytical methods were adopted as per APHA (1998), Trivedi and Goel (1986), Khanna and Bhutiani (2004).

Results and Discussion

The average value of physico-chemical and biological parameters of samples analyzed from three sampling site are given in Tables 88.1 and 88.2 and shown in Figures 88.1–88.9. The average values of temperature ranged between 13.33±1.69-17.00±2.00 °C. A more or less similar trend has been observed by Khanna and Bhutiani (2003) in the river Ganga at Haridwar and Iqbal *et al.* (2006) in the River Soan (Punjab), Pakistan. Velocity ranged between 0.82±0.13-1.05±0.19 m/s. Total solids, total dissolved solids and total suspended solids ranged between 1690.00±50.38-1866.00±48.63 mg/l, 203.00±7.02-292.33±9.50 mg/l, 1487.00±46.05-1573.66±40.79 mg/l respectively. The similar trend has been observed by Khanna and Bhutiani (2007) in River Suswa, Logankumar *et al.* (1989) in the river Bhavani at Sirumugai, Tamilnadu and Abbasi *et al.* (1996) in Punmurpuzha river of Kerala. pH value has been recorded between 7.50±0.00-7.83±0.16. Gasim *et al.* (2005) reported similar trend in the Bebar River, Pahang, Malaysia. Dissolved oxygen and biological oxygen demand varied between 8.91±0.46-7.56±0.26 mg/l and 2.42±0.24-3.37±0.27 mg/l respectively. The similar trend has been observed by Khanna *et al.* (2007) in Song River and Khaiwal *et al.* (2003) reported in the River Yamuna. Chemical oxygen demand was found between 2.56±0.24-3.36±0.24 mg/l. Hardness, calcium and magnesium varied between 206.00±10.27-233.66±15.08 mg/l, 86.16±4.00-104.20±4.17 mg/l and

26.62±1.54-34.44±2.25 mg/l respectively. A similar trend has been found by Singh *et al.* (2006) in the river Ganga. Kulkarni *et al.* (2002) in the Khushavati river at Quepem, Goa. Free CO_2 ranged between 0.88±0.25- 1.90±0.14 mg/l. A similar trend has been found by Khanna *et al.* (2006). Chloride was found between 17.98±1.70-22.24±1.71 mg/l. Alkalinity varied between 160.00±7.05-186.66±7.60 mg/l. Khanna and Singh (2000) found similar trend in river Suswa in Raiwala.

Table 88.1: Average Value of Physical Parameters of Stream Nalhota from January to June 2008

Parameter	*January*	*February*	*March*	*April*	*May*	*June*	*Average*
Temperature (°C)	13.33 ±1.69	14.33 ±2.72	13.66 ±2.33	14.00 ±2.51	16.00 ±1.52	17.00 ±2.00	14.72 ±0.59
Velocity (m/sec)	0.82 ±0.13	0.83 ±0.13	0.88 ±0.13	0.92 ±0.13	0.93 ±0.14	1.05 ±0.19	0.90 ±0.03
Total solid (mg/l)	1690.00 ±50.08	1720.00 ±5.00	1745.66 ±2.35	1779.66 ±43.64	1820.33 ±46.24	1866.00 ±8.63	1770.18 ±6.77
TDS (mg/l)	203.00 ±7.02	222.33 ±10.73	241.00 ±9.02	257.00 ±11.85	280.66 ±12.01	292.33 ±9.50	249.38 ±13.99
TSS (mg/l)	1487.00 ±6.05	1497.66 ±34.72	1504.66 46.72	1522.66 ±35.41	1539.66 ±46.84	1573.66 40.79	1520.88 ±13.07

Table 88.2: Average Value of Chemical Parameters of Stream Nalhota from January to June 2008

Parameter	*January*	*February*	*March*	*April*	*May*	*June*	*Average*
pH	7.50 ±0.00	7.66 ±0.16	7.66 ±0.16	7.66 ±0.16	7.66 ±0.16	7.83 ±0.16	7.66 ±0.04
Free CO_2 (mg/l)	0.88 ±0.25	1.17 ±0.l4	1.46 ±0.14	1.61 ±0.14	1.61 ±0.l4	1.90 ±0.14	1.43 ±0.14
D.O. (mg/l)	8.91 ±0.46	8.51 ±0.46	8.23 ±0.35	7.96 ±0.35	7.83 ±0.35	7.56 ±0.26	8.16 ±0.10
Hardness (mg/l)	206.00 ±10.27	211.33 ±.9.62	214.66 ±10.46	217.33 ±.9.96	222.00 ±11.72	230.66 ±15.08	216.99 ±3.52
Calcium (mg/l)	86.16 ±4.00	88.83 ±4.04	90.17 ±4.00	94.18 ±4.17	96.18 ±1.62	104.20 ±4.17	93.28 ±2.64
Magnesium (mg/l)	26.62 ±1.54	27.27 ±1.48	29.23 ±2.06	30.53 ±1.31	32.32 ±2.18	34.44 ±2.25	30.06 ±1.22
BOD (mg/l)	242 ±0.24	256 ±0.35	283 ±0.46	2.97 ±0.26	3.24 ±0.23	3.37 ±0.27	2.89 ±0.15
COD (mg/l)	256 ±0.24	2.72 ±0.27	2.88 ±0.24	3.04 ±0.24	3.20 ±0.24	3.36 ±0.24	2.96 ±0.11
Chloride (mg/l)	17.98 ±1.70	19.40 ±1.70	19.88 ±1.69	21.30 ±1.69	21.77 ±1.70	22.24 ±1.71	20.42 ±0.66
Alkalinity (mg/l)	160.00 ±7.05	170.00 ±7.07	178.33 ±7.05	180.00 ±7.00	186.66 ±7.90	186.66 ±7.60	176.94 ±12.68

Maximum values of the studied parameters were observed during summer except DO and minimum values of these parameters obtained during winter, while higher value of DO was observed during winter which may be due to super saturation of oxygen at lower ambient temperature and less

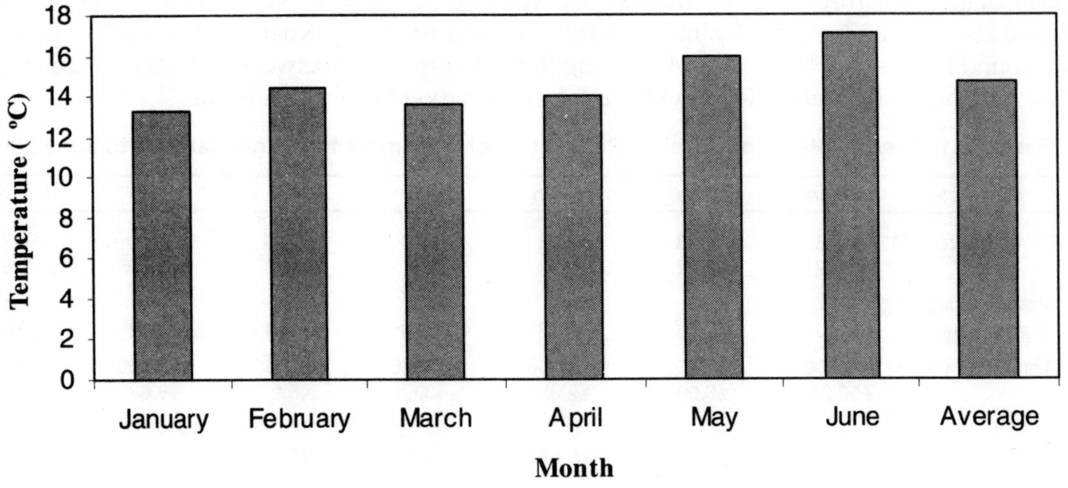

Figure 88.1: Showing Monthly Fluctuation in Water Temperature (ºC) of Stream Nalhota

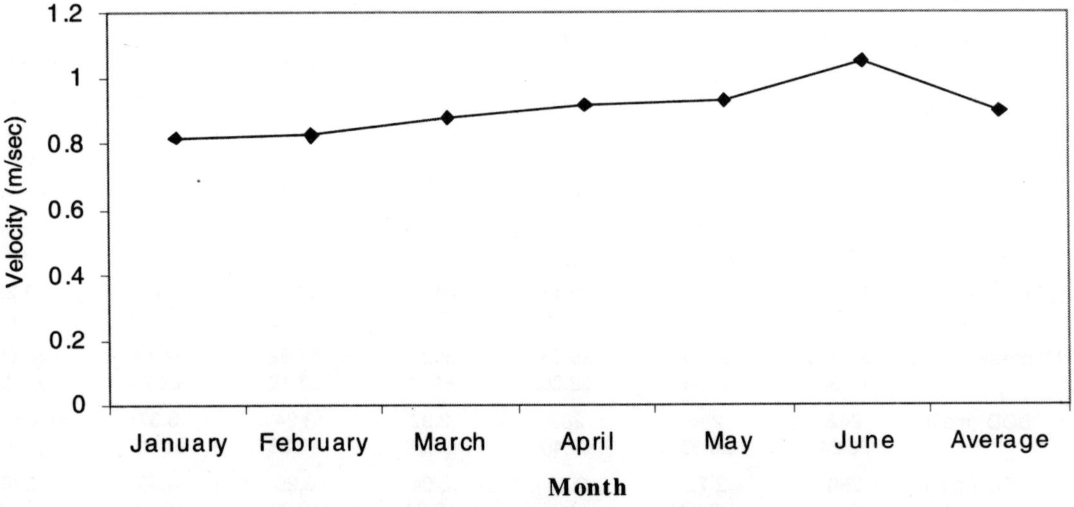

Figure 88.2: Showing Monthly Fluctuation in Water Velocity (m/sec) of Stream Nalhota

activities of localities and tourists in winter. After winter period as temperature of the environment rises activities of tourists are also increased which gradually deteriorates the water quality of stream Nalhota. The values of studied parameters are obtained higher at sampling site C than sampling site B which in turn has higher values of studied parameters than sampling site A which may be due to the

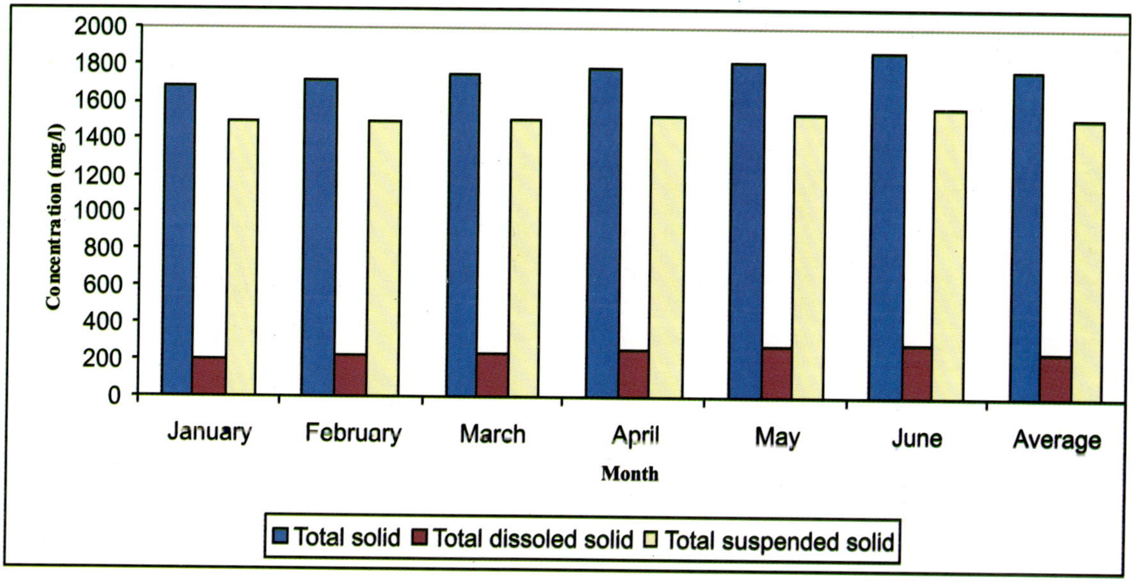

Figure 88.3: Showing Monthly Fluctuation in Total Solid, TDS and TSS (mg/l) of Stream Nalhota

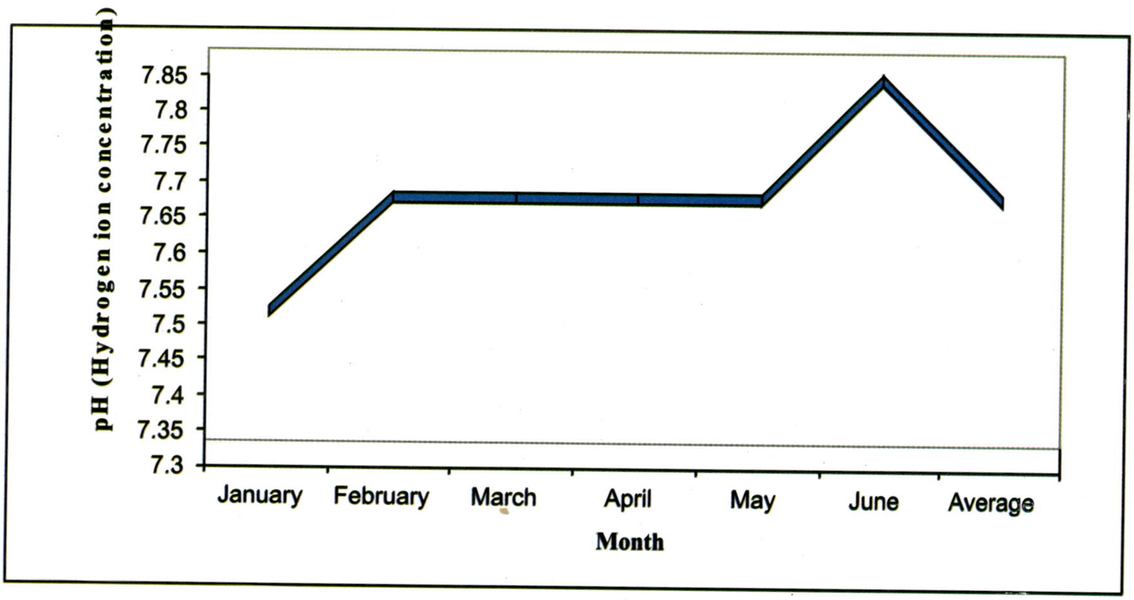

Figure 88.4: Showing Monthly Fluctuation in pH of Stream Nalhota

higher bathing and washing activities of tourists and localities on site C and also due to the small sewage drains of village Guchu pani at this sampling site. The value of studied parameter of site B was obtained higher than site A, because there are many bathing point for tourists and small restaurants at this site. Water quality of stream Nalhota with respect to pH is alkaline. The value of BOD as well as

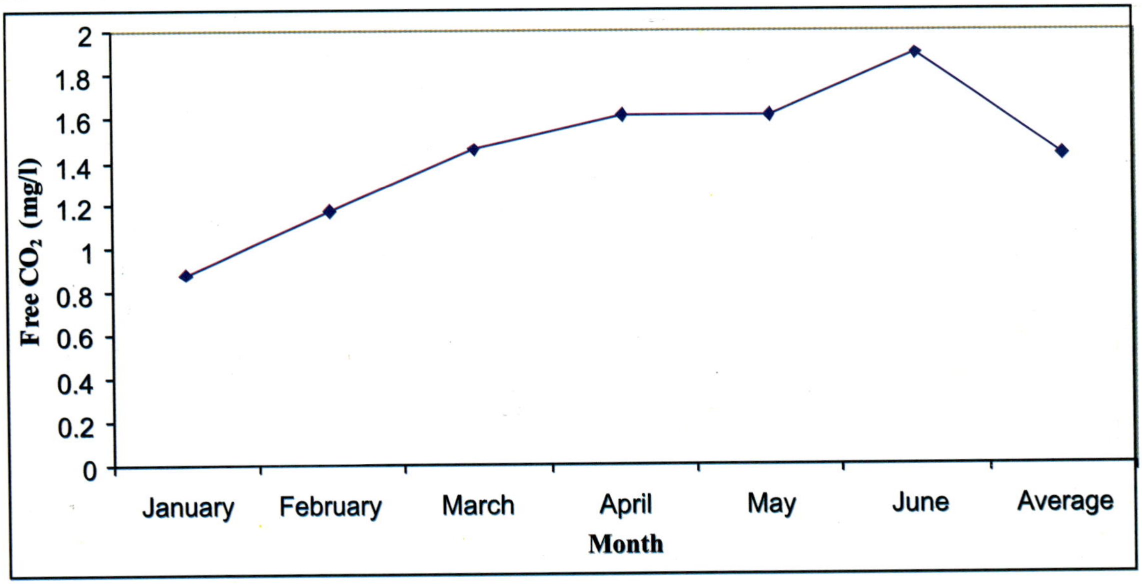

Figure 88.5: Showing Monthly Fluctuation in Free CO_2 (mg/l) of Stream Nalhota

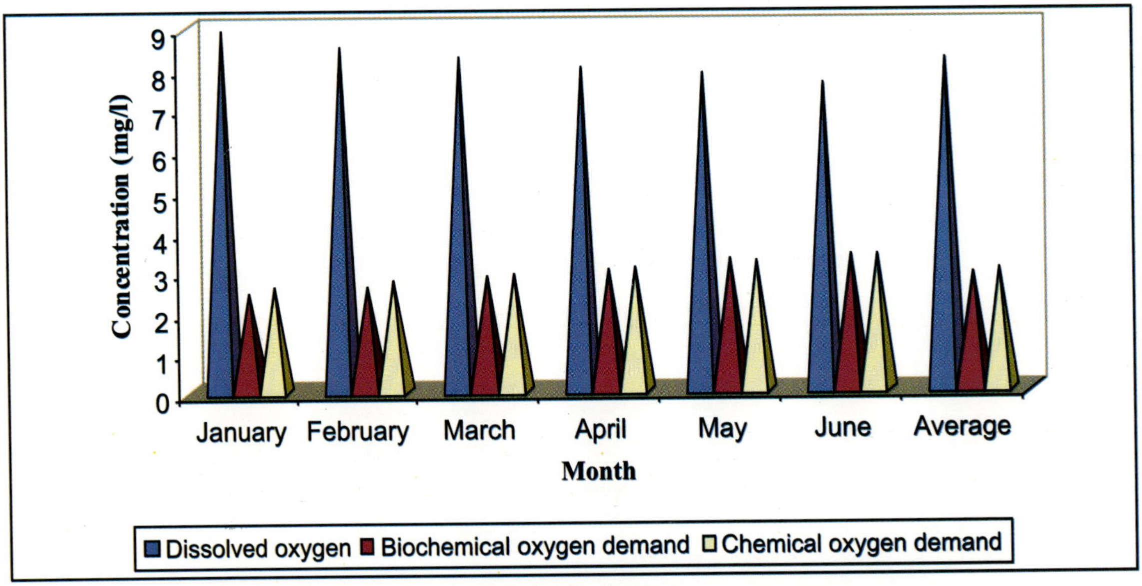

Figure 88.6: Showing Monthly Fluctuation in DO, BOD, COD (mg/l) of Stream Nalhota

COD was recorded highest at site C in all the season while the average value of both the parameters was obtained highest in the month of June due to higher activities of tourists in summer tourism. The average value of pH, DO, COD, alkalinity, Hardness and Chloride were obtained with in the prescribed tolerance limits of drinking purpose (BIS-1991) where the average values of BOD (2.42±0.24-3.37±0.27

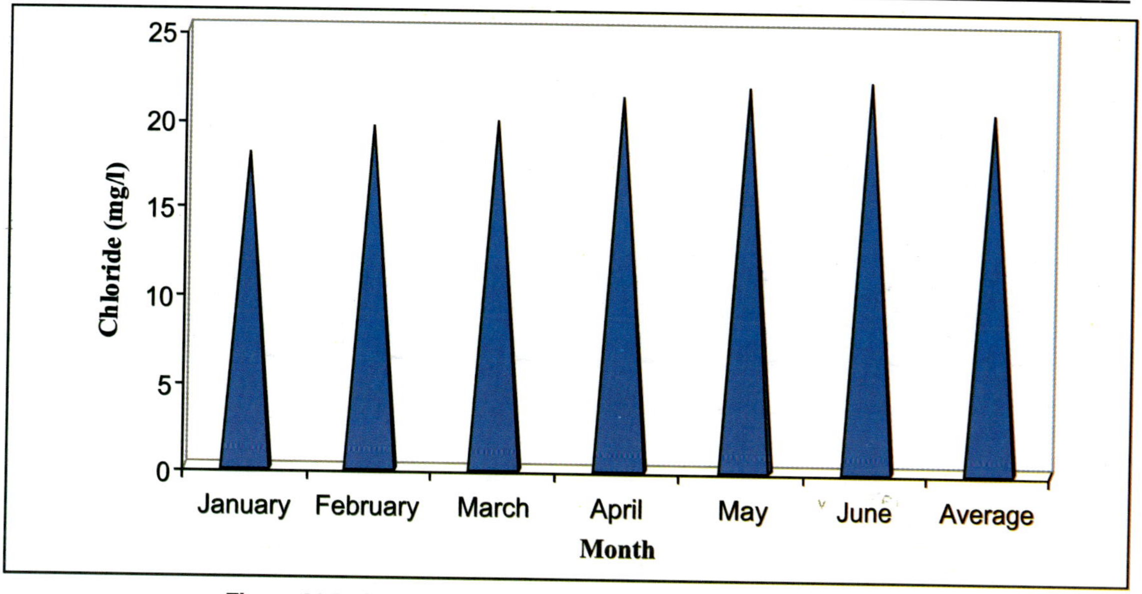

Figure 88.7: Showing Monthly Fluctuation in Chloride (mg/l) of Stream Nalhota

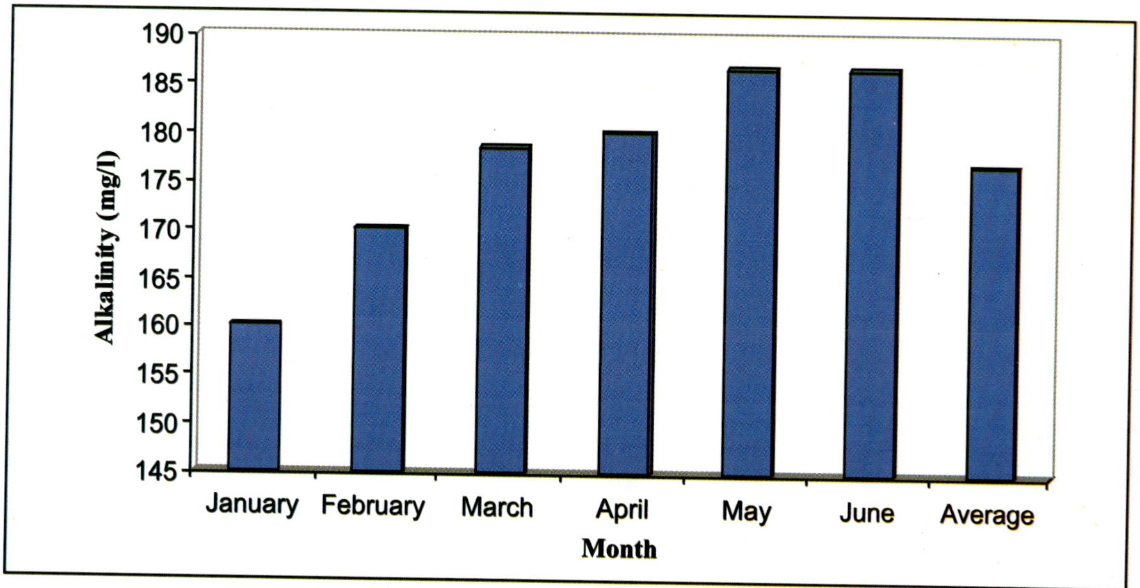

Figure 88.8: Showing Monthly Fluctuation in Alkalinity (mg/l) of Stream Nalhota

mg/l) and TS 1(690.00±50.38-1866.00±48.63) mg/l was observed beyond the standard limits. The observed value of BOD beyond the standard limits clearly indicates the presence of bacteria, which may cause diseases like gastroenteritis, dysentery, cholera and other fatal diseases. Indian standards of drinking water (BIS-1991) are given in Table 88.3.

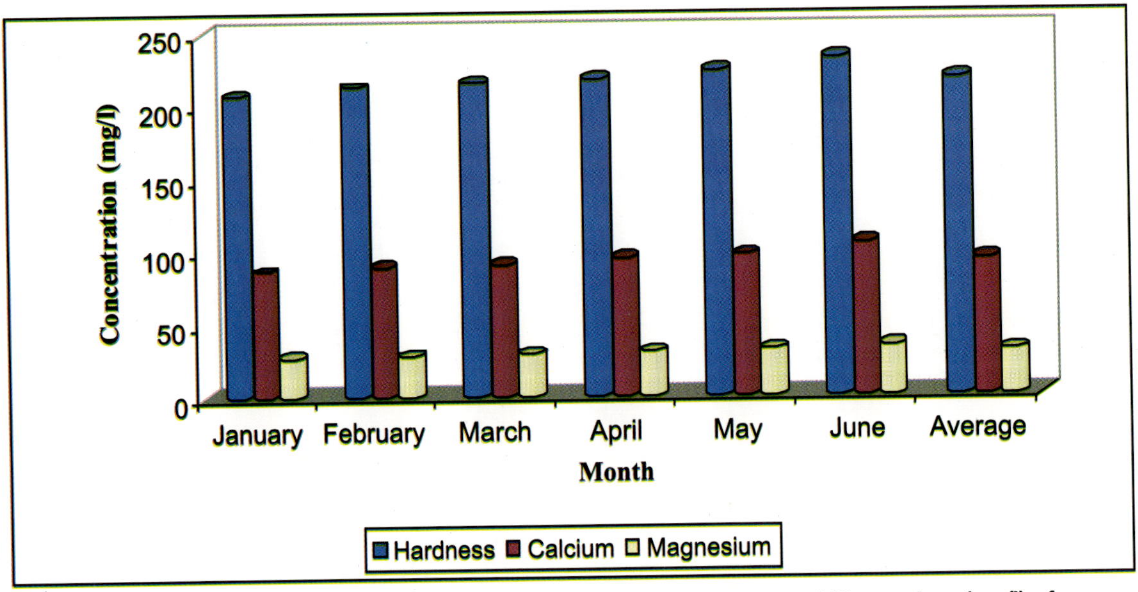

Figure 88.9: Showing Monthly Fluctuation in Hardness, Calcium and Magnesium (mg/l) of Stream Nalhota

Table 88.3: Standard Limits of Drinking Water in Terms of Physico-chemical Parameters

Parameters	*Desirable Limit*
pH	6.5–8.5
TS (mg/l)	500
DO (mg/l)	>6.0
BOD	2.0
Hardness (mg/l)	300.0
Calcium hardness (mg/l)	75.0–200.0
Magnesium hardness (mg/l)	30.0–100.0
Alkalinity (mg/l)	<200.0
Chloride (mg/l)	250.0

The correlation coefficients among the different parameters are presented in Table 88.3. The analysis show the high degree positive correlation between temperature and TS, temperature and TDS, temperature and Hardness, temperature and calcium, temperature and BOD, temperature and COD, velocity and TS, velocity and TDS, velocity and CO_2, velocity and Hardness, velocity and Calcium, velocity and Magnesium, velocity and BOD, velocity and COD, TS and TSS, TS and TDS, TS and CO_2, TS and Hardness, TS and Calcium, TS and Magnesium, TS and BOD, TS and COD, TS and Chloride, TS and Alkalinity. TSS and TDS, TSS and CO_2, TSS and Hardness, TSS and Calcium, TSS and Magnesium, TSS and BOD, TSS and COD, TSS and Chloride, TSS and Alkalinity, TDS and CO_2, TDS and Hardness, TDS and Calcium, TDS and Magnesium, TDS and BOD, TDS and COD, TDS and Chloride, pH and Hardness, pH and Calcium, CO_2 and DO, Hardness and Calcium, Hardness and Magnesium, Hardness and BOD, Hardness and COD, Hardness and Chloride, Hardness and Alkalinity, Calcium and Magnesium, Calcium and BOD, Calcium and COD, Calcium and Chloride, Magnesium and BOD, Magnesium and COD, Magnesium and Chloride, Magnesium and Alkalinity, BOD and COD, BOD and Chloride, BOD and Alkalinity, COD and Chloride, COD and Alkalinity, Chloride and Alkalinity.

Table 88.4: Correlation Matrix of Physico-chemical Parameters of Stream Nalhota during January to June 2008

	Temperature	*Velocity*	*TS*	*TSS*	*TDS*	*pH*	*CO_2*	*DO*	*Hardness*	*Calcium*	*Magnesium*	*BOD*	*COD*	*Chloride*	*Alkalinity*
Temperature		0.86	0.91	0.87	0.93	0.80	0.76	–0.81	0.91	0.91	0.88	0.86	0.87	0.81	0.76
Velocity			0.96	0.92	0.98	0.81	0.92	–0.92	0.97	0.98	0.97	0.93	0.94	0.89	0.03
TS				0.98	0.98	0.85	0.94	–0.97	0.99	0.98	0.99	0.98	0.99	0.96	0.92
TSS					0.95	0.02	0.96	–0.96	0.97	0.94	0.98	0.99	0.99	0.98	0.96
TDS						0.06	0.91	–0.93	0.98	0.99	0.90	0.95	0.96	0.91	0.85
pH							.0.89	–0.86	0.90	0.88	0.82	0.80	0.84	0.82	0.80
CO_2								0.94	–0.98	–0.97	–0.96	–0.95	–0.96’	–0.93	–0.89
DO									–0.96	–0.93	–0.96	–0.97	–0.98	–0.99	–0.97
Hardness										0.98	0.98	0.96	0.98	0.94	0.90
Calcium											0.97	0.94	0.96	0.91	0.85
Magnesium												0.99	0.99	0.95	0.91
BOD													0.99	0.97	0.95
COD														0.98	0.95
Chloride															0.96
Alkalinity															

The analysis showed high degree negative correlation between temperature and DO, Velocity and DO, TS and DO, TDS and DO, TSS and DO, pH and DO, pH and CO_2, CO_2 and Hardness, CO_2 and Biological Oxygen Demand, CO_2 and Chemical Oxygen Demand, CO_2 and Chloride, CO_2 and Calcium, CO_2 and Magnesium, CO_2 and Alkalinity, DO and Biological Oxygen Demand, DO and Chemical Oxygen Demand, DO and Chloride, DO and Hardness, DO and Calcium, DO and Magnesium and DO and Alkalinity. The statistical analysis of present study showed the correlation coefficient among the parameters and was negative 23 times and positive 73 times. This study indicated that positive correlation dominated in the present stream.

References

Abbasi, S.A., Arya, D.S., Ahmed, A.S. and Abbasi, N., 1996. Water quality of a typical river Punnurpuzha of Kerala. *Pollution Research*, 15(2): 163–166.

APHA, AWWA, WPCF, 1998. *Standard Methods for Examination of Water and Wastewater*, 20th edn. American Public Health Association, 1015, Fifteen Street, New Washington, 15: 1–1134.

Gasim, Muhammad Barzani, Ismail, B.S., Ekhwan Toriman, Sujaul Islam Mir and Tan Choon, C., 2007. A physico-chemical assessment of the Bebar river, Pahang, Malaysia. *Global Journal of Environmental Research*, 1(1): 07–11.

Iqbal, Furhan, Ali Muhammad and Kanwal, N., 2006. Limnological study of River Soan (Punjab), Pakistan. *Agriculturae Conspecius Scientificus*, 7(2): 65–73.

Khaiwal, Ravindra; Ameena, Meenakshi; Monika, Rani and Anubha, K., 2003. Seasonal variations in physico-chemical characteristics of River Yamuna in Haryana and its ecological best designated use. *J. Environ. Monit.*, 5: 419–426.

Khanna, D.R. and Bhutiani, R., 2003. Limnological characteristics of the river Ganga Haridwar (Uttaranchal). *U.P. J. Zool.*, 23(3): 179–183.

Khanna, D.R. and Bhutiani, R., 2004. *Water Analysis at a Glance.* ASEA Publications, pp. 1–115.

Khanna, D.R. and Bhutiani, R., 2007. *Limnological Modeling: A Case Study of River Suswa.* Daya Publishing House, pp. 1–297.

Khanna, D.R. and Singh, R.K., 2000. Seasonal fluctuation in plankton of Suswa river at Raiwala (Dehradun). *Env. Cons. J.*, 1(2–3): 89–92.

Khanna, D.R., Pathak, S.K., Bhutiani, R. and Chandra, K.S., 2006. Study of water quality of river Suswa near Raiwala, Uttaranchal. *Env. Cons. J.*, 7(3): 79–84.

Khanna, D.R., Singh, Vikas, Bhutiani, R., Kumar, S.C., Matta, Gagan and Kumar, Dheeraj, 2007. A study of biotic and abiotic factors of Song River at Dehradun, Uttarakhand. *Env. Cons. J.*, 8(3): 117–126.

Kulkarni, R.R., Sharma, R.N. and Burkari, M., 2002. Diurnal variations of physico-chemical aspects of pollution in Khushavati River at Quepem, Goa. *J. Aqua. Biol.*, 17: 27–28.

Logankumar, K., Sivakurnar, A.A., Logasamy, S. and Aruchami, M., 1989. Seasonal variation in the nutrients of the river Bhavani at Sirumugai, Coimbatore District, Tamil Nadu. *Ind. J. Ecol.*, 16(1): 60–63.

Santra, S.C., 2004. *Environmental Science*. New Central Book Agency (P) Ltd.

Singh, J.P., Singh, Shakun and Khanna, D.R., 2006. Water quality status of River Ganges in respect of physico- chemical and microbial characteristics at Anupshahar, Bulandshahar. *Env. Cons. J.*, 7(2–3): 29–34.

Trivedi, R.K. and Goel, P.K., 1986. *Chemical and Biological Methods of Water Pollution Studies*. Environmental Publications, Karad, pp. 1–251.

Water and Wastewater Management in 2 Vols (2011) *Pages 588–594*
Editors: **D.R. Khanna, R. Bhutiani & Gagan Matta**
Published by: **DAYA PUBLISHING HOUSE, NEW DELHI**

Chapter 89

Occurrence of Small Free-living Amoebae from Natural Water Resources

Tabrez Ahmad, Newton Paul and A.K. Sharma
Protozoology Research Laboratory, Department of Zoology, University of Lucknow

ABSTRACT

Small free-living amoebae are also called as amphizoic amoebae, because of their capability to exist both as free-living as well as opportunistically pathogenic. Some of the species of *Naegleria* and *Acanthamoeba* are known to cause fatal disease (Primary Amoebic Meningoencephalitis= PAM and Granulomatous Amoebic Encephalitis= GAE) of central nervous system affecting brain and amoebic keratitis affecting eyes. The purpose of our study was to find out whether small free-living amoebae were also found in natural resource of water such as river and pond from Lucknow region. A total 68 samples of water were collected from six different sites of Gomti River and four different ponds from Lucknow city. Total 38 samples were found to be positive for small free-living amoebae. The most common amoebae present were species of *Naegleria, Schizopyrenus, Acanthamoeba, Hartmennella* and *Vannella*. Out of total 47 water samples from Gomti River, 27 samples were positive for amoebae and out of 21 samples of ponds examined 11 were found to be positive.

The occurrence of these amoebae in natural water sources such as river and pond in Lucknow region poses a threat to human being for meddling with water without taking adequate care. Since the pathogenic strains of amoebae are known to infect human being *via* nasal route during swimming and/or *via* cut in the skin/body *via* haematogenous route. Exposure of eyes with contaminated water may cause amoebic keratitis in human being. A great awareness among masses is essential to educate about these new amoebic disease, their possible preventive measures and occurrence of these pathogenic amoebae from natural water resources.

Keywords: *Free-living, Amphizoic, Naegleria, Acanthamoeba, PAM, GAE and natural water (river and pond).*

Introduction

Small free-living amoebae are widely distributed in the human environment (soil, water and air). These amoebae have been isolated from various natural water sources such as rivers, pond, lake *etc.* (Kasprzak and Mazur, 1974 and Jonckheere, 1981), from thermal acidic stream (Sheehan *et al.*, 2003), from artificial well (Shenoy *et al.*, 2002) and from domestic tap water (Kilvington *et al.*, 2004, Pandey and Sharma, 2006).

These amoebae are called "amphizoic" because of their capability to exist both as free-living as well as opportunistically pathogenic (Page, 1974). People probably have more frequent contact with these organisms particularly with their resistant cysts and temporary amoebo-flagellate stage (Martinez and Visvesvara, 1997). The species of *Naegleria, Acanthamoeba* and *Balamuthia* have been identified as opportunistic pathogens of human beings and other domestic animals and are known to cause spectrum of infections in immunocompromised individuals including persons with AIDS (Denny *et al.*, 1997 and Sison *et al.*, 1995).

Occurrence of Primary Amoebic Meningoencephalitis (PAM) has also been reported from India. Pan and Ghosh (1971) from Kolkata, first reported the occurrence of PAM in two children. Third case was reported by Bedi *et al.* (1972) from Udaipur, Rajasthan. One more case has been reported, in an infant of five months old from Mangalore, South India (Shenoy *et al.*, 2002).

Recently, Jain *et al.* (2002) from Chandigarh also reported survival of a patient after the infection of *Naegleria*. Recently, Kaushal *et al.* (2008) reported one more case of primary amoebic meningoencephalitis from Ludhiana, who had the history of taking bath in the village pond. Till date a total of nine cases have been reported between 1971- 2008 from India. More than 200 cases of PAM have been reported so far abroad and most of them are from America. This may be because of the greater awareness about these diseases rather than the incidence. The aim of our investigation was to detect these amphizoic amoebae from aquatic environment used by human being and to know the distribution of the pathogenic and non-pathogenic free-living amoebae in natural water resources.

Materials and Method

The sampling sites included Gomti River at different points and pond of different localities in ten parts of Lucknow city. Using sterile 500 ml screw-cap bottle, 200 ml surface water samples were collected at proposed collecting sites. Water temperature and pH were recorded at the sampling sites. The water samples were filtered through a sterile filter paper using sterile conical funnel. The residue was collected in the inner side of the cone of filter paper and the 1.00 cm cone of this filter paper was cut by sterilized scissors and was placed in the center of non-nutrient agar petri-dish plates already pre-seeded with *Escherichia coli* (as food for amoebae) keeping the inner side of paper downward. These plates were incubated at 37.00 °C for 8-10 days or longer for the growth of amoebae for their detailed study and biological characterization.

Amoeba Identification

Amoebae were identified on the morphological basis of trophozoites, cysts and flagellates following Singh (1985) and the latest classification of protozoa (Levine *et al.*, 1980).

Results and Discussion

All the samples were collected throughout the year. Total 68 samples were collected from different points of Gomti River and various ponds of Lucknow city and surrounding areas. The pond water

The Experimental Design of Water Sampling Sites

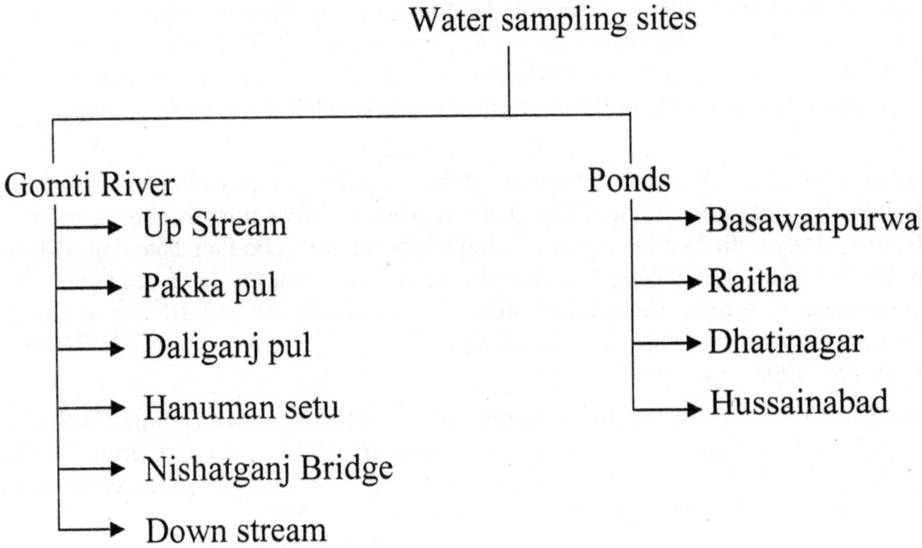

All samples were collected throughout the year

samples were collected from the villages of Bakshi-Ka-Talab area of Lucknow. Out of total 68 water samples, 38 samples were found to be positive for small free-living amoebae (57 per cent). Out of total 47 samples from river, 27 were positive (57 per cent) and out of 21 samples of pond examined, 11 samples were found to be positive for amoebae (52 per cent) (Figures 89.1 and 89.2). The water temperature recorded was in the range of 29.00 °C-37.00 °C and pH from 5.00 to 7.00. *Acanthamoeba* strains and other free-living amoebae could be recovered from pH range of 5.00 to 7.00 and temperature range (27.00 °C to 35.00 °C) of water, where as *Naegleria* strains were found in the water temperature from 28.00 °C to 35.00 °C and at the pH range of 6.50-7.00.

In our investigation, the strains of free-living amoebae isolated from different water samples were identified as different strains and species of *Naegleria, Acanthamoeba, Schizopyrenus* and *Hartmannella* (Plates 89.1 and 89.2).

The identification of the amoebic isolates was done following their detailed biological characterization (studying the morphology of cysts and trophozoite, locomotion, nuclear division and pathogenicity) using the method of Singh, 1985 (Table 89.1 and Plates 89.1 and 89.2). In pathogenicity test using *Albino* mice (weight approximately 12 gm) the strain (R-4) of *Naegleria fowleri* was found to be highly pathogenic. The occurrence of small free-living amoebae from water sources has also been reported from different parts of India and abroad (Singh and Das, 1972a; Gogate *et al.*, 1984; Pandey and Sharma, 2006). Human infection by *Naegleria* and *Acanthamoeba* has been reported world wide (Nacapunchai *et al.*, 1999). The environmental sources of amoebic isolates range from air to water and to soil (John and Howard, 1995).

Our results are inconformity with that of specific characteristic of amoebae isolates by other workers (Pandey and Sharma, 2006; Singh and Das, 1972a, Martinez *et al.*, 1997 and Rohr *et al.*, 1998).

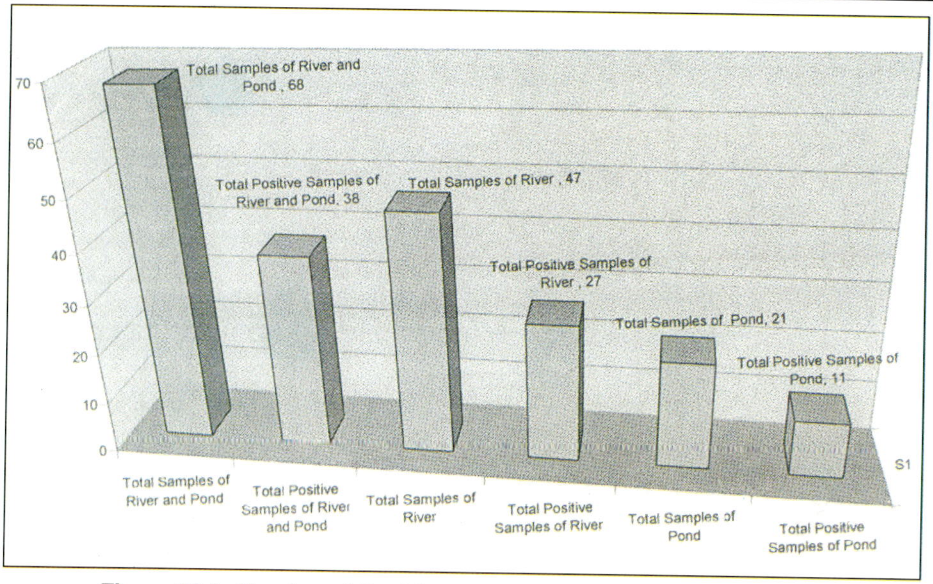

Figure 89.1: Number of Positive Samples from River and Ponds in Ten Different Sites of Lucknow

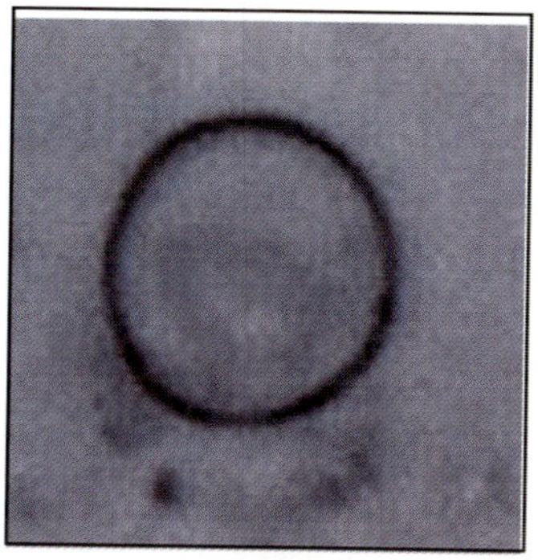

Cyst

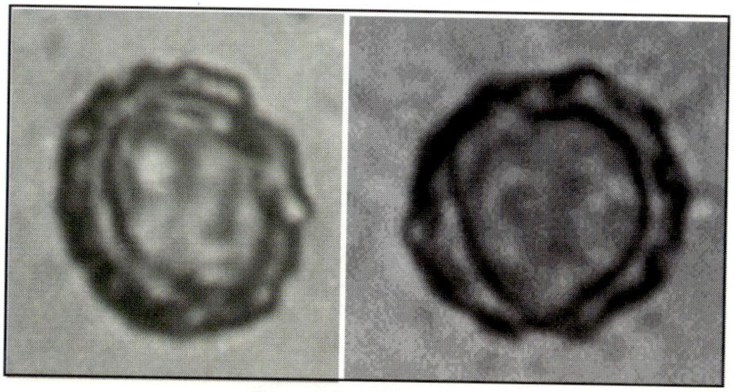

Trophozoites

***Naegleria* spp.**

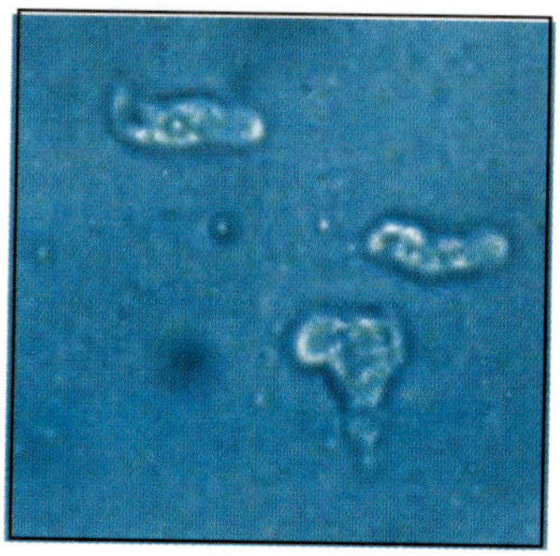

Cyst

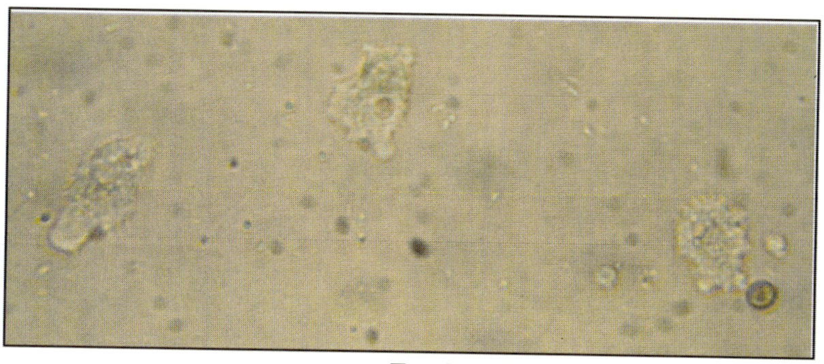

Trophozoites

***Acanthamoeba* spp.**

Plate 89.1

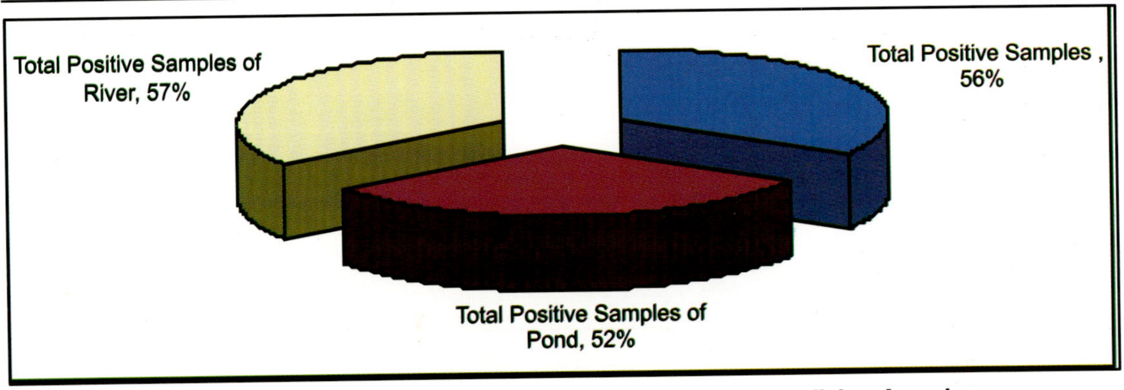

Figure 89.2: Percentage of Positive Water Samples for Free-living Amoebae from River and Pond of Lucknow

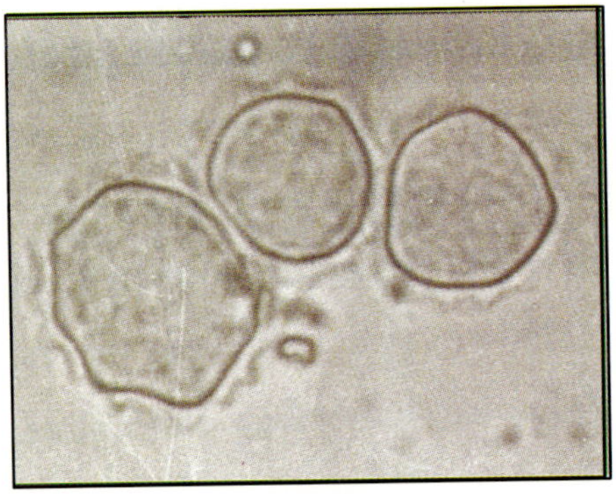

Cyst

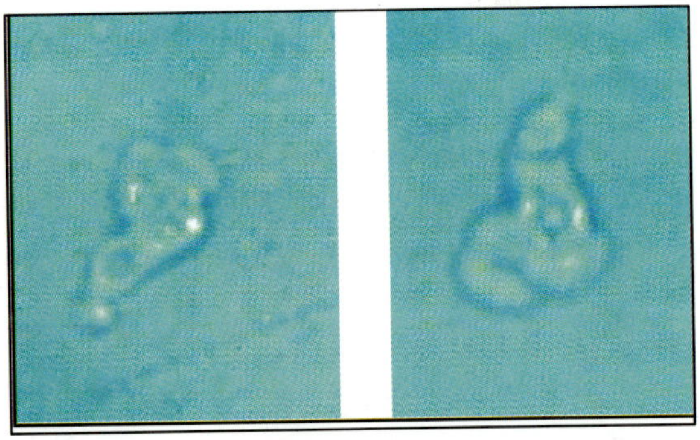

Trophozoites

***Schizopyrenus* spp.**

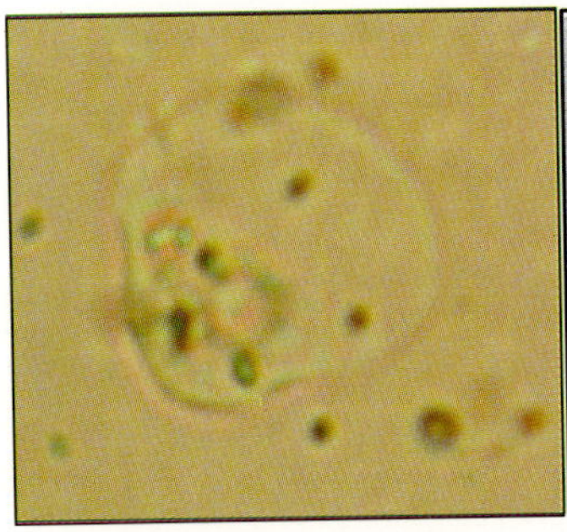

Trophozoites *Vanella*

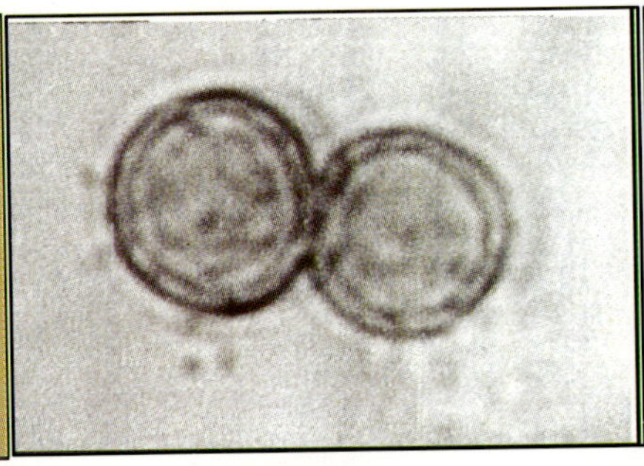

Cyst

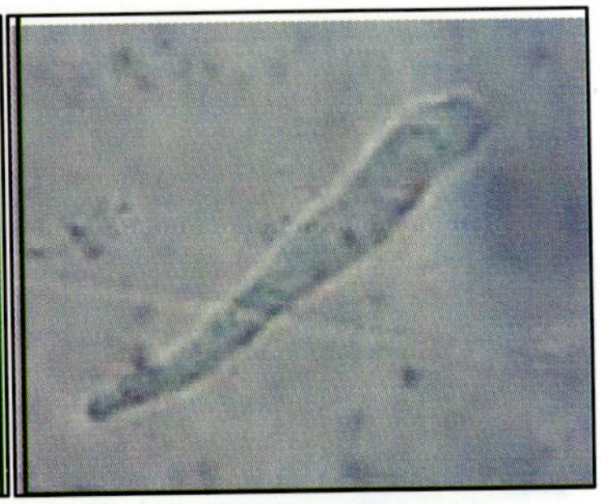

Trophozoites

***Hartmennella* spp.**

Plate 89.2

Conclusion

These findings serve as additional evidence for the presence of either pathogenic or non-pathogenic free-living amoebae responsible for human diseases under natural conditions, therefore, these results may have profound implication with regard surveillance of water system for amoebae and especially for the prevention of diseases, either directly or indirectly. A greater awareness among masses is essential to educate about the new amoebic diseases and occurrence of these pathogenic amoebae in natural water sources like rivers, ponds, lake *etc.*

Table 89.1: Isolation of Amphizoic Amoebae from Water Resources (River and Pond) in Lucknow

Site Name	*Site Code*	*Water Sample*		*Amoeba*
River and Pond		*Total Sample*	*Positive Sample*	
Up-stream	R-1	9	6	A,N,H
Pakka pul	R-2	5	3	A,N,Vn
Daliganj pul	R-3	6	3	A,V,H
Hanuman setu	R-4	12	7	A,N,Scz
Nishat ganj Bridge	R-5	9	5	A,N,H,Vn
Down–stream	R-6	6	3	N,A
Basawan Purwa pond	P-1	5	3	A,N,H
Raith pond	P-2	7	4	A,N,Vn
Dhatingara pond	P-3	5	2	A,N,Scz
Hussainabad pond	P-4	4	2	A,N,Scz
Total		68	38	A,N,Scz,H,V

Note: N: *Naegleria*; Scz: *Schizopyrenus*; A: *Acanthamoeba*; Vn: *Vvanella*; H: *Hartmannella.*

Acknowledgement

We are thankful to the Head, Deptt. of Zoology, Prof. (Mrs.) M. Srivastava for providing necessary laboratory facilities to carry out this work.

References

Bedi, H.K., Devapura, J.C. and Bomb, B.S., 1972. Primar amoebic meningoencephalitis. *J. Indian Med. Assoc.*, 58: 13–14.

De Jonckheere, J.F., 1981. *Naegleria australiensis* sp. nov., another pathogenic *Naegleria* from water. *Protistologica*, 17: 423–429.

Denny, C.F., Iragui, V.J., Uber-Zak, L.D., Karpinski, N.C., Zeigler, E.J., Visvesvara, G.S. and Reed, S.L., 1997. Amoebic Meningoencephalitis caused by *Balamuthia mandrillaris*: Case report and review. *Clin. Infect. Dis.*, 25: 1354–1380.

Gogate, A., Singh, B.N., Deohar, L.P., and Jhala, H.I., 1984. Primary amoebic meningoencephalitis caused by *Acanthamoeba*. *J. Postgrad. Med.*, 30: 125–128.

Jain, R., Prabhakar, S., Modi, M., Bhatia, R. and Sehgal, R., 2002. *Naegleria* meningitis: A rare survival. *Neurology India*, 50: 470–472.

John, D.T. and Howard, M.J., 1995. Techinque for isolating thermotolerent and pathogenic free-living amoebae. *Folia Parasitol.*, 43: 267–271.

Kasprzak, W. and Mazur, T., 1974. Studies on some pathogenic strains of free-living amoebae isolated from lakes in Poland. *Ann. Soc. Belge. Med. Trop.*, 54: 351–357.

Kaushal, V., Chhina, D.K., Ram S., Singh G., Kaushal, R.K. and Kumar, R., 2008. Primary amoebic meningoencephalitis due to *Naegleria fowleri*. *JAPI*, 56: 459–462.

Kilvington, S., Gray, T., Dart, J., Morlwt, N., Beeching, J.R., Frazer, D.G. and Matheson, M., 2004. *Acanthamoeba keratitis* the role of domestic tap water contamination in the United Kingdom. *Invest. Opthalmol. Vis. Sci.*, 45: 165–169.

Levine, N.D., Corless, J.O., Cox, F.E.G., Deroux, G., Grain, J., Honigberg, B.M., Leedale, G.F., oeblick, A.R., Lom, J., Lynn, D., Merinfeld, E.G., Page, F.C., Polijansky, G., Sprague, V. Vavra, J. and Wallance, F.C., 1980. A newly revised classification of the protozoa. *J. Protozool.*, 27: 37–58.

Martinez, A.J. and Visvesvara, G.S., 1997. Free-living amphizoic and opportunistic amoebae (review). *Brain Pathology*, 7: 583–598.

Nacapunchai, D., Lamom, C., Ruangsittichai, C. and Sriwichai, P., 1999. Isolation of free living amoebae from Soil and water resources in Thailand. *J. Trop. Med. Parasitol.*, p. 22–25.

Page, F.C., 1974. A further study of taxonomic criteria for limax amoebae with description of new species and a key to the genera. *Arch. Protistenk.*, 116:149–184.

Pan, N.R. and Ghosh, T.N., 1971. Primary amoebic meningoencephalitis in two Indian children. *J. Indian Med. Assoc.*, 56: 134–137.

Pandey, R. and Sharma, A.K., 2006. Isolation of pathogenic *Naegleria fowleri* and non cyst forming free-living amoebae *Vanella mira* and *Flabellula calkinsi* from domestic water supply U.P. India. *Lucknow Journal of Science*, 3–1: 27–32.

Rohr, U., Weber, S., Michel, R., Selenka, F. and Wilhelm, M., 1998. Comparison of free-living amoeba in hot water system of hospitals with isolates from moist sanitary areas by identifying genera and determining temperature tolerance. *Appl. Environ. Microbiol.*, 64: 1822–1824.

Sheehan, K.B., Michael, F.J. and Henson, J.M., 2003. Detection of *Naegleria* sp. in a thermal acidic stream in Yellowstone National Park. *The Journal of Eukaryotic Microbiology*, 50(4): 263–265.

Shenoy, S., Godwin W., Prashanth, H.V., Vidyalakshmi, K., Dhanashree, B. and Bharath, R., 2002. Primary meningoencephalitis by *Naegleria fowleri*; first reported case from Mangalore, South India. *J. Clin. Microbiol.*, 40(1): 309–310.

Singh, B.N. and Das, S.R., 1972a. Occurrence of pathogenic *Naegleria aerobia*, *Hartmannella culbertsoni* and *H. rhysodes* in sewage sludge samples of Lucknow. *Curr. Sci.*, 41: 277–281.

Singh, B.N., 1985. Free-living soil amoebae as human pathogen. *Proc. Indian Nat. Sci. Acad.*, B51(3): 297–312.

Sison, J.P., Kemper, C.A., Loveless, M., McShane, D. and Visvesvara, G.S., 1995. Disseminated *Acanthamoeba* infection in patients with AIDS: Case report and review. *Clin. Infect. Dis.*, 20: 1207–1216.

Water and Wastewater Management in 2 Vols (2011) *Pages 595–610*
Editors: **D.R. Khanna, R. Bhutiani & Gagan Matta**
Published by: **DAYA PUBLISHING HOUSE, NEW DELHI**

Chapter 90

Water Quality Analysis of River Panv Dhoi in Reference to its Physico-chemical Parameters and Heavy Metals

D.R. Khanna[1], J. Ashraf[2], Beena Chauhan[2], R. Bhutiani[1], Gagan Matta[1] and V. Singh[1]
[1]*Department of Zoology, Gurukul Kangri University, Haridwar*
[2]*Department of Zoology, M.S. College, Saharanpur, U.P.*

ABSTRACT

This paper deals with the analysis of different water parameters of River Panv Dhoi which flows through Saharanpur district. It is a stream fed river and a tributary of Hindon. The sample collection was usually completed during morning hrs. between 8:00 AM to 10:00 AM. The parameters like Temperature, Turbidity, Conductivity, Total Solids (TS), Total Suspended Solids (TSS), Total Dissolved Solids (TDS), pH, Dissolved Oxygen (DO), Free Carbon dioxide (CO_2), Biochemical Oxygen Demand (BOD), Chemical Oxygen Demand (COD), Alkalinity, Total Hardness, Chloride (Cl^-) and heavy metals like Lead, Zinc, Mercury and copper were analyzed.

Keywords: Water, River, Panv Dhoi, Saharanpur, Heavy metals.

Introduction

Water is the most precious commodity of life. It is not only the basic need for sustaining human life but also vital to all the segments of economic development. Water is being adversely affected, qualitatively and quantitatively by all kinds of human activities on land, in air and/or in water.

River Panv Dhoi flows through Saharanpur district. It is a streamfed river and a tributary of Hindon. This river originates near Shanklapuri Shiv Mandir of Panwarka, than it goes to Saharanpur.

It is about 15 km in length and then it mixes with river Dhamola (Figure 90.1). Metallic scraps, metallic dust, oil, paper and wooden products, besides domestic garbage appear to be the main items of the sewage. Main drain carries away some effluent from the factories and also from the residential colonies. This drain carries domestic sewage, which is poured into river Panv Dhoi. Therefore, it can be well considered that it carries a variety of pollutants of equally different in physico-chemical nature. A lot of work has been done to evaluate the impact of human interferences on different water bodies by Khanna *et al.* (I 997), Khanna *et al.* (2000), Khanna and Bhutiani (2003) and Khanna and Chugh (2004) and many more. It is very much essential to monitor the water quality of different water bodies present to assess their suitability for different purposes. Therefore the physico-chemical parameters of river Panv Dhoi were analysed for different seasons during 2002–2003.

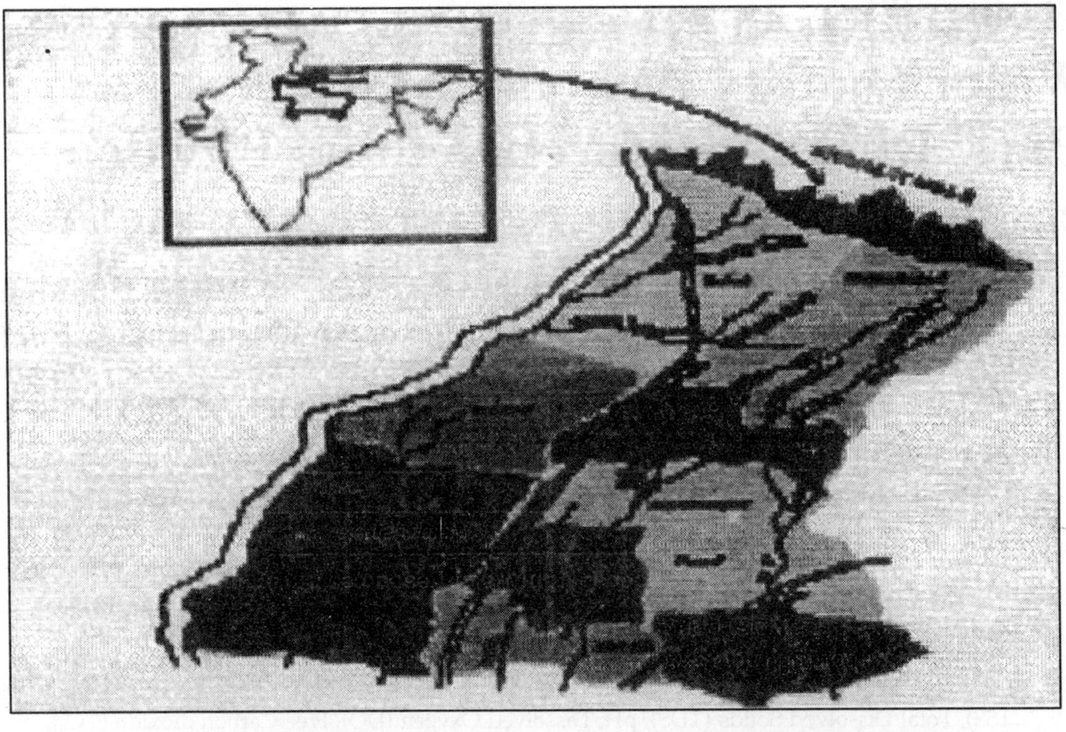

Figure 90.1: Map of District Saharanpur Showing the Situation of Panv Dhoi River

Materials and Method

The samples for different parameters were analysed with the help of the procedure described by Welch (1948), APHA(1980), Mathur (1982), Ross (1983), Trivedi and Goel (1984) and Khanna (1993). The water samples were collected from five different sampling sites Shanklapuri Shiv Mandir (A), Makhraj Ka Pul (B), Laldas Ka Baada (C), Jogyan Pul (D) and Near Dhamola (E). The sample collection was usually completed during morning hrs. between 8:00 A.M. to 10:00 A.M. The parameters like Temperature, Turbidity, Conductivity, Total Solids (TS), Total Suspended Solids (TSS), Total Dissolved Solids (TDS), pH, Dissolved Oxygen (DO), Free Carbon dioxide (CO_2), Biochemical Oxygen Demand (BOD), Chemical Oxygen Demand (COD), Alkalinity, Total Hardness, Chloride (Cl^-) and heavy metals like Lead, Zinc, Mercury and copper were analyzed.

Results and Discussion

In an aquatic ecosystem the physico-chemical parameters are the main important factors responsible for the biotic healthiness of the system. They show their effect individually and also collectively. The values of different physico-chemical parameters in different seasons were tabulated in the Tables 90.1–90.10 and are graphically represented in Figures 90.2–90.11.

The temperature is one of the most important parameter in an aquatic environment. In the present study, a difference in the fluctuation of water temperature was observed 10.4±1.5 (minimum) in the winter season and 28.0±0.5 (maximum) in monsoon season. Annual average values of temperature varied between 15 .6°C±4.5 to 25.0°C±4.4. Minimum annual average value of temperature was observed at sampling station A and maximum was observed at sampling station E. The water temperature showed an upward trend from winter season to summer season followed by downward from monsoon season onwards. A more or less similar trend has been observed in the river Yamuna by Chakrabarty *et al.* (1959) and in the Kallayi (John, 1976). Badola and Singh (1981) also reported similar trend in the river Alaknanda. Same trend of temperature was observed by Khanna *et al.* (2001) in river Ganga at Haridwar.

The present study showed conductivity fluctuation of 65.10 μmhos/cm±2.98 in winter season to 336.78 μmhos/cm±0.58 in monsoon season. Annual average values of conductivity varied between 142.07 μmhos/cm±104.39 to 191.44 μmhos/cm±127.32. Minimum annual average value of conductivity was observed at sampling station E and maximum was observed at sampling station A. Identical results were observed by Raina *et al.* (1982) and Mittal and Sengar (1990) from various Indian rivers. Khanna *et al.* (2003) also observed similar trend of conductivity in Ganga river at Bulandshahar.

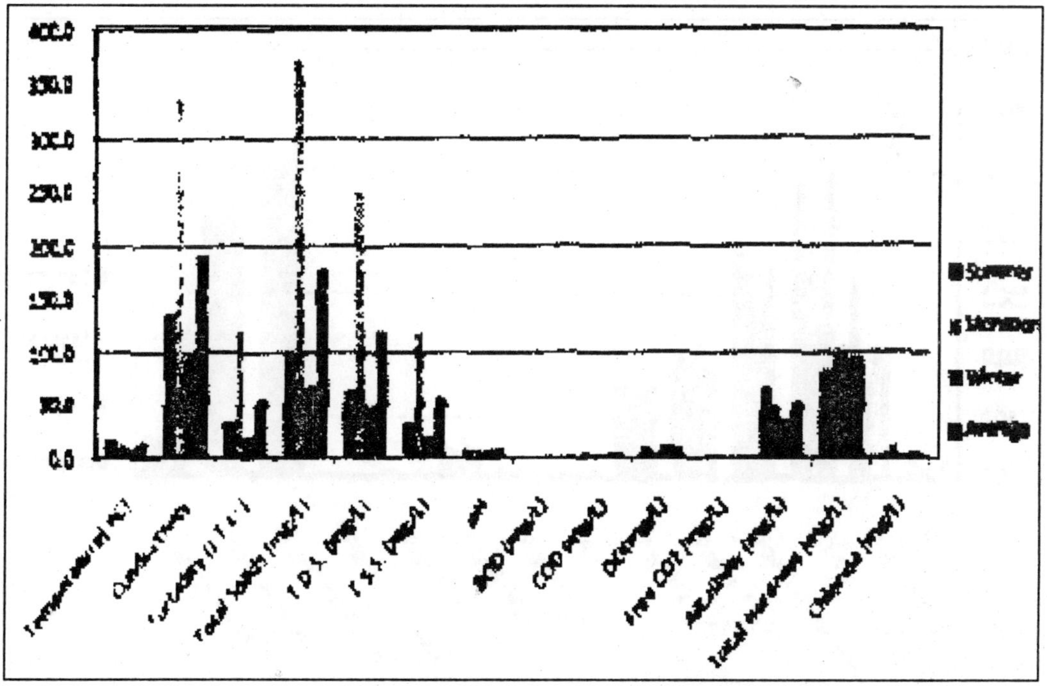

Figure 90.2: Seasonal Variation in Physico-chemical Parameters of River Panv Dhoi at Sampling Station A (2002–2003)

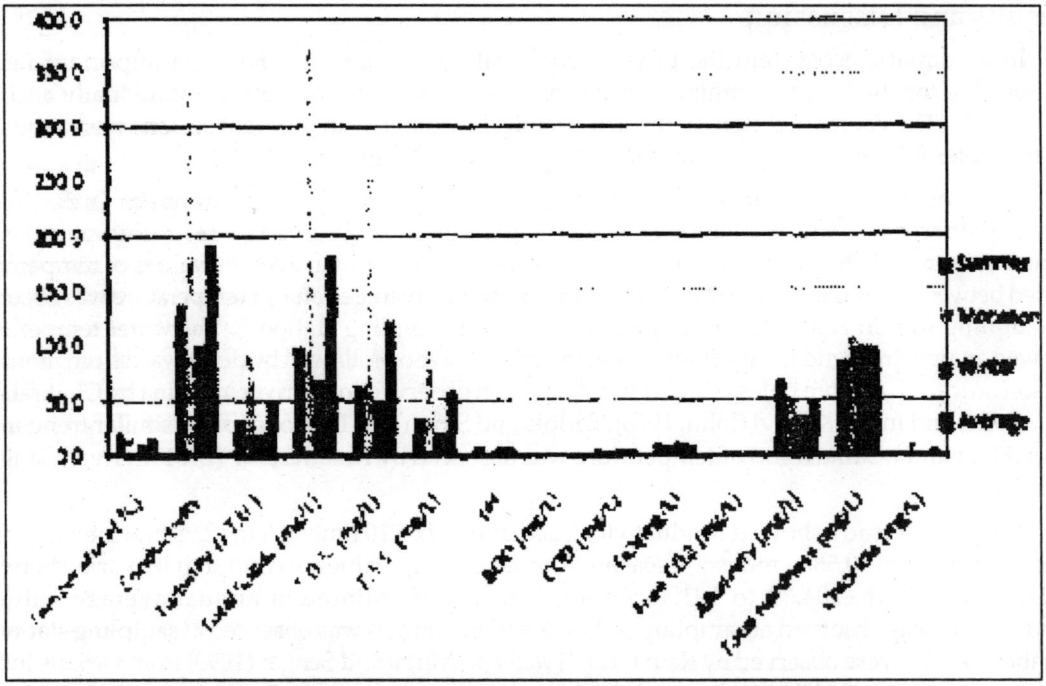

Figure 90.3: Seasonal Variation in Physico-chemical Parameters of River Panv Dhoi at Sampling Station B (2002–2003)

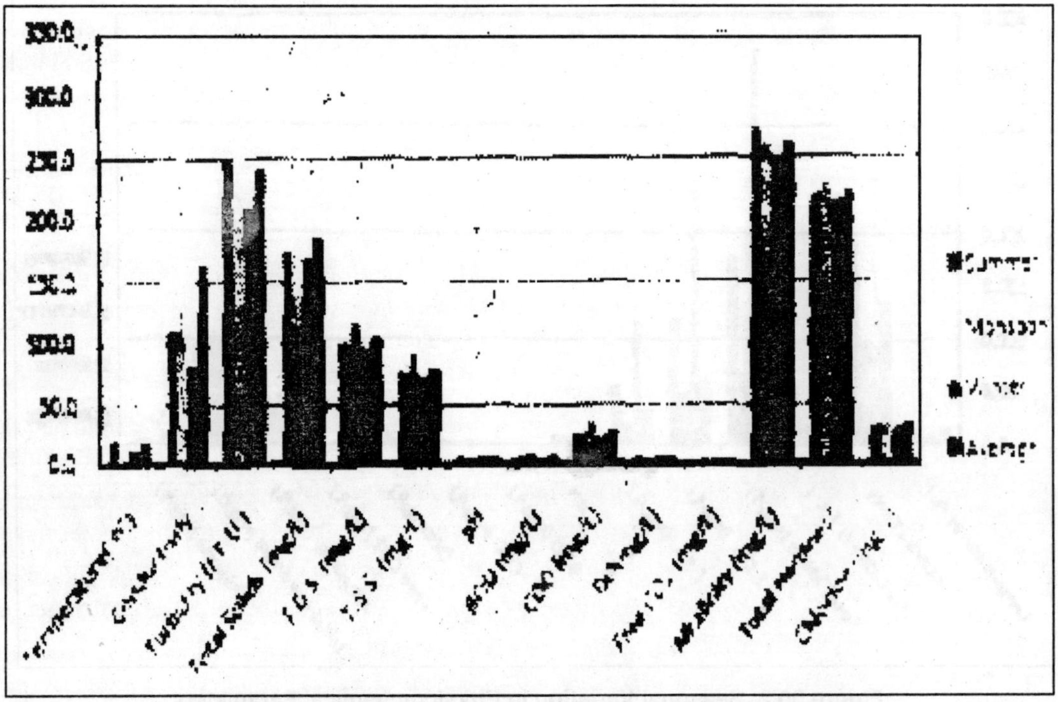

Figure 90.4: Seasonal Variation in Physico-chemical Parameters of River Panv Dhoi at Sampling Station C (2002–2003)

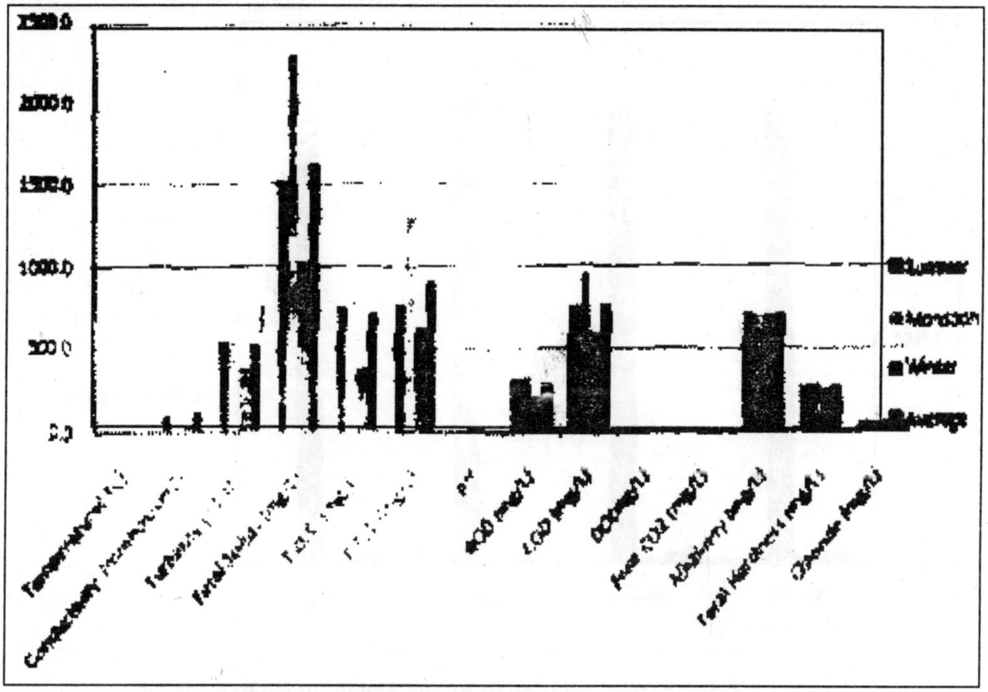

Figure 90.5: Seasonal Variation in Physico-chemical Parameters of River Panv Dhoi at Sampling Station D (2002–2003)

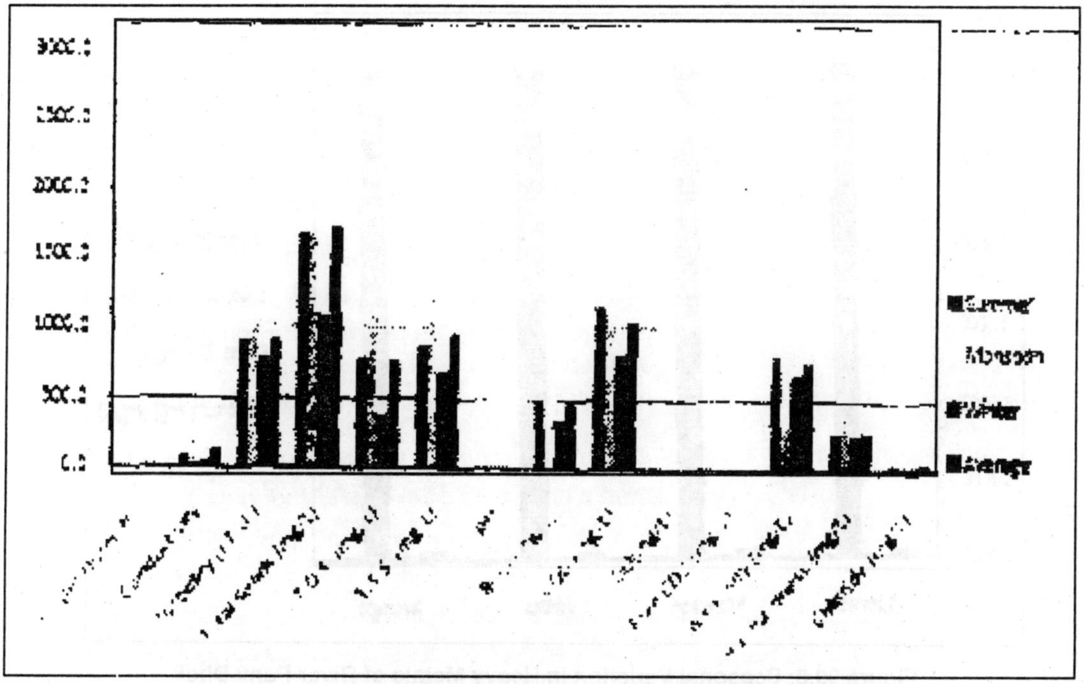

Figure 90.6: Seasonal Variation in Physico-chemical Parameters of River Panv Dhoi at Sampling Station E (2002–2003)

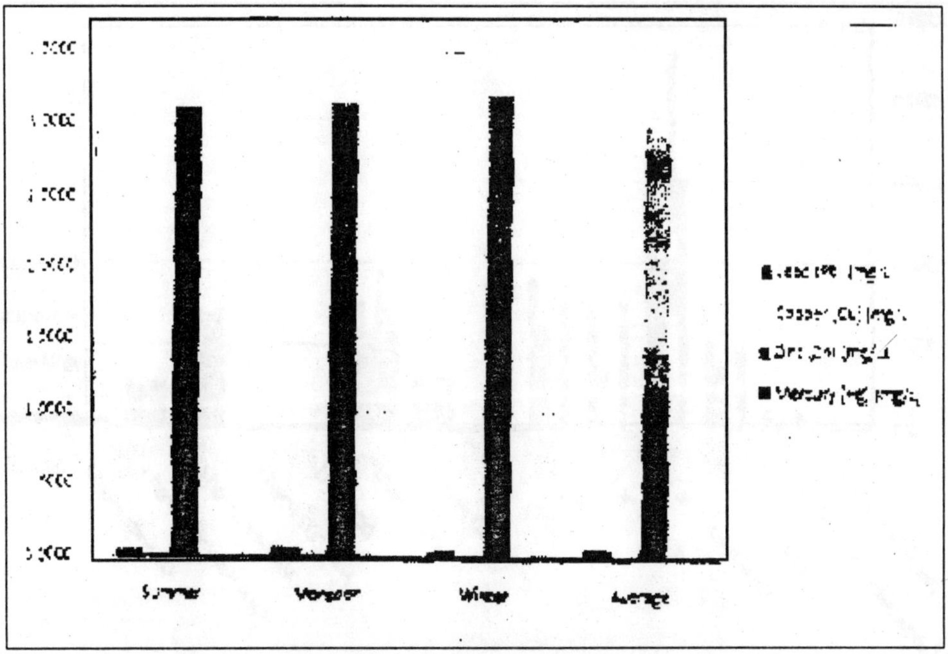

Figure 90.7: Seasonal Variation in Heavy Metals of River Panv Dhoi at Sampling Station A (2002–2003)

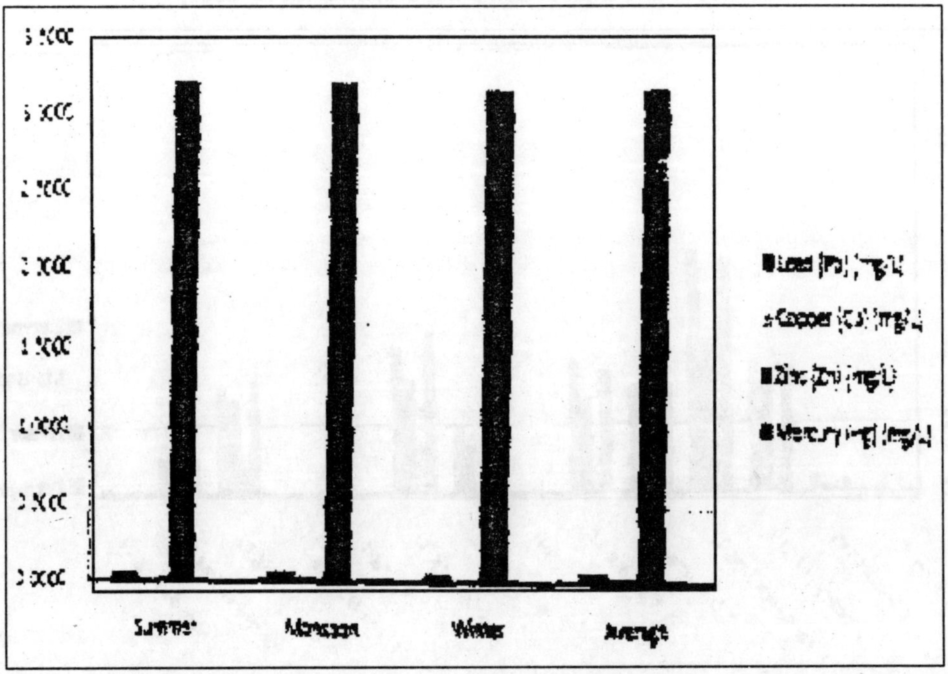

Figure 90.8: Seasonal Variation in Heavy Metals of River Panv Dhoi at Sampling Station B (2002–2003)

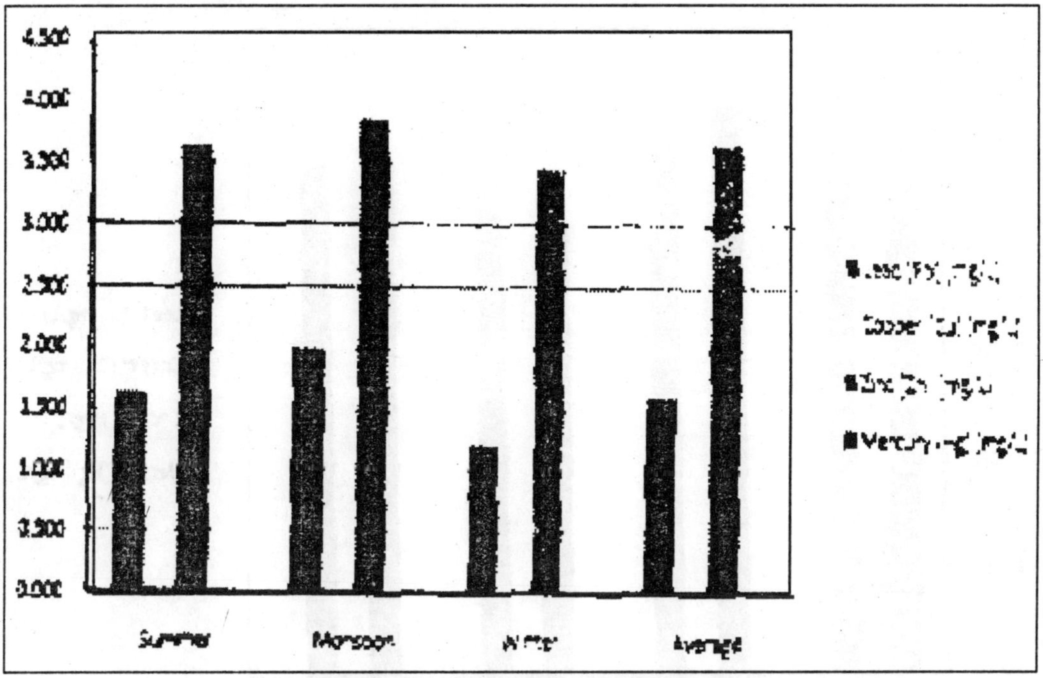

Figure 90.9: Seasonal Variation in Heavy Metals of River Panv Dhoi at Sampling Station C (2002–2003)

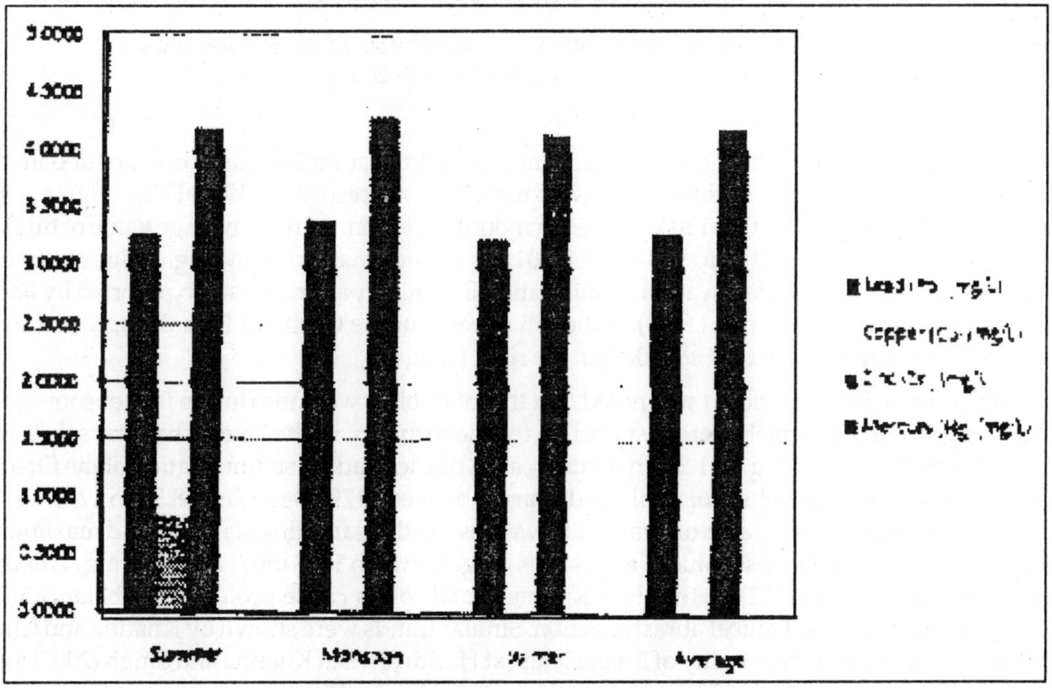

Figure 90.10: Seasonal Variation in Heavy Metals of River Panv Dhoi at Sampling Station D (2002–2003)

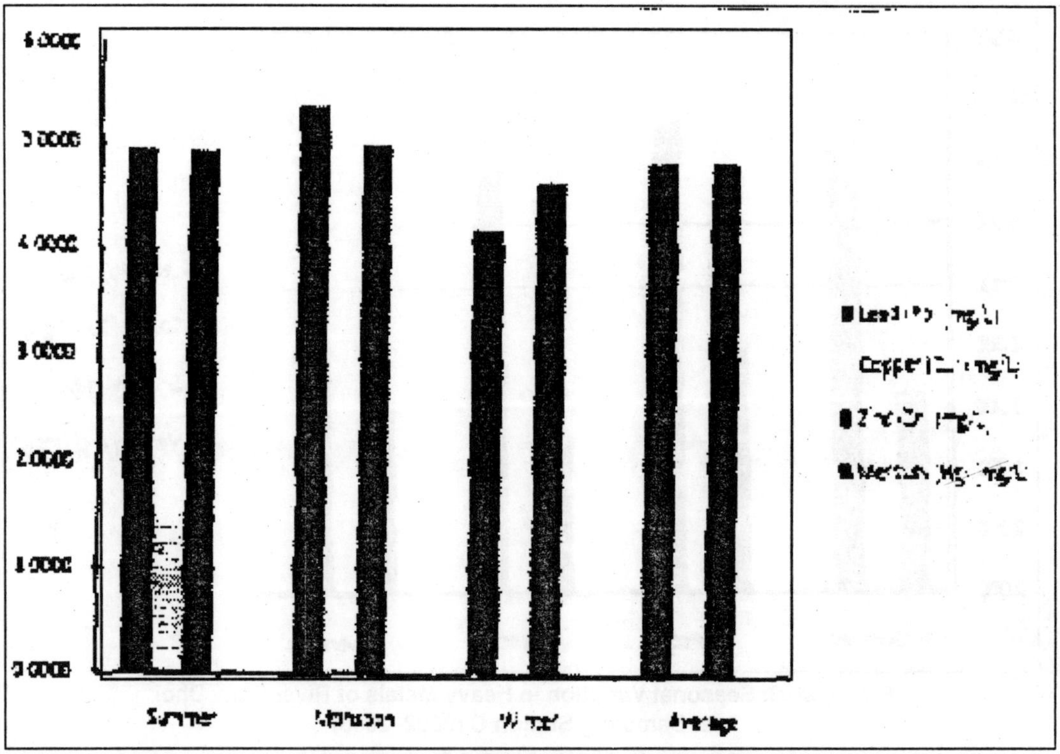

Figure 90.11: Seasonal Variation in Heavy Metals of River Panv Dhoi at Sampling Station E (2002–2003)

The water of the river Panv Dhoi becomes start turbid from summer season onward and in rainy season the water was highly turbid. The lowest turbidity was recorded (19.78 JTU±1.00) in winter season and highest (1000.00 JTU±150.21) noted in monsoon season. Annual average value of turbidity varied between 49.58 JTU±40.44 to 908.46 JTU±101.05 in which maximum average value observed at sampling station E and minimum at sampling station B. Similar pattern was also reported by Badola and Singh (1981), Dobriyal *et al.* (1983) in the hill streams of the Garhwal Himalaya, Khanna *et al.* (1997) in river Ganga, and Seth *et al.* (2000) in the river Ganga.

In the present investigation it was noted that the total solids were maximum in monsoon season (2400.00 mg/l±226.63), which were responsible for the turbidity in the river. The total solids were recorded minimum (68.44 mg/l±1.36) in winter season due to gradual sedimentation of the filterable residue. Annual average values of total solids varied between 179.53 mg/l±168.53 to 1726.94 mg/l±651.22. Minimum average value of total solids was observed at sampling station A and maximum at sampling station E. The highest value of total solids range between 1873 mg/l and 3573 mg/l reported by Kumar and Sharma (2002) in the river Krishni. Total solids cause ecological imbalance in the aquatic ecosystem by mechanical abrasive action. Similar trends were shown by Khanna and Chugh (2004) during study of water quality of River Ganga at Haridwar. But Khanna and Singh (2000) found that total solids were maximum in summer in the water of Suswa River.

Table 90.1: Seasonal Variation in Physico-chemical Parameters of River Panv Dhoi at Sampling Station A (2002–2003)

Physico-chemical Parameters	*2002–2003*			
	Summer	*Monsoon*	*Winter*	*Average*
Temperature (°C)	18.0±1.3	18.4±1.3	10.4±1.5	15.6±4.5
Conductivity (μmhos/cm)	138.56±0.40	336.78±0.58	99.000.56	191.44±127.32
Turbidity (JTU)	33.22±0.36	2120.28±0.37	19.83±0.45	57.78±54.54
Total Solids (mg/l)	96.70±1.35	373.45±1.26	68.44±1.36	179.53±168.53
TDS (mg/l)	64.13±1.41	252.56±1.48	48.33±1.57	121.67±113.62
TSS (mg/l)	32.57±0.04	120.89±0.08	20.11±0.11	57.85±54.94
pH	7.30±0.31	7.57±0.29	7.07±0.26	7.32±0.25
BOD (mg/l)	2.30±0.27	3.08±0.35	1.39±0.17	2.26±0.82
COD (mg/l)	5.02±0.29	7.36±1.09	4.69±0.18	5.69±1.45
DO (mg/l)	9.52±0.34	7.34±1.11	11.78±0.25	9.55±2.18
Free CO_2 (mg/l)	2.02±0.29	3.98±0.19	0.07±0.01	2.02±1.95
Alkalinity (mg/l)	66.67±1.34	48.39±1.22	35.00±1.38	50.02±15.76
Total Hardness (mg/l)	83.78±0.56	107.20±.71	96.00±0.85	95.66±11.71
Chloride (mg/l)	2.96±1.57	11.58±1.90	1.97±1.66	5.51±5.29

Table 90.2: Seasonal Variation in Physico-chemical Parameters of River Panv Dhoi at Sampling Station B (2002–2003)

Physico-chemical Parameters	*2002–2003*			
	Summer	*Monsoon*	*Winter*	*Average*
Temperature (°C)	19.8±1.6	20.2±1.2	12.6±0.7	17.6±4.2
Conductivity (μmhos/cm)	138.15±1.00	336.61±31.12	99.58±7.04	191.44±127.18
Turbidity (JTU)	33.34±1.92	95.62±0.63	19.78±1.00	49.58±40.44
Total Solids (mg/l)	97.75±1.82	377.55±14.51	70.47±1.21	181.92±159.96
TDS (mg/l)	65.17±1.74	254.57±10.79	49.35±4.98	123.03±114.19
TSS (mg/l)	32.58±1.98	122.98±2.03	21.12±1.23	58.88±55.79
pH	7.35±0.90	7.59±0.28	7.09±0.05	7.34±0.25
BOD (mg/l)	2.88±0.62	3.18±0.58	1.57±0.35	2.54±0.88
COD (mg/l)	5.52±0.98	6.36±1.67	4.98±0.79	5.82±0.69
DO (mg/l)	9.65±0.72	8.18±1.02	11.01±0.58	9.61±1.41
Free CO_2 (mg/l)	2.08±0.06	4.20±0.12	1.24±0.97	2.56±1.52
Alkalinity (mg/l)	67.67±1.72	49.08±1.21	35.94±0.76	50.89±15.94
Total Hardness (mg/l)	84.01±1.67	108.38±2.61	96.94±0.76	96.44±12.19
Chloride (mg/l)	3.15±0.19	12.05±0.99	2.37±1.22	5.85±5.37

Table 90.3: Seasonal Variation in Physico-chemical Parameters of River Panv Dhoi at Sampling Station C (2002–2003)

Physico-chemical Parameters	*2002–2003*			
	Summer	*Monsoon*	*Winter*	*Average*
Temperature (°C)	22.1±0.7	24.2±0.9	14.2±0.6	20.2±5.2
Conductivity (μmhos/cm)	111.26±6.27	300.24±15.25	85.20±1.79	165.56±117.35
Turbidity (JTU)	250.00±1.34	270.00±9.50	210.00±4.56	243.33±30.55
Total Solids (mg/l)	176.22±13.57	213.32±9.28	170.26±2.87	186.60±23.33
TDS (mg/l)	99.20±5.42	120.28±14.26	98.26±6.21	105.24±13.10
TSS (mg/l)	77.02±4.31	93.04±2.74	74.00±1.28	81.53±10.23
pH	7.48±0.18	7.62±0.20	7.40±0.10	7.50±0.09
BOD (mg/l)	8.44±0.63	11.16±0.29	5.67±0.12	8.42±2.74
COD (mg/l)	26.30±1.26	35.40±1.04	22.65±0.44	28.11±6.56
DO (mg/l)	7.85±0.27	6.56±0.45	8.27±0.89	7.56±0.89
Free CO_2 (mg/l)	2.21±0.27	4.12±0.39	1.75±0.34	2.69±1.25
Alkalinity (mg/l)	272.98±9.02	261.55±2.54	249.00±1.04	261.17±11.99
Total Hardness (mg/l)	221.00±1.39	229.32±4.58	215.51±7.21	221.94±6.96
Chloride (mg/l)	32.72±1.85	39.39±1.09	31.26±0.77	34.45±4.33

Table 90.4: Seasonal Variation in Physico-chemical Parameters of River Panv Dhoi at Sampling Station D (2002–2003)

Physico-chemical Parameters	*2002–2003*			
	Summer	*Monsoon*	*Winter*	*Average*
Temperature (°C)	26.0±0.6	26.9±0.6	17.0±0.7	23.4±5.5
Conductivity (μmhos/cm)	101.90±1.02	274.78±6.84	70.20±2.22	148.96±110.11
Turbidity (JTU)	550.80±9.95	650.70±17.12	400.24±6.27	533.93±12.08
Total Solids (mg/l)	1536.42±41.23	2315.28±112.21	1030.50±79.62	1627.40±647.20
TDS (mg/l)	755.62±22.32	1015.28±14.21	390.50±7.96	720.46±313.86
TSS (mg/l)	760.80±65.83	1300.00±12.04	640.00±29.32	906.93±347.60
pH	8.80±0.24	8.40±0.15	8.00±0.10	8.42±0.40
BOD (mg/l)	315.00±6.94	325.00±10.06	205.00±14.28	281.66±66.58
COD (mg/l)	750.00±45.21	950.00±40.75	605.00±20.29	766.33±173.22
DO (mg/l)	4.90±0.098	4.00±0.34	5.10±0.45	4.66±0.58
Free CO_2 (mg/l)	3.74±1.22	5.38±0.46	2.67±0.94	3.93±1.36
Alkalinity (mg/l)	715.00±47.68	710.25±22.98	708.00±38.12	711.08±3.57
Total Hardness (mg/l)	272.50±6.74	294.00±2.34	261.00±41.08	275.83±16.75
Chloride (mg/l)	58.37±14.78	60.00±8.96	53.12±21.49	57.16±3.59

Table 90.5: Seasonal Variation in Physico-chemical Parameters of River Panv Dhoi at Sampling Station E (2002–2003)

Physico-chemical Parameters	*2002–2003*			
	Summer	*Monsoon*	*Winter*	*Average*
Temperature (°C)	27.0±0.7	28.0±0.5	20.0±0.6	25.0±4.4
Conductivity (μmhos/cm)	100.20±4.27	260.90±8.42	65.10±2.98	142.07±104.39
Turbidity (JTU)	925.22±69.32	1000.00±150.21	800.00±22.34	908.46±101.05
Total Solids (mg/l)	1680.82±167.34	2400.00±226.63	1100.00±175.00	1726.94±651.22
TDS (mg/l)	800.00±26.82	1100.00±55.00	400.00±12.05	766.66±351.18
TSS (mg/l)	880.82±14.99	1300.00±87.73	700.00±63.29	960.26±307.78
pH	8.85±0.18	8.88±0.19	8.22±0.10	8.65±0.37
BOD (mg/l)	600.00±35.20	530.00±15.98	375.00±14.71	468.33±80.96
COD (mg/l)	1160.00±0.89	1240.00±145.26	825.00±68.74	1075.00±220.17
DO (mg/l)	2.02±0.24	0.07±0.01	2.96±0.19	1.68±1.47
Free CO_2 (mg/l)	4.68±0.64	6.24±0.96	3.12±0.72	4.74±1.56
Alkalinity (mg/l)	820.00±55.27	815.00±45.72	710.00±41.38	781.66±62.11
Total Hardness (mg/l)	276.40±18.45	390.82±26.71	262.00±22.94	279.74±17.27
Chloride (mg/l)	60.54±2.50	68.32±4.56	50.17±5.04	59.67±3.16

Table 90.6: Seasonal Variation in Heavy Metals of River Panv Dhoi at Sampling Station A (2002–2003)

Heavy Metals	*2002–2003*			
	Summer	*Monsoon*	*Winter*	*Average*
Lead (Pb) (mg/l)	0.0680±0.0072	0.0788±0.0070	0.0638±0.0072	0.0702±0.0076
Copper (Cu) (mg/l)	0.0695±0.0067	0.0740±0.0069	0.0214±0.0022	0.0560±0.0292
Zinc (Zn) (mg/l)	3.1224±0.0467	3.1488±0.0513	3.2242±0.0536	3.1551±0.0528
Mercury (Hg) (mg/l)	0.0000±0.0000	0.0000±0.0000	0.0000±0.0000	0.0000±0.0000

Table 90.7: Seasonal Variation in Heavy Metals of River Panv Dhoi at Sampling Station B (2002–2003)

Heavy Metals	*2002–2003*			
	Summer	*Monsoon*	*Winter*	*Average*
Lead (Pb) (mg/l)	0.0697±0.0077	0.0790±0.0072	0.0647±0.0065	0.0711±0.0073
Copper (Cu) (mg/l)	0.0739±0.0284	0.0614±0.0301	0.0324±0.0324	0.0626±0.0263
Zinc (Zn) (mg/l)	3.225±0.0265	3.2280±0.0251	3.1810±0.0231	3.2109±0.0259
Mercury (Hg) (mg/l)	0.0000±0.0000	0.0000±0.0000	0.0000±0.0000	0.0000±0.0000

Table 90.8: Seasonal Variation in Heavy Metals of River Panv Dhoi at Sampling Station C (2002–2003)

Heavy Metals	*2002–2003*			
	Summer	*Monsoon*	*Winter*	*Average*
Lead (Pb) (mg/l)	1.645±0.3830	1.982±0.4136	1.224±0.2184	1.6170±0.3797
Copper (Cu) (mg/l)	0.0872±0.0181	0.0985±0.0186	0.0649±0.0161	0.0835±0.017
Zinc (Zn) (mg/l)	3.6527±0.2056	3.8928±0.2132	3.4812±0.2001	3.6755±0.2067
Mercury (Hg) (mg/l)	0.0000±0.0000	0.0000±0.0000	0.0000±0.0000	0.0000±0.0000

Table 90.9: Seasonal Variation in Heavy Metals of River Panv Dhoi at Sampling Station D (2002–2003)

Heavy Metals	*2002–2003*			
	Summer	*Monsoon*	*Winter*	*Average*
Lead (Pb) (mg/l)	3.2755±0.0796	3.3972±0.08120	3.2527±0.0612	3.3084±0.0776
Copper (Cu) (mg/l)	0.8475±0.1761	0.9976±0.1810	0.6474±0.161	0.8308±0.1756
Zinc (Zn) (mg/l)	4.2186±0.01230	4.3270±0.0112	4.1874±0.0149	4.2443±0.0732
Mercury (Hg) (mg/l)	0.0001±0.0000	0.0002±0.0000	0.0000±0.0000	0.0000±0.0000

Table 90.10: Seasonal Variation in Heavy Metals of River Panv Dhoi at Sampling Station E (2002–2003)

Heavy Metals	*2002–2003*			
	Summer	*Monsoon*	*Winter*	*Average*
Lead (Pb) (mg/l)	4.9492±0.6260	5.3975±0.7123	4.1889±0.5600	4.8452±0.6109
Copper (Cu) (mg/l)	1.5737±0.3631	1.9774±0.3996	1.2324±0.2998	1.5945±0.3729
Zinc (Zn) (mg/l)	4.9384±0.1796	5.0012±0.1814	4.6692±0.1616	4.8695±0.1763
Mercury (Hg) (mg/l)	0.0002±0.0004	0.0009±0.0007	0.0004±0.0005	0.0005±0.0006

Minimum values of total dissolved solids were obtained during winter season (48.3 3 mg/l±1.57) and maximum values during monsoon season (1100.00 mg/l±55.00). Annual average value of total dissolved solids varied between 105.24 mg/l±13.10 to 766.66 mg/l±351.18, in which minimum average value was obtained from sampling station C and maximum from sampling station E. Same study was also done by Khanna and Chugh (2004) in the river Ganga at Haridwar.

Minimum values of total suspended solids were obtained during winter season (20.11 mg/l±0.11) and maximum values during monsoon season (1300.00 mg/l±120.04). Annual average value of total suspended solids varied between 57.85 mg/l±54.94 to 960.26 mg/l±307.78, in which minimum average value was obtained from sampling station A and maximum from sampling station E. Same study was also done by Khanna and Chugh (2004) in the river Ganga at Haridwar.

The Panv Dhoi river at Saharanpur showed high pH value (8.88±0.19) in monsoon season which might be due to high chemical load in the river. The minimum pH values (7.07±0.26) were observed in winter season. The annual average values of pH varied between 7.32±0.25 to 8.65±0.37. Highest annual average value was recorded at sampling station E and minimum at station A. It was recorded during this study that pH of the river Panv Dhoi is slightly alkaline in nature. Identical results were reported by Sangu and Sharma (1985) in the river Yamuna and Khanna *et al.* (1999) in Ganga river. Khanna *et al.* (2001) in the river Ganga at Haridwar and Khanna *et al.* (2003) in the river Ganga at Bulandshahar have also shown the alkaline nature of river.

The biochemical oxygen demand was observed maximum (530 mg/l±15.98) in monsoon season and minimum (1.39 mg/l±.17) in winter season. The annual average value of biochemical oxygen demand ranged between 2.26 mg/l±0.82 to 468.33 mg/l±80.98. The minimum average value was found at sampling station A and maximum at sampling station E. Highest annual average value of bio-chemical oxygen demand at sampling station E may be due to drainage of several small sewage drains into the river and runoff of sludgy, silted sewage during months of rainy season. Khanna and Singh (2000) noticed peak values during summer in Suswa river and Khanna *et al.* (1997) observed peak values in monsoon season in river Ganga.

Chemical Oxygen Demand (COD) represents chemically oxidizable load of organic matter in water. It was noted highest (1240.00 mg/l±145.26) in monsoon season and minimum (4.69 mg/l±0.18) in winter season. The annual average value of COD ranged between 5.62 mg/l±0.69 to 1075.00 mg/l±220.17 whereas least average value was found at sampling station B and maximum at sampling station E. Similar trends of COD have shown by Khanna *et al.* (2002, 2003) in the river Ganga and Khanna and Singh (2000) in Suswa River at Raiwala.

Maximum dissolved oxygen was recorded (11.78 mg/l±0.25) in the winter season. The minimum value of dissolved oxygen (0.07 mg/l±0.01) was observed in monsoon season. The annual average value of dissolved oxygen ranged between 1.68 mg/l±1.47 to 9.61 mg/l±1.41, whereas the minimum annual average value of DO was observed at sampling station E and maximum was observed at sampling station B. The dissolved oxygen reduced gradually from summer onward due to turbidity which retarded the photosynthetic activity of aquatic flora. The temperature showed an inverse relationship with the DO almost throughout the study. The cause of maximum dissolved oxygen in winter is due to reduced rate of decomposition by decreased microbial activity at low temperature (Strommer and Smock, 1989). Chopra *et al.* (1990), Gopal and Sah (1993) and Sharma (1999) also have got the same result and have opined that low temperature in winter increases the oxygen retaining capacity of water and solubility of O_2 in water. This trend was also observed by Badola and Singh (1981) in the river Alaknanda. Khanna (1993, 2001) and Khanna and Chugh (2004) has also reported the same trends in the river Ganga at Haridwar. Free carbon dioxide was observed maximum (624 mg/l±0.96) in monsoon season due to higher turbidity and water temperature, but was recorded minimum (0.07 mg/l±0.01) in winter season. Annual average value of free carbon dioxide varied between 2.02 mg/l±1.95 to 4.74 mg/l±1.56 in which maximum average value obtained from sampling point E and minimum observed at sampling point A. Pahwa and Mehrotra (1966) and Ray *et al.* (1966) have reported that the Ganga river contains maximum free carbon dioxide in rainy season at Allahabad. Khanna *et al.* (1997) and Seth *et al.* (2000) have also reported the same trends in the river Ganga at Haridwar but Khanna and Singh (2000) observed maximum free carbon dioxide during summer in Suswa River at Raiwala, Dehradun.

Alkalinity of water is a measure of weak acid present in it and of the cations balanced against them (Sverdrup *et al.*, 1942). The highest concentration (820.00 mg/l±55.27) was observed in summer season and lowest (35.00 mg/l±1.38) in winter season. The annual average value of alkalinity varied between 50.02 mg/l±15.76 to 781.66 mg/l±62.11 in which maximum average value was obtained from sampling station E and minimum from sampling station A. Similar observation was also obtained by Khanna and Chugh (2004), Holden and Green (1960), Tallying and Rzoska (1967) and Abdin (1948).

Maximum values of total hardness were recorded in monsoon season (390.82 mg/l±26.71) and minimum was recorded in summer season (83.78 mg/l±0.56). The annual average values of total hardness ranged from 95.66 mg/l±11.71 to 279.74 mg/l±17.27 in which maximum average value is recorded from sampling station E and minimum at sampling station A. Khanna *et al.* (1993, 2003) and Mishra (2003) observed hardness in river Ganga at Haridwar and found more or less similar trends in their study.

The highest and lowest values of chloride were found in monsoon and winter season, 68.32 mg/l±4.56 and 1.97 mg/l±1.66 respectively. The annual average value of chloride varied between 5.51 mg/l±5.29 to 59.67 mg/l±3.16 in which maximum average value was recorded from the station E and minimum was at sampling station A. Chlorides are present in sewage, sewage effluents and farm drainage. Significant levels of chloride were shown by many rivers like Yamuna (Sengar *et al.*, 1985); Tungbhadra (Reddy and Venkateswarlu, 1987); Jhelum (Raina *et al.*, 1984) and Kshipra (Mishra and Saxena, 1984). CPCB (2003) reported the value of chloride in between 14 to 51 mg/l during Ganga monitoring from Bithur, Kanpur to Sangam, Allahabad.

The minimum value 0.0638 mg/l±0.0072 of lead was found from sampling station A and maximum 5.3975 mg/l±0.7123 from sampling station E. The annual average of lead (Pb) ranged between 0.0702 mg/l±0.0076 to 4.8452 mg/l±0.6109. The minimum value 0.0214 mg/l±0.0022 of copper was found from sampling station A and maximum 1.9774 mg/l±0.3996 from sampling station E. The annual average value of copper varied between 0.0550 mg/l±0.0292 to 1.5945 mg/l±0.3729 in which minimum average value of copper was obtained at sampling station A and maximum average value at sampling station E.

Hg is not required even in small amount by any organisms. Virtually all metals, including the essential metal micronutrients, are toxic if exposure levels are sufficient high. The increased circulation of toxic metals in recent times resulted in the unavoidable build up of such toxic substances in the human food chain. The minimum value 3.1224 mg/l±0.0467 of zinc was found from sampling station A and maximum 5.0012 mg/l±0.1814 from sampling station E. The annual average value of Zinc ranged between 3.1651 mg/l±0.0528 to 4.8695 mg/l±0.1763 in which minimum average value of Zinc was obtained from sampling station A and maximum was recorded at sampling station E. The annual average value of Mercury varied between 0.0000±0.0000 to 0.0005 mg/l±0.0006. The minimum was found as nil at sampling station A, B, C and maximum (0.0009 mg/l±0.0007) at sampling station E. The concentration of these metals at sampling station C, D and E gave a highly misleading picture of the degree of metal pollution. Khanna *et al.* (2003) also reported heavy metals in water of Ganga river at Bulandshahar.

References

Abdin, G., 1948. Physical and chemical investigations relating to algal growth in the river Nile, Cairo. *Bull. Inst., Egypt*, 29: 20–24.

Abiotic factors on the Phytoplanktonic population of a pond. *Env. Cons. J.*, 1(1): 41–46.

APHA, 1980. *Standard Methods for the Examination of Water and Wastewater*. American Public Health Association, 1015 Fifteen Street, NW Washington, 15: 1–1134.

Badola, S.P. and Singh, H.R., 1981. Fish and fisheries of river Alaknanda. *Proc. Nat Acad. Sci.*, 15(B): 133–142.

C.P.C.B., 2003. *A Report on Colour Problem of River Ganga.* Central Pollution Control Board, Zonal Kanpur, pp. 1–9.

Chakarbarty, R.D., Ray, P. and Singh, S.B., 1959. A quantitative survey of plankton and physiological conditions of the river Yamuna at Allahabad in 1954–1955. *Indian J. Fish.*, 6(1): 186–203.

Chopra, A.K., Madhwal, B.P. and Singh, H.R., 1990. Abiotic variable and Primary productivity of river Yamuna at Naugaon, Uttar Kashi, Garhwal. *Indian J. Ecol.*, 17(1): 61–64.

Dobriyal, A.K., Singh, H.R. and Bisht, K.L., 1983. Diurnal variation in hydrobiological parameters of two Hill streams of Garhwal Himalaya, India, Uttar Pradesh. *J. Zool.*, 3: 30–34.

Gopal, B. and Sah, M., 1993. Conservation and management of rivers in India; Case study of the Yamuna. *Environmental Conservation*, 20(2): 243–253.

Holden, J.M. and Green, J., 1960. Hydrobiology and plankton of the river Sokoto. *J. Anim. Ecol.*, 29(1): 65–84.

John, V., 1976. Hydrobiological studies on the river Kallayi in Kerala. *Indian J. Fish.*, 23: 72–85.

Khanna, D.R. and Bhutiani, Rakesh, 2003. Ecological status of Sitapur Pond at Haridwar (U.A.) India. *Indian J. Environ. and Ecoplan.*, 7(1): 175–178.

Khanna, D.R. and Chugh, T., 2004. *Microbial Ecology: A Study of River Ganga at Haridwar*, p. 1–273.

Khanna, D.R. and Singh, R.K., 2000. Seasonal fluctuations in the plankton of Suswa River at Raiwala (Dehradun). *Env. Cons. J.*, (2 and 3): 89–92.

Khanna, D.R., 1993. *Ecology and Pollution of Ganga River.* Ashish Publishing House, Delhi, pp. 1–241.

Khanna, D.R., Badola, S.P. and Dobriyal, A.K., 1993. Plankton ecology of the river Ganga at Chandighat, Haridwar. In: *Advance in Limnology*, (Ed.) H.R. Singh. Narendra Publishing House, New Delhi, p. 171–174.

Khanna, D.R., Badola, S.P. and Malik, D.S., 1997. Population of green algae in relation to physico-chemical factors of the river Ganga at Lal Ji Wala, Haridwar, U.P. *J. Zool.*, 17(3): 237–240.

Khanna, D.R., Bhutiani, Rakesh and Trivedi, Manoj, 2002. Impact of paper mill effluent on some water parameters of Hindon river at Saharanpur. *Him. J. Env. Zool.*, 16(1): 125–128.

Khanna, D.R., Gautam, A. and Sarkar, P., 2001. Water quality of bathing ghats of river Ganga at Haridwar. In: *National Conference on Status of Indian Environment, Pre-conference Proceeding*, ASEA, Rishikesh, p. 38.

Khanna, D.R., Malik, D.S., Seth, T.R. and Rupendra, 1999. Correlation between abiotic factors and planktonic population in river Ganga at Rishikesh (U.P.). In: *Eco. Sys. and Env.*, (Eds.) D.R. Khanna, A. Gautam and A. Goutam. ASEA, Rishikesh, pp. 69–76.

Khanna, D.R., Singh, Shakun, Gautam, A. and Singh, J.P., 2003. Assessment of water quality of river Ganga in District Bulandshahar (U.P.). *India. J. Nat. Conservation*, 15(1): 167–175.

Khanna, D.R., Gautam, Ashutosh, Chugh, Tarun and Sarkar, Praveen, 2000. Impact of abiotic factors on the phytoplanktonic population of a pond. *Env. Con. J.*, 1(1): 41–46.

Kumar, N. and Sharma, R.C., 2002. Water quality of river Krishni with reference to physico-chemical parameters. *J. Natcon,* 14(2): 273–297.

Mathur, R.P., 1982. *Water and Wastewater Testing.* Nem Chand and Bros. Publishers, Roorkee, pp. 1–54.

Mishra, N. and Saxena, A.B., 1984. The effect of sewage with special reference to aquatic insects in the river Krishna (India). *Int. J. Environ. Studies,* 23: 191–208.

Mishra, S. and Joshi, B.D., 2003. Assessment of water quality with few selected parameters of River Ganga at Haridwar. *Him. J. Env. Zool.,* 17(2): 113–122.

Mittal, S. and Sengar, R.M.S., 1990. Phytoplanktonic diversity in relation to certain physico-chemical characteristics of river water. *Proc. Nat. Acad. Sci.,* India, 60(B): 3.

Pahwa, D.V. and Mehrotra, S.N., 1966. Observation on fluctuations in abundance of plankton in relation to certain hydrological conditions of river Ganga. *Proc. Nat. Acad. Sci.,* 36B(2): 157–189.

Raina, R., Subia, B.A. and Zutshi, D.P., 1982. Water quality and plankton of Jhelum River. *Int. J. Ecol. Environ. Sci.,* 8: 11–17.

Raina, V., Saha, A.R. and Ahmed, S.R., 1984. Pollution studies on river Jhelum–1: An assessment of water quality. *Indian J. Env. Hlth.,* 26: 187–210.

Ray, P., Singh, S.B. and Sehgal, K.L., 1966. A study of some aspects of the river Ganga and Yamuna at Allahabad (U.P.) in 1958–59. *Proc. Nat. Acad. Sci., India,* 36 B(3): 235–272.

Reddy, P.M. and Venkateshwarlu, V., 1987. Assessment of water quality and pollution in river Tungbhadra near Kurnool (AP.). *J. Environ. Biol.,* 8(2): 109–119.

Ross, Fedrick C., 1983. *Introductory Microbiology.* Charles E. Merill Publishing Company. A Bell and Howell Company, Columbus, Ohoio, pp. 1–615.

Sangu, R.P.S., and Sharma, K.D., 1985. Studies on water pollution on Yamuna river at Agra. *Indian J. Environ. Hlth.,* 27(3): 257–261.

Sengar, R.M.S., Sharma, K.D. and Pathak, P.D., 1985. Studies on distribution of algal flora in polluted and non-polluted regions in Yamuna river at Agra (U.P.). *J. Indian Bot. Soc.,* 64: 365–376.

Seth, T.R., Khanna, D.R., Gautam, A., Chugh, T. and Sarkar, P., 2000. Temporal trends of phytoplanktonic diversity in the river Ganga at Haridwar. *Him. J. Env. Zool.,* 14: 129–134.

Sharma, A., 1999. Limnological studies of Ban Ganga and distributional pattern of stream bottom fauna. *Ph.D. Thesis,* University of Jammu.

Strommer, J.L. and Smock, L.A., 1989. Vertical distribution and abundance of invertebrates within, the sandy substrate of low gradient head water stream. *Freshwater Biology,* 22: 263–274.

Sverdrup, H.H., Johnson, M.W. and Fleming, R.H., 1942. *The Oceans: Their Physics, and Chemistry and General Biology.* Prentice Hall Inc., New York.

Tailing, J.F. and Rzoska, J., 1967. The development of plankton in relation to hydrobiological regime in Blue Nile. *J. Ecol.,* 55: 636–662.

Trivedi, R.K. and Goel, P.K., 1984. *Chemical and Biological Methods for Water Pollution Studies.* Environmental Publications, Karad, pp. 1–251.

Welch, P.S., 1948. *Limnological Methods.* The Blakiston Co., Philadelphia, pp. 1–381.

Water and Wastewater Management in 2 Vols (2011) *Pages 611–621*
Editors: D.R. Khanna, R. Bhutiani & Gagan Matta
Published by: DAYA PUBLISHING HOUSE, NEW DELHI

Chapter 91

Riffles Inhabiting Benthic Macro-invertebrate Communities in a Spring-fed Tributary of River Alaknanda, Garhwal Himalaya

Manju Prakash Gusain, Bhaskranand Semalty, Jitendra Singh Rana, Anupama Negi, Lakhpat Singh and O.P. Gusain
Freshwater Biology Unit, P.O. Box No. 60, Department of Zoology, HNB Garhwal University, Srinagar–Garhwal, Uttarakhand

ABSTRACT

Bachchan Gad is a spring-fed tributary of River Alaknanda in the Garhwal region of Uttarakhand. Benthic macroinvertebrates in the riffles of Bachchan Gad was investigated during March 2008 to February 2009 at four sampling sites between 590 m to 1600 m above sea level. Preliminary observations revealed that riffle inhabiting benthic macroinvertebrates communities in the Bachchan Gad comprised of 36 insect genera belonging to 8 orders and 29 families along with 1 genus each of crustacean and annelida. Dominated by *Ephemeropteran, Trichopteran* and *Coleopteran* genera the density was highest (1300±142 Individual m^{-2}) during winter at confluence, the downstream site, it was recorded lowest (80±15.2 Individual m^{-2}) during monsoon at Pata, the headwater site.

Keywords: *Benthic macroinvertebrate, Springfed, River Alaknanda.*

Introduction

Fast or swiftly flowing stream is usually dominated by riffles or fast moving water habitats. These are the habitats preferred by many bottom dwelling macro invertebrates as the water current is relatively faster, the depth vary from moderate to shallow and the substrate is coarser. These areas generally alternate with deeper pools with finer deposits and slower water (Moss, 1998).

Located almost centrally in the long Himalayan sweep the Garhwal Himalaya is endowed with rich water resources in the form of rivers, streams, rivulets and lakes. The major rivers namely Bhagirathi, Alaknanda, Yamuna and some of their tributaries originate in the greater Himalayas, there by owing their origin to the glaciers. In addition, a large number of streams originate as spring in the lesser Himalaya. These small spring-fed fluvial systems are locally known as Gad. A number of invertebrate organism associated with the solid-liquid interface inhabit these water bodies these are referred as zoobenthos, benthic invertebrates and macro invertebrate benthos. The macroinvertebrates of freshwater are dominated by insects and their diversity is greatest in running water.

Aquatic insects play potentially major qualitative role in the processing and turnover of nutrients in freshwater ecosystems (Merrit *et al.*, 1984). Being involved in the recycling of organic detritus and serving as the main food source for stream fishes, estimates of the benthic production are of considerable importance for understanding the stream fish ecology. The benthic macroinvertebrate have been frequently studied to evaluate stream conditions (Hynes, 1970) and *in situ* assessment of environmental stress in aquatic communities has been increasingly utilized in biomonitoring programmes and toxicity test procedure (Vuori, 1995). Macroinvertebrate responses have increasingly received attention as a part of evaluation programmes of restoration works (Friberg *et al.*, 1998; Lassonen *et al.*, 1998). Benthic macroinvertebrates have been the subject of investigations in many parts of the world since the beginning was made by Needham, and Christenson (1927), Shelford and Eddy (1929) and Hynes (1984). However such studies in Garhwal Himalaya have been fewer and fragmentary. Some noteworthy work in this regard are by Badola and Singh (1981), Dobriyal (1985), Nautiyal (1986), Sharma (1986), Negi (1989), Negi and Singh (1990), Kumar (1991), Gusain (1994a), Gusain (1994b), Singh *et al.* (1994), Kumar *et al.* (1999), Kishor *et al.* (1998), Singh (2002), Sharma *et al.* (2004) and Gusain and Gusain (2005). Understandably, there is relatively little or no information available on macroinvertebrates in the remote tributary habitats of the river systems in Garhwal Himalaya. Data collected during the present study gave an insight into the riffle inhabiting benthic macro-invertebrates in Bachchan Gad, a spring-fed stream in Garhwal Himalaya.

Materials and Method

Study Area

Bachchan Gad a spring-fed tributary of River Alaknanda in the Rudraprayag district of Uttarakhand originates upstream to Pata at an altitude of 1600 m above sea level (Figure 91.1). This approximately 12 km long perennial stream receive water from several streams, underground seepages and surface runoff before meeting the River Alaknanda, about 21 km (590 m above sea level) upstream to Srinagar near the NH 58 on route to the shrine resort of Badrinath. The investigated area is located between latitude 30°11′45.08″ to 30°15′32.77″N and 78°54′58.87″ to 78°59′38.64″E.

After a preliminary survey, depending on the accessibility, four sampling sites namely Pata (S_1), Kandai (S_2), Khankra (S_3) and Confluence (S_4) were selected (Table 91.1). The vegetation type influenced by altitudinal limits, in the Bachchan Gad catchment constituted of sub montane (Tropical Forest) and montane (Temperate Forests) (Meher-Homji, 1978). Terraced agriculture land along with open mixed forest is abundant in the riparian zone of the catchment.

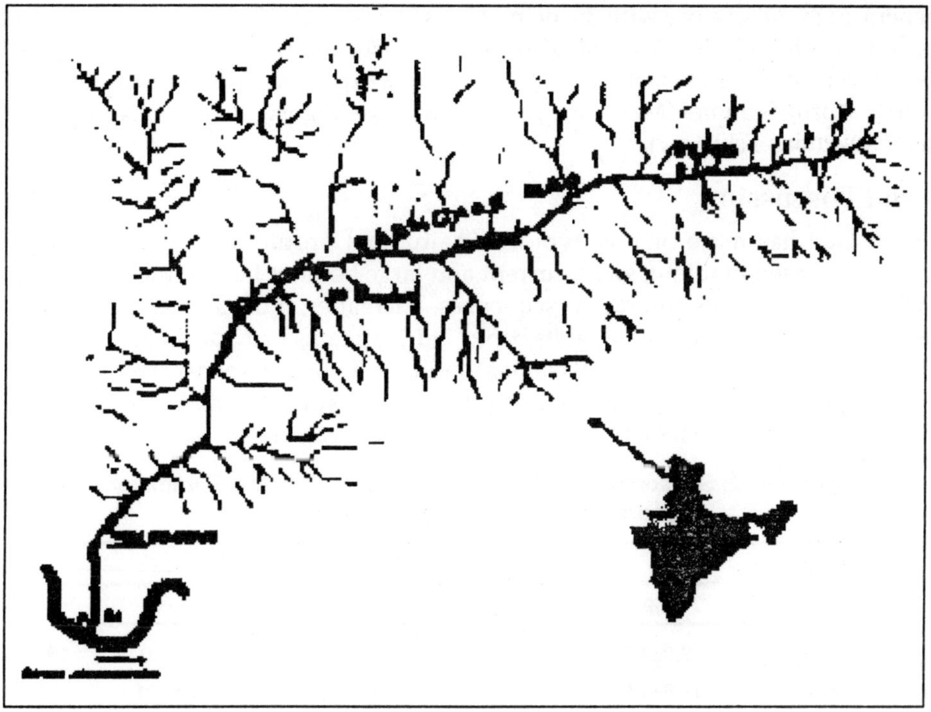

Figure 91.1: Location Map of Study Area Bachchan Gad

Monthly sampling for analyzing the selected physicochemical parameters and benthic macro invertebrates was conducted during March 2008 to February 2009. Air and water temperature were recorded with the help of a centigrade thermometer. The mean water velocity was measured using float drift method. pH of water was estimated by the Control Dynamic pH meter.

Table 91.1: Sampling Sites on Bachchan Gad

Sampling Sites	*Altitude (m above sea level)*	*Substratum Characteristics*
Pata	1600	Pebbles, cobbles (dominant) and boulders; Cascade and riffles
Kandai	820	Pebbles, cobbles (dominant) and boulders; Cascade, riffles and pools.
Khankra	610	Small pebbles (dominant), cobbles and boulders; Riffles and pools
Confluence	590	Large pebbles (dominant), cobbles and boulders; Riffles and pools

Turbidity was measured with the help of Nephelometer. Dissolved oxygen, free CO_2, total alkalinity, total solid and total dissolved solid, were estimated by following standard methods outlined in Trivedi and Goel (1984) and APHA (1995). The 'MAC' conductivity meter was used for measuring the conductivity of water.

Benthic macro-invertebrates from the riffles were collected from one square feet area by lifting stones and sieving substratum using modified Surber's square foot sampler (Welch, 1952). Samples were preserved in 4 per cent formalin. Identification was carried out to the lowest recognizable level usually genera, in the laboratory with the help of keys by Usinger (1950), Ward and Whipple (1959), Needham and Needham (1962), Macan(1979), Tonapi (1980) and Edington and Hildrew (1995). The data collected were analyzed seasonally as the study area experiences five seasons in a year namely Spring (March–April), Summer (May–June), Monsoon (July–August), Autumn (September–October) and Winter (November–February).

Results and Discussion

Bachchan Gad has more or less, typical longitudinal profile, with the headwater reaches characterized by a steep gradient, swift, current and large boulder bed. In the middle reaches, the stream has intermittent pools and rapids sequences. Thereafter, it usually tends to widen near the mouth, depositing the finer sediment on its banks. Such streams provide ample opportunities to a large number of bottom dwelling faunal elements especially the macro invertebrates. According to Moss (1998) stones, the pockets between them and the interstices of gravel and, sand provide a complex architecture on stream bottom for habitation by the benthic organisms.

Table 91.2: Seasonal Variation in the Physico-chemical Parameters Recorded in the Riffles of Bachchan Gad

Parameters	*Seasons*				
	Spring	*Summer*	*Monsoon*	*Autumn*	*Winter*
Air temperature (°C)	27.5±4.93	30±3.35	28.7±1.76	24.0±2.0	13.5±2.19
Water temperature (°C)	17.6±1.92	20.8±1.96	20.2±1.45	17.5±0.91	10±0.81
Mean depth (m)	0.18±0.04	0.23±0.06	0.31±0.09	0.24±0.09	0.20±0.05
Mean width (m)	0.63±0.14	2.06±1.20	2.55±1.47	1.76±1.22	0.76±0.23
Mean velocity (ms^{-1})	0.59±0.16	0.76±0.07	0.80±0.08	0.55±0.13	0.44±0.13
Turbidity (NTU)	4.66±3.08	2.6±0.38	6.5±2.12	1.25±0.35	1.9±0.141
Conductivity (S cm^{-1})	140±12.83	1.53±17.77	131±21.05	115±15.29	93.25±3.12
Dissolved Oxygen (mgl^{-1})	9.4±1.22	7.16±0.06	8.05±0.46	7.5±0.27	8.9±0.63
Free CO_2 (mgl^{-1})	1.50±0.20	0.92±0.41	0.99±0.66	1.37±0.36	1.81±0.23
Alkalinity (a) P.A (mgl^{-1})	0	0	0	0	0
(b) T.A (mgl^{-1})	77.5±5.36	71±5.24	70±5.77	70±4.59	69±12.12
pH	7.85±0.14	8.8±0.23	9.5±0.26	8.37±0.28	7.87±0.29
Total Solids (gl^{-1})	0.24±0.05	0.16±0.01	0.14±0.02	0.145±0.03	0.105±0.02
Total Dissolved Solids (gl^{-1})	0.14±0.01	0.11±0.018	0.10±0.008	0.11±0.02	0.07±0.01

Several factors are known to influence the distribution of benthic invertebrates, but the most important factors which affect the density and diversity of benthic macro invertebrates in aquatic ecosystem are water temperature, water velocity, substrate composition, hydromedian depth and turbidity. However, the spring-fed streams often vary in physico-chemical and biological characteristics from other types of streams (Covich, 1988). Most importantly, they tend to maintain a more constant aquatic environment than the streams supplied by runoff. In the present study similar environmental regime was evident all along the stream from headwater to the downstream region. Seasonal variation in the physicochemical parameters in the riffles of Bachchan Gad is depicted in Table 91.2. A total of 36 insects genera belonging to 29 families and 8 orders along with one genus each of annelida and crustacean were recorded from the riffles of Bachchan Gad (Table 91.3). Out of the 36 insect genera recorded 11 belonged to trichoptera, 9 to ephemeroptera, 7 to diptera, 3 to odonata, 2 to plecoptera and coleoptera each and I genus each to hemiptera and neuroptera.

Table 91.3: Checklist of Benthic Macro-invertebrates Inbabiting the Riffles of Bachchan Gad

Phylum	*Order*	*Family*	*Genera*
		Arthropoda	
		Class-Insecta	
	Ephemeroptera	Heptageniidae	*Hevtagenia*
			Epeorus
			Ironodes
			Cinygma
		Baetidae	*Baetis*
		Leptophlebiidae	*Leptophlebia*
		Ephemerellidae	*Ephemerella*
		Caenidae	*Caenis*
		Ephemeridae	*Ephemera*
	Odonata	Agrioniidae	*Enallagama*
		Lestidae	*Lestes*
		Gomphidae	*Ophiogomphus*
	Plecoptera	Perlidae	*Perla*
			Isogenus
	Hemiptera	Pleadae	*Plea*
	Neuroptera	Corydalidae	*Corydalus*
	Coleoptera	Psephenidae	*Psephenus*
		Elmidae	*Promoresia*
	Trichoptera	Hydropsychidae	*Hydropsyche*
		Philopotamidae	*Philopotamus*
		Rhycophilidae	*Rhyacophila*
		Glossosomatidae	*Glossosoma*
			Algapetus
		Brachycentridae	*Brachycentrus*
		Limnephilidae	*Limnephilus*
			Phacoptertyx
		Leptoceridae	*Leptocella*
		Polyentropodidae	*Polycentropus*
		Phryganeidae	*Phryganea*
	Diptera	Blepharoceridae	*Bibiocephala*
		Simuliidae	*Simulium*
		Athericidae	*Atherix*
		Chironomidae	*Chironomus*
		Tabanidae	*Tabanus*
		Tipulidae	*Antocha*
			Elliptera
Class-Crustacea			Crab
Annelida			Leech

The density of benthic macroinvertebrates in the riffles of Bachchan Gad showed an increase in the downstream region (Table 91.4). The density was found to be maximum (1300±142 Individual m^{-2}) during winter at confluence, the downstream region and it was recorded minimum (80±15.2 Individual m^{-2}) during monsoon at Pata, the headwater region.

Table 91.4: Density (Individual m^{-2}) of Benthic Macro-invertebrates Inhabiting the Riffles of Bachchan Gad

Sampling Sites	*Density (Individual m^{-2})*	
	Minimum	*Maximum*
Pata	80±15.2 (MO)	765±44.63 (WI)
Kandai	490±58.88 (SU)	825±67.9 (AU)
Khankra	394±27.2 (SP)	990±136.0 (SU)
Confluence	320±24.7 (MO)	1300±142.0 (WI)

SP: Spring; SU: Summer; MO: Monsoon; AU: Autumn; WI: Winter.

The abundance of macroinvertebrates during the winter may be explained as due to the relatively low velocity of water current, high dissolved oxygen and low turbidity of stream water. The macro invertebrates density declined in the month of summer and was recorded lowest during monsoon. The reason for such decline during monsoon may be attributed to the disturbance in substratum, increased water level, depth of stream, current velocity and turbidity (Crayton and Sommerfeld, 1979). During rainy season the spate causes disturbance in the stream by overturning the stones, underneath which the insects live. Also, the organisms living on the upper surface of stones are seriously affected due to scouring and stress (Malmqvist and Otto, 1987). Some reduction in number can also be attributed to the difficulties in sampling during high water but much was due to losses caused by wash out (Hynes, 1970). Ward (1976) stated that sedimentation decrease substrate heterogeneity fills interstices with silt, may severely reduce algal population which directly affects the zoobenthos.

Site wise distribution revealed that 30 genera were recorded from middle region (S_2), while the downstream region (S_3 and S_4) recorded 25 genera each (Figures 91.2–91.5). Among the benthic macro invertebrates inhabiting the riffles of Bachchan Gad, 18 genera were recorded at all the sampling sites, whereas 6 genera were absent at only one site, 4 were absent at 2 sites and 10 genera were recorded only at one site. Among the genera *Epeorus, Leptophlebia, Psephenus, Brachycentrus, Corydalis* and Leech were found at three sites. *Ephemera, Ophiogampus* and *Perla* were absent at S_3 and S_4 in the downstream region, while *Plea* was absent at S_2 and S_4. *Phacopteryx, Phrygenia, Antocha* and *Elliptera* were found only at S_2, and *Tabanus, Lestes, Polycentropus* and Crab were recorded only at S_1 *i.e.*, the headwater region.

Also, *Agapetus* and *Cinygma* were found only at S_4, the downstream region. Ephemeroptera was the dominant macroinvertebrate order in the riffles of Bachchan Gad followed by Trichoptera and Diptera. Similar composition of benthic invertebrates has been observed by Suren (1994), while investigating the invertebrate communities of the streams of Western Nepal. The taxonomic composition of the fauna was similar to that reported by Rundle *et al.* (1993) and Ormerod *et al.* (1994) in Central and Northeast Nepal.

The present study though preliminary still revealed that the riffle inhabiting benthic macro invertebrates in spring-fed stream are diverse and varied seasonally along an altitudinal gradient.

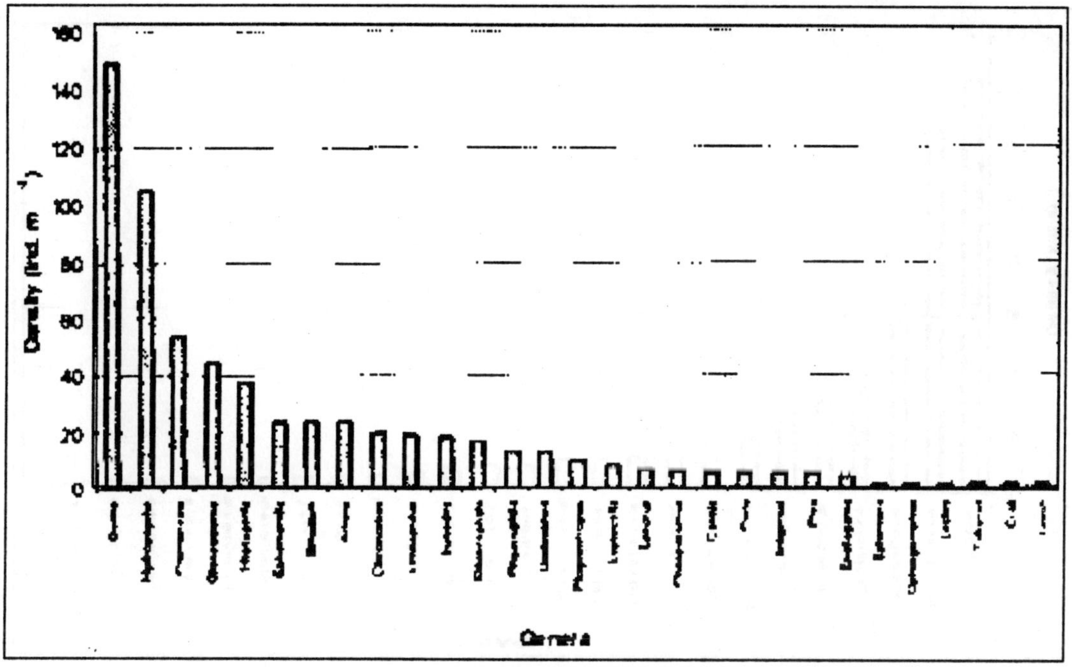

Figure 91.2: Density (Individual m^{-2}) of Benthic Macro-invertebrates in the Riffles of Bachchan Gad at S$_1$

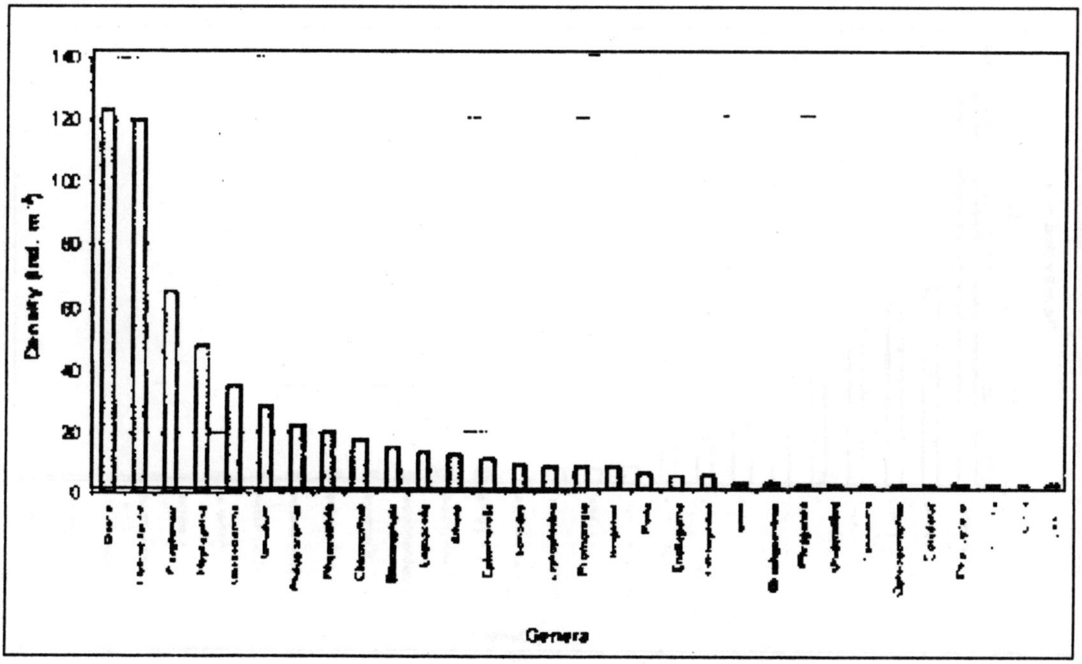

Figure 91.3: Density (Individual m^{-2}) of Benthic Macro-invertebrates in the Riffles of Bachchan Gad at S$_2$

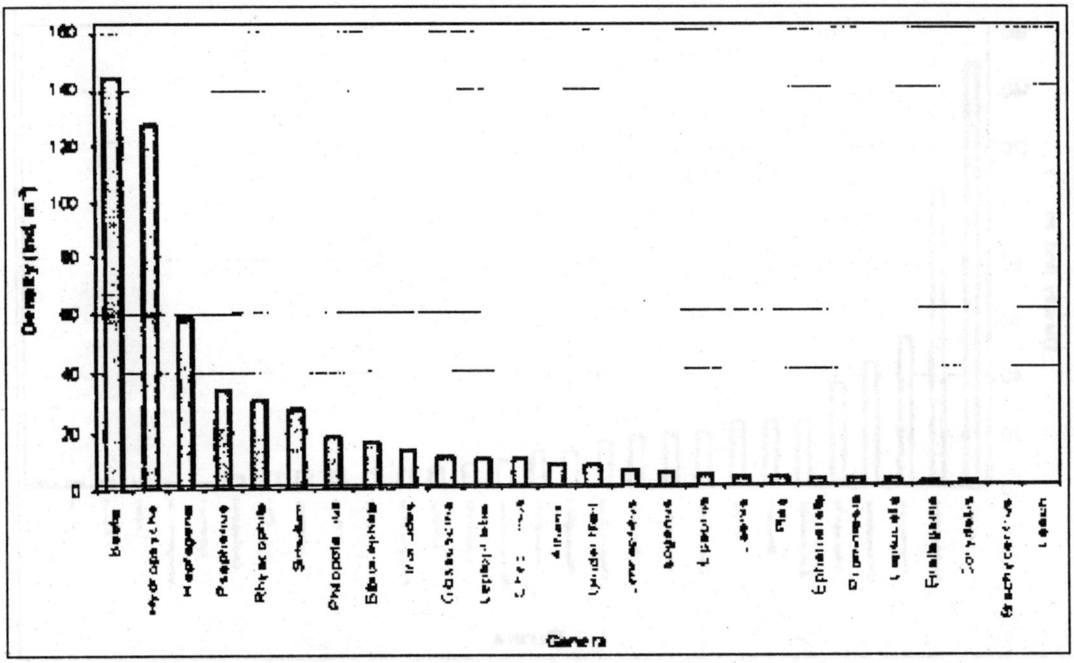

Figure 91.4: Density (Individual m^{-2}) of Benthic Macro-invertebrates in the Riffles of Bachchan Gad at S_3

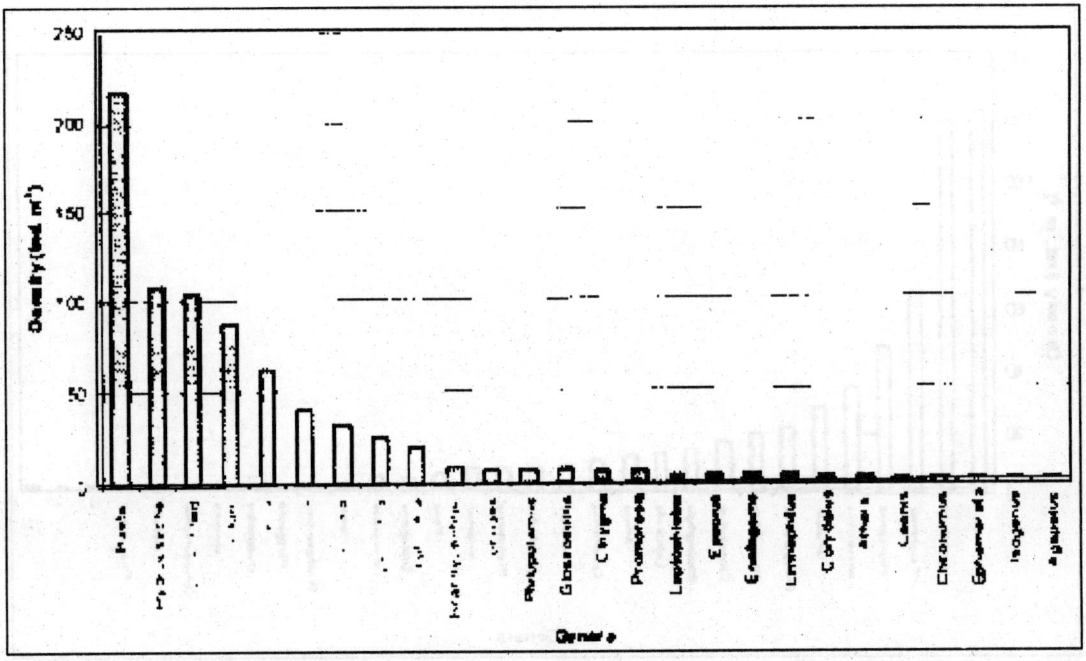

Figure 91.5: Density (Individual m^{-2}) of Benthic Macro-invertebrates in the Riffles of Bachchan Gad at S_4

Acknowledgement

The authors are thankful to Professor J.P. Bhatt, Head, Department of Zoology, HNB Garhwal University, Srinagar-Garhwal (Uttarakhand) for providing the laboratory facilities. Financial assistance extended by UCOST, Dehradun is thankfully acknowledged.

References

APHA, 1995. *Standard Methods for the Examination of Water and Wastewater* (19th edn). Amer. Pub. Health Assoc., AWWA, WEF Washington D.C.

Badola, S.P. and Singh H.R., 1981. Hydrobiology of the river Alaknanda of Garhwal Himalaya. *Indian J. Ecol.*, 8: 269–276.

Covich, A.P., 1988. Geological and historical comparisons of neotropical streams. Biotic diversity and detrital processing in highly variable habitats. *J. North American Benthological Soc.*, 7(4): 361–386.

Crayton, W.M. and Sommerfield, M.R., 1979. Composition and abundance of phytoplankton in tributaries of the lower Colorado River, Canyon region. *Hydrobiologia*, 661: 81–93.

Dobriyal, A.K., 1985. Ecology of limnofauna in the small streams and their importance to the village life in Garhwal Himalaya. *U.P. J. Zool.*, 5: 139–144.

Edington, J.M. and Hildrew, A.G., 1995 Caseless caddish larvae of the British Isles (A key with ecological notes). *FBA Sci. Pub. No.* 53, London.

Friberg, N., Kronvang, B., Hansen, H.O. and Svendsen, L.M., 1998. Long-term, habitat-specific response of a macro-invertebrate community to river restoration. *Aquatic Consev. Mar. Freshwut. Ecosyst.*, 8: 87–99.

Gusain, M.P., 1994. Himalayan Mahseer: Ecological perspectives migration routes, River Alaknanda. In: *Mahseer: The Game Fish*, (Ed.) P. Nautiyal. Jagdamba Prakashan, Dehradun for Rachna,,Srinagar (Garhwal), pp. 123–135.

Gusain, O.P. and Gusain, M.P., 2005. Caddis (Trichoptera) larvae of a spring-fed stream in Garhwal Himalaya. In: *Biological Diversity in Freshwater Environments*, (Eds.) P. Nautiyal, J.P. Bhatt, O.P. Gusain and A.K. Dobriyal. Transmedia, Srinagar, Uttrakhand, pp. 249–261.

Hynes, H.B.N., 1970. *The Ecology of Running Waters.* Univ. Toronto Press, Toronto, Ont., pp. 555.

Hynes, H.B.N., 1984. The relationships between the taxonomy and ecology of aquatic insects. In: (Eds.) *The Ecology of Aquatic Insects*, (Eds.) V.B. Resh and O.M. Rosenberg. Praeger New York, p. 9–23.

Kishor, B., Bhatt, J.P., Rawat, V.S. and Nautiyal, P., 1998. Stream regulation: Variations in the density of benthic macro-invertebrate fauna of Ganga in lateral canals at Haridwar. *Journal of Hill Research*, 11(1): 62–67.

Kumar, N., Kumar, B. and Dobriyal, A.K., 1999. Geomorphic influences on the distribution of benthic diversity in two hillstreams of Garhwal Himalaya. In: *Proc. Indo–US Workshop*, Punjab Univ., Chandigarh, pp. 53–66.

Kumar, N., 1991. Ecological studies on the macrozoobenthic communities of some hillstream of the Alaknanda river system. *D.Phil. Thesis*, HNB Garhwal Univ., Srinagar.

Lassonen, P., Muotka, T. and Kivijarvi, F., 1998. Recovery of macro-invertebrate communities from stream habitat restoration. *Aquatic Cons.. Mar. Freshwat. Ecosyst.*, 8: 101–113.

Macan, T.T., 1979. A key to the nymphs of the British species of Ephemeroptera with notes on their ecology. *Freshwater Biol. Assoc. Sci. Pub.*, 20(3): 5–80.

Malmqvist, B. and Otto, C., 1987. The influence of substrate stability on the composition of stream benthos an experimental study. *Oikos*, 48: 33–38.

Meher-Homji, V.M., 1978. The term subtropical in phytogeography: Facts and fallacies. In: *Environmental Physiology and Ecology of Plants*, (Eds.) D.N. Sen and R.P. Bansal. Bishen Singh Mahendra Pal Singh, Dehradun.

Merritt, R.W., Cummins, K.W. and Burton, T.M., 1984. The role of aquatic insects in the processing and cycling of nutrients. In: *The Ecology of Aquatic Insects*, (Eds.) V.B. Resh and O.M. Rosenberg. Praeger New York, pp. 358–400.

Moss, B., 1998. *Freshwater, Man and Median: Past to Future*, 3rd edn. School of Biological Sciences University of Liverpool, UK, pp. 1–499.

Nautiyal, P., 1986. Studies on the riverine of torrential waters in the Indian upland of the Garhwal region. III. Flora and fauna survey. *Trop. Ecol.*, 27: 157–165.

Needham, J.G. and Christenson, R.O., 1927. Economic insects in some streams of Northern Utah. *Utah Agr. Expt. Sta. Bull.*, 201: 1–36

Needham, J.G. and Needham, P.R., 1962. *A Guide to the Study of Freshwater Biology*, 5th edn. Holden Day Inc., San Francisco.

Negi, M., 1989. Ecological studies on the benthic communities of the river Alaknanda of Garhwal Himalaya. *D.Phil Thesis*, H.N.B. Garhwal University.

Negi, Manju and Singh, H.R., 1990. Substratum as determining factor for bottom fauna in the river Alakananda. *Proc. Indian Nat. Sci. Acad.*, B56(5 and 6): 417–423.

Ormetod, S.J., Rundle, S.D., Wilkinson, S.M., Daly, G.P., Dale, K.M. and Juttner, I., 1994. Altitudinal trends in the diatoms, bryophytes, macro invertebrates and fish of a Nepalese river system. *Freshwater Biol.*, 32: 309–322.

Rundle, S.D., Jenkins, A. and Ormerod, S.J., 1993. Macro-invertebrate communities in stream in the Himalaya, Nepal. *Freshwater Biol.*, 30: 169–180.

Sharma, R.C., 1986. Effect of physico-chemical factors on benthic fauna of Bhagirathi river, Garhwal Himalaya. *Indian J. Ecol.*, 13(1): 133–137.

Sharma, R.C., Bhanot, Geetu and Singh, Deepak, 2004. Aquatic macro invertebrate diversity in Nanda Devi Biosphere Reserve, India. *The Environmentalists*, 24: 211–221.

Shelford, V.E. and Eddy, S., 1929. Methods for the study of stream communities. *Ecology*, 10(4): 382–391.

Singh, H.R., Nautiyal, P., Dobriyal, A.K., Pokhriyal, R.C., Negi, M., Baduni, V., Nautiyal, R., Agarwal, N.K., Nautiyal, Preeti and Gautam, A., 1994. Water quality of river Ganga (Garhwal Himalaya). *Acta Hydro-biologica*, 36(1): 3–15.

Singh, J., 2002. A study of macroinvertebrates benthos substratum relationship along the altitudinal gradient in a Himalayan stream, Takoli gad. Unpublished *D.Phil. Thesis*, H.N.B. Garhwal University, Srinagar.

Suren, A.M., 1994. Macroinvertebrate communities of streams in Western Nepal: Effect of altitude and land use. *Freshwater Biol.*, (32): 323–336.

Tonapi, G.T., 1980. *Freshwater Animals of India: An Ecological Approach.* Oxford and IBH Publ. Co., New Delhi, pp. 341.

Trivedi, R.K. and Goel, P.K., 1984. *Chemical and Biological Methods for Water Pollution Studies.* Environ. Publ., Karad (India), pp. 251.

Usinger, R.L., (Ed.), 1950. *Aquatic Insects of California.* University of Califorina, Press Berkeley, C.A. pp. 508.

Vuori, K.M., 1995. Species and pollution-specific response of translocated hydropsychid larvae (Tricoptera, Hydropsychidae) to run off from acid sulphate soils in the river Kyronjoki, Western Finland. *Freshwat Biol.*, 33: 305–318.

Ward, H.B. and Whipple, G.C., 1959. *Freshwater Biology*, 2nd edn. (Ed.) W.T. Edmondson. John Willey and Sons, Inc., New York.

Ward, J.V., 1976. Comparative limnology of differentially regulated, sections of a Colorado mountain river *Arch. Hydrobiologia*, 78(3): 319–342.

Welch, P.S., 1952. *Limnology*, 2nd edn. McGraw-Hill Book Comp., New York, pp. 538.

Water and Wastewater Management in 2 Vols (2011) *Pages 622–626*
Editors: D.R. Khanna, R. Bhutiani & Gagan Matta
Published by: DAYA PUBLISHING HOUSE, NEW DELHI

Chapter 92

Water Quality of Natural Springs in Garhwal Himalayas

Avanish Chauhan[1], Namita Tewari[2] and Gagan Matta[1]
[1]*Department of Zoology and Environmental Sciences, Gurukula Kangri University, Haridwar, Uttarakhand*
[2]*Department of Environmental Science, Uttaranchal College of Science and Technology, Dehradun, Uttarakhand*

ABSTRACT

Instead of having plenty of water in the Garhwal Himalayas in the form of Glaciers and snow fed rivers, a large population is facing the problem of easy availability of freshwater natural springs. This chapter deals with the study of natural water springs in Garhwal region. The parameters studied were alkalinity, acidity, DO, BOD, free CO_2, nitrate, H_2S, chlorides, hardness, inorganic phosphates, temperature, pH and coliform number. The study elucidates that the water quality of selected natural water springs is suitable for drinking purpose.

Keywords: Water quality, Natural springs, Garhwal, Physico-chemical, Biological.

Introduction

Natural springs are the major source of potable water in most of villages in Garhwal Himalaya and are naturally emerging from earth. In hills, middle and upper dense vegetation of broad leaved trees *viz. Quercus leucotrichophora* (Banj Oak), *Quercus floribunda* (Moru Oak), *Rhododendron arboretum* (Buarns) *etc.*, absorbs rainwater during monsoon. This water is slowly released over the year by these broad leaved plants. Released water percolates in land and forms numerous channels which flows over the year and called as natural springs. Water is the most essential commodity for the entire living system on the earth (Shivashankara and Shanmila, 2004). Till now a very less amount of studies has been carried out on the springs of Garhwal region. Rare are the studies that have been conducted on the riverine ecology in the torrential reaches of the Indian uplands, to which the major rivers, the Ganga and Brahamputra owe the existence (Nautiyal, 1986). Khanna and Bhutiani (2004) worked on fish and their ecology of river Ganga at Gohri Ghat, Garhwal. The main objective of the present study, which is the first of its kind in this region, was to assess the current situation and state of spring in Garhwal region. The present study was conducted in order to obtain an overall picture of the prevailing ecological conditions and thus the water quality, in the spring fed Alkananda.

Materials and Method

The physico-chemical and biological parameters of the water were recorded monthly at different stations followed APHA (1998) and Khanna and Bhutiani (2008). Total four sites were selected for the study as following:

Site-1 (Kothar Dhara)

Water emerges after passing through dense vegetation at this sampling site.

Site-2 (Sweeth Bridge)

This spring lies adjacent to Government Medical College, emerges directly from earth and is surrounded by cemented pool which remains covered by slabs.

Site-3 (Bhola Mahadev)

This spring is situated at Srikot. Spring is surrounded by cemented pool.

Site-4 (Barkot Spring)

The sampling site lies 3.0 km away from Chauras campus in north. Water was flowing in cemented channels.

Results and Discussion

Table 92.1 represents results of various physico-chemical and biological parameters and is graphically presented in Figures 92.1 and 92.2 at different sampling sites in natural springs.

The alkalinity of water samples was recorded as 180.31±4.10, 178.12±7.43, 170.25±5.50 and 230.06±7.23 at site 1, 2, 3 and 4, respectively, whereas the permissible limit as 200–600 mg/l. D.O. was recorded as 2.30±0.66, 2.9±0.37, 1.3±0.28 and 6.4±0.36 mg/l at sites 1, 2, 3 and 4, respectively. The acceptable limit varies from 4-6 mg/l. BOD of water sample was recorded as 0.83±0.16, 1.6±0.16 and 0.60±0.12 and 1.0±0.26 mg/l at sites 1, 2, 3 and 4, respectively. The acceptable limits of BOD are 2–3 mg/l, above this limit water is considered to be not fit for drinking purpose. Nitrate of water sample was recorded as 0.58±0.08, 0.38±0.06 and 0.32 mg/l at sites 1, 2 and 4, respectively. H_2S was reported as 5.90±0.40, 5.30±0.35, 5.30±1.08 and 3.76±0.63 at site 1, 2, 3 and 4, respectively. Chlorides of water samples were recorded as 34.17±2.11, 15.12±1.14 and 35.36±1.89 and 11.79±1.04 mg/l at site 1, 2, 3 and 4, respectively. Free CO_2 recorded as 46.1±2.33, 34.7±2.13 and72.6±3.83 and 37.31±2.21 mg/l at site 1, 2, 3 and 4, respectively. Hardness of water samples were recorded as 176±3.71, 141±10.12, 247±11.89 and 263±5.92 mg/l at sites 1, 2, 3 and 4. Inorganic phosphate in given water sample recorded as 0.06±0.22, 0.06±0.01 and 0.09±0.01 mg/l at sites 1, 2 and 4, respectively. pH was recorded as 6.90±0.65, 7.30±0.41, 7.26±0.12 and 7.29±0.14 at sites 1, 2, 3 and 4, respectively. Temperature was recorded as 23.8±0.39, 23.7±0.86, 23.9±2.69 and 24.9±0.79 at sites 1, 2, 3 and 4, respectively. Coliform in given water samples was recorded as 0.50–22.0, 7.0– 63.0, 0.5–10.0 and 32.00 to 450/100 ml at sites

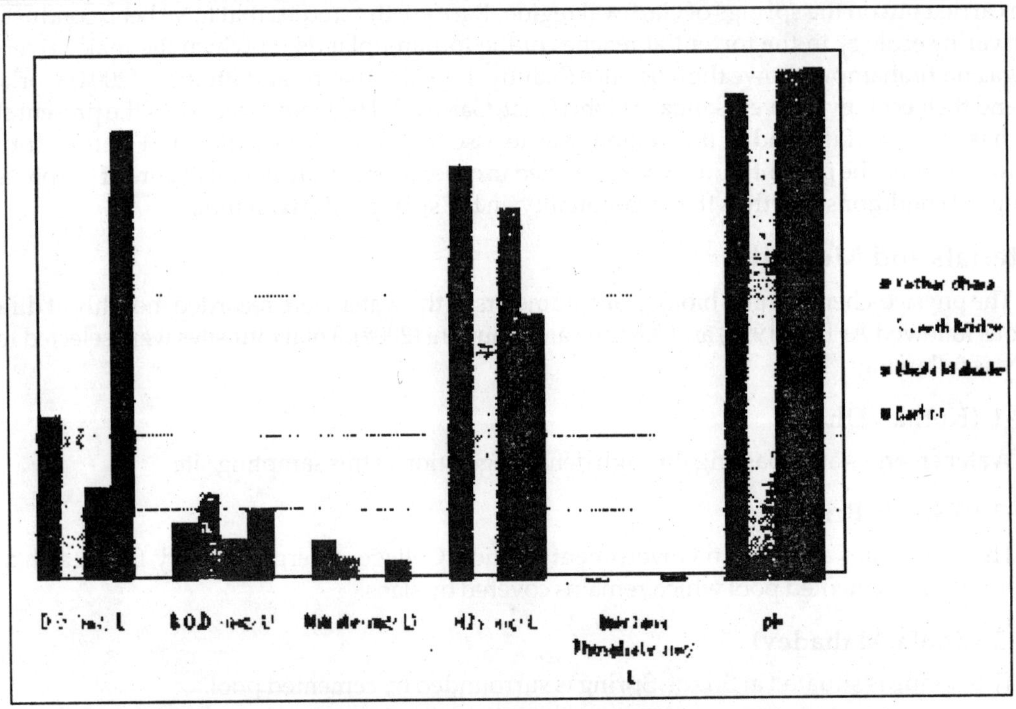

Figure 92.1: The Graphical Representation of Physico-chemical and Biological Parameters Observed in Natural Springs of Garhwal Region

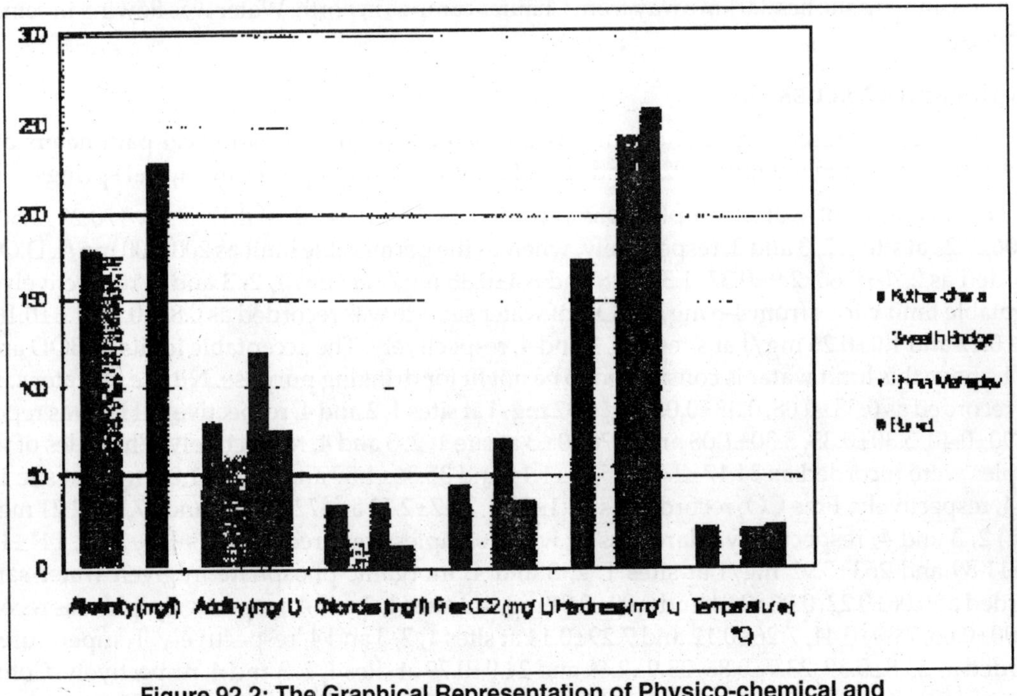

Figure 92.2: The Graphical Representation of Physico-chemical and Biological Parameters Observed in Natural Springs of Garhwal Region

1, 2, 3 and 4, respectively. This test is indicator of fecal contamination. The permissible value of coliform in the sample is very much disputed. WHO recommends the coliform count 0/100 ml for drinking water while the Indian standards are very much flexible and permissible value of coliform is 50–500/100 ml.

Table 92.1: The Value of Physico-Chemical and Biological Parameters Observed in Natural Springs of Garhwal Region

Parameters	*Sampling Sites*				
	Kothar Dhara	*Sweeth Bridge*	*Bhola Mahadev*	*Barkot Spring*	*Permissible Limit (PL)*
Alkalinity (mg/l)	180.31±4.10	178.12±7.43	170.25±5.51	230.06±7.23	200–600*
Acidity (mg/l)	81±1.84	63±1.79	121±3.74	76±2.75	N.R.
DO (mg/l)	2.3±0.66	2.9±0.37	1.3±0.28	6.4±0.36	4.0–6.0*
BOD (mg/l)	0.83±0.16	1.6±0.16	0.6±0.12	1.0±0.26	2.0–3.0*
Nitrate (mg/l)	0.58±0.08	0.38±0.06	NT	0.32±0.04	45–100*
H_2S (mg/l)	5.9±0.40	5.3±0.35	5.3±1.08	3.76±0.63	N.R.
Chlorides (mg/l)	34.17±2.11	15.12±1.14	35.36±1.89	11.79±1.04	250–1000*
Free CO_2 (mg/l)	46.1±2.33	34.7±2.13	72.6±3.83	37.3±2.21	N.R.
Hardness (mg/l)	176±3.71	141±10.12	247±11.89	263±5.92	300–600*
Inorganic phosphate (mg/l)	0.06±0.22	0.06±0.01	NT	0.09±0.01	0.046–0.068*
Temperature (°C)	23.8±0.39	23.7±0.86	23.9±2.69	24.9±0.79	N.R.
pH	6.90±0.65	7.30±0.41	7.26±0.12	7.29±0.14	65–7.5*
Coliform no. (Per 100 ml.)	L=05 U=22	L=7.0 U=63	L=0.5 U=10.0	L=32.0 U=450	50–500*

*Public Health Engineering Department, Govt. of West Bengal.

NT: Not Traceable; L: Lower Limit; U: Upper Limit.

Groundwater is rich in carbonic acid and dissolved oxygen usually possesses a high solubilizing potential towards soil or rocks that contain appreciable amount of mineral calcite, gypsum and dolomite and consequently hardness level may increased. That's why the, values of conductivity, TDS and DO were observed beyond the limit of drinking purpose (Singh *et al.*, 2007). Negi *et al.* (2008) have reported water temperature to range between 14.62 to 20.25 °C at Ganga River. It is a well established fact that the dissolved oxygen budget of a river is a direct indicator of its biological state, as was also suggested by Lamb (1985). The temperature may not be considered as most important factor in case of pure water due to the eurythermic nature of aquatic biota, but in polluted waters, the temperature has serious impact on dissolved oxygen and BOD (Palharya and Malviya, 1988). Phosphate-phosphorus is one of the most important limiting nutrients of primary concern to aquatic ecology (Datta *et al.*, 1988). The world average for nitrate in unpolluted freshwaters as reported by Reid (1961) is 0.30 mg/l.

References

APHA, 1998. *Standard Methods for Examination of Water and Wastewater*, 20th edn. American Public Health Association; 1015, Fifteen Street, New Washington, 15: 1–1134.

Datta, N.C., Bandhopadhyay, B.K., Majumdar, A. and Ahuja, D., 1988. Hydrological profile of Hoogly sector (Bally to Bandel) of river Ganga. In: *Ecology and Pollution of Indian Rivers*, (Ed.) R.K. Trivedy. Ashish Publishing House, New Delhi, pp. 113–130.

Khanna, D.R. and Bhutiani, R., 2008. *Laboratory Manual for Water and Wastewater Analysis.* Daya Publishing House, New Delhi, pp. 1–189.

Lamb, J.C., 1985. *Water Quality and its Control.* John Wiley and Sons Inc., New York, pp. 384.

Nautiyal, P., 1986. Studies on the riverine ecology of torrential waters in the Indian uplands of the Garhwal region. III. Floristic and faunistic survey. *Tropical Ecology*, 27: 157–165.

Negi, R.K., Negi, T. and Joshi, P.C., 2008. Study on physico-chemical parameters of Hinval freshwater stream and Ganga river water at Shivpuri in the Garhwal region. *Journal of Environment and Biosciences*, 22(2): 203–212.

Palharya, J.P. and Malviya, S., 1988. Pollution of Narmada river at Hosangabad in Madhya Pradesh and suggested measure for control. In: *Ecology and Pollution of Indian Rivers*, (Ed.) R.K. Trivedy. Ashish Publishing House, New Delhi, pp. 55–86.

Reid, G.K., 1961. *Ecology of Inland Waters and Estuaries.* Reinhold Publication Corporation, New York.

Shivashankar, G.P. and Sharmila, A.P., 2004. Coliform bacteria and nitrate contamination of wells from septic tank systems in urban area. *Nature Environment and Pollution Technology*, 3(3): 261–264.

Singh, J.P., Singh, S.K. and Singh, S., 2007. Impact of stored effluents of slaughter houses on groundwater quality of hand pumps situated nearby slaughter house at Khurja town, District Bulandshahar (U.P.), India. *Environment Conservation Journal*, 8(1 and 2): 31–36.

Water and Wastewater Management in 2 Vols (2011) *Pages* **627–633**
Editors: **D.R. Khanna, R. Bhutiani & Gagan Matta**
Published by: **DAYA PUBLISHING HOUSE, NEW DELHI**

Chapter 93

Snowtrout Fishery in Garhwal Himalaya: Causes of Depletion and Strategy for Propagation

N.K. Agarwal and Harpal Singh
Fish Reproduction and Conservation Biology Research Laboratory,
Department of Zoology, HNB Garhwal University Campus Badshahithaul,
Tehri Garhwal, Uttarakhand

ABSTRACT

Snowtrout (*Schizothorax* sp.) is an important group of fishes in Indian uplands and is prone to decline due to several anthropogenic activities and natural disasters. However, the natural behaviour of the group is also one of the important constraints for its self-propagation in the nature. It requires attention for the conservation and propagation through *in-situ* as well as ex-*situ* measures. The *in-situ* refers to aquaculture under controlled conditions for propagation of the group by developing artificial breeding programmes for high seed requirement, better management of incubation patterns and rational management of water bodies including development of sanctuaries, where the fishing would be banned. The artificial breeding and snowtrout seed production technique in flow through hatchery can be adopted at commercial level for mass scale snowtrout seed production. Sometimes unavailability of either sex of the mature brooder at the time of artificial fertilization in hatchery is a major constraint in the artificial breeding programmes. Cryopreservation of gametes has emerged as a promising and a very useful technique to facilitate artificial breeding in several fishes. The cryopreservation of milt of snowtrout finds its role right here. Initial attempts for developing cryopreservation protocol for snowtrout milt are very much promising. More concerted efforts for commercializing the reproductive techniques including cryopreservation of gametes will certainly be helpful for the strategic propagation of snowtrout in cold-water bodies.

Keywords: *Snowtrout, Habitat destruction, Breeding, In-situ conservation, Ex-situ conservation, Cryopreservation.*

Introduction

India is blessed with the vast and varied fish germplasm resource distributed widely in various aquatic bodies. With an estimate, the coldwater fish germplasm resource contributes around 3.32 per cent to the total fish germplasm resource of the country (Das and Pandey, 1999). More than 725 freshwater fish species have been reported to occur in India and 68 species dwell in the Garhwal Himalayas. Among these fishes, some species are on decline. Likewise, the schizothoracids, the dominant group of fishes in Garhwal Himalayas, are rapidly decreasing since past 3-4 decades due to various factors including over exploitation, habitat destruction besides having its own specific genetic reasons. It requires urgent consideration for its conservation and propagation by making scientific strategies and sustainable exploitation.

In view of the global slogan of "seed, feed and breed" for intensification of fish farming and organized pisciculture, fish sperm bank has gained current importance (Gjedrem, 1981). Therefore, the preservation of gametes has emerged as a promising and a very useful technique to facilitate artificial breeding. In order to meet high fish seed requirements, pressure on harvesting of brooders from wild stocks, in the absence of availability of the farm-raised bloodstock, is on the increase. At the same time, fish farming is in a state of expansion. This situation further emphasizes the need for the Cryopreservation of gametes for propagation of fishery in captivity as well as in the wild.

Materials and Method

The snowtrout were surveyed for their occurrence in river Ganga from Chilla barrage to Rishikesh, in river Bhagirathi from Tehri to Harsil, in river Bhilangana from Tehri to Ghansali, in river Alaknanda from Devprayag to Chamoli, in river Mandakini from Rudraparyag to Agustmuni. The fishermen and the local inhabitants were interviewed for their past knowledge regarding the fish catch experiences. The breeding and feeding ground of the fishes were identified on the basis of presence of eggs, fry and brooders.

Results and Discussion

The fishery in Garhwal Himalaya is observed mainly in the nature of capture fishery. The survey conducted during the present study has noticed that snowtrout fishery contributes a large (nearly eighty per cent of total fish catch), for subsisting fish food resources to the local denizens of the villages settled nearby to rivers and streams. The old aged experienced fisherman largely reported that now they have to make more efforts to catch fish as their nets often found without fish when they do fishing. It is not like as such those of 30-40 years back when they fished out their nets from rivers contained number of fish. They remember it as a past story. The fish catch per unit effort is declining year after year resulting in to overall decline in capture fishery of the region. This has also well documented in several previous reports by number of workers (Singh *et al.*, 1987; Agarwal *et al.*, 2005)

Factors Affecting Snowtrout Population

With the rapid ever-increasing demand of fish as food, the aquatic ecosystems are under constant pressure of man-induced stresses to the determent of the aquatic flora and fauna (Jhingran, 1991). Though the decline of snowtrout species in the Garhwal Himalayan region is very often related to more than one proximate factor, the various causes of decline in the population of snowtrout has been observed and grouped as –Anthropogenic or man induced factors and Natural or Genetic factors of the population.

Anthropogenic or Man Induced Factors

The anthropogenic or man induced factors responsible for the decline of the snowtrout population are:

Habitat Destruction

The feeding and breeding grounds of snowtrout including many others were found adversely affected by heavy silt load in the river Bhagirathi and Bhilangana due to developmental activities such as road construction *etc.* from the catchment areas. Collection of stones and boulders from riverbanks as building material resulted into the destruction of breeding grounds of snowtrout and mahseer. Recent emergence of reservoirs due to hydroelectric projects (Maneri Bhali and Tehri Project) has dramatically changed the breeding and feeding habitats of the fish. Singh *et al.* (1992), Kirchhoffer and Hefti (1996) and Shrestha (1997) have also reported that siltation in the feeding and breeding grounds and damming of rivers for power generation are the reasons of habitat degeneration for local fish communities. The migration routes of the important native fishes such as mahseer (*Tor putitora* and *T. tor*) and snowtrouts (*Schizothoraichthys progastus*) have also been blocked by the construction of dam infrastructures at Tehri and Maneri in Bhagirathi River and Veer Bhadra Barrage at Rishikesh in River Ganga.

The recent emergence of Tehri dam reservoir has resulted into the large destruction of natural habitat of snowtrout species (*viz. Schizothorax richardsonii, Schizothorax plagiostomus, Schizothorax curvifrons*) that are basically bottom feeder and lithophil spawner, thrive in the snowfed river habitat of clear, shallow water of stony substratum with a average depth from 1 to 3 meters, and river flow not less than 0.5 meter per sec (Singh *et al.*, 1985; Singh and Agarwal, 1986; Agarwal, 1996, Agarwal, 2001). The impoundments of the river Bhagirathi and Bhilangana for Tehri dam reservoir has caused a loss of riverine habitat of 44 km stretch between Tehri and Chinyalisaur and 25 km stretch between Tehri and Ghansali. The impoundments of these two river for Tehri Hydro-Electric project has affected the snowtrout population in this area up to the extent that snowtrout population has totally disappeared from the reservoir and shifted upstream to Chinyalisour and Ghansali in the river Bhagirathi and Bhilangana respectively due to its highly sensitive nature to river bed, being bottom feeder and lithophil spawner (High water column and steep banks of the reservoir do not support feeding and breeding behaviour of the snowtrout).

Over-exploitation

Increasing dependence of human population for nutrition on fishes has caused diminishing of the resources in Uttaranchal. The fish resource has also been found over-exploited for extraordinary economic benefits and has caused increasing vulnerability of the snowtrout population as evidenced by decline in catch of snowtrout in per unit effort in the Bhagirathi, Bhilangana and Alaknanda Valleys.

Mass-destruction

Use of dynamites and electric shocks by the persons involved in road construction and hydroelectric projects are very often for catching fishes. Some people also used ichthyotoxic plants (*viz. Xenthoxylum armatum*, common name Timuru) and chemical poisons (*viz.* bleaching powder) to catch fish with fewer efforts in the shallow portion of the river. These methods not only kill adults but Juveniles also and cause severe damage to the micro habitat by affecting the survival of micro-organism including periphyton and benthic invertebrates that serve as food of the fish. Thus cumulative effect of it leads to decline in the fish population. Dehadrai *et al.* (1994), Shrestha, (1997) and Ponniah *et al.*

(1998) also opined that the unscientific fishing techniques causes mass-destruction of brooders, small sized fishes, fingerlings and fry, which have distinct impact on capture fisheries in subsequent years.

Uncontrolled Introduction of Exotics

Many exotic fish species have been introduced and well established in Indian waters due to their fast growing capability. However, these species have initiated competition with indigenous fish species at different levels and are becoming dominant in Indian waters (Pullin, 1994; Singh and Pandey, 1995). Common carp, introduced in the Kashmir has almost exterminated the indigenous schizothoracids of the Valley. Though in the Garhwal Himalayan rivers, common carps was not introduced till the year 2006. But recent survey to Tehri dam reservoir has shown the presence of substantial number of common carp in the catch along with mahseer population. Although snowtrout from the reservoir has shifted itself to upstream due to its highly sensitive nature with the river bed but there is a further threat to riverine population of snowtrout from the exotic common carp. If common carp established to themselves in the reservoir and somehow migrate to the downstream (passive migration with downstream water into the river after dam) or upstream (where river entering into the reservoir at Chinyalisour and Ghansali), may give tough competition to the indigenous local population of snowtrout and may dominate over them in coming years.

Another exotic fish species brown trout and rainbow trout have been introduced in the coldwater of Garhwal Himalaya. The seeds of trout are being raised in Varagana, Talwari and Gangori hatcheries and stocked in the adjacent high altitude streams by state fishery department. These exotic trout being carnivore in nature may cause danger to the survival of native herbivore snowtrout by feeding upon their hatchlings and small fry. Therefore one must be aware of the potential hazards of the introduction of the exotics and this should be done under careful monitoring.

Genetic Factors

In addition to above-mentioned factors, snowtrouts have some own genetic constrains for its vast and speedy propagation in the nature. Important ones are–

Asynchronization of Gonadal Maturation

It is one of the major constraints that impede breeding in schizothoracids. Males often show testicular recrudescence earlier during the season, therefore causing earlier ripeness of the males than females when they are not yet mature and ready for spawning (Agarwal, 1996, 2001). Similarly, during the later periods of the breeding season, the males become spent while the females are still capable of producing eggs. This asynchronization in the maturation of the sexes causes lesser chances of fertilization thereby less production.

Low Egg Production Potential

The egg production potential (fecundity) of snowtrouts is comparatively very less than other plain counterparts as well as other coldwater fishes (Agarwal *et al.*,1988; Agarwal, 1996; Thapliyal, 2002). However this may be due to the struggle for existence in adverse coldwater environment, but it has also caused low production of the group.

Strategies for Snowtrout Protection and Propagation

The further decline of the fish germplasm resource may be prevented by devising all the possible measures of conservation and rehabilitation (Pavolov, 1993; Penczak, 1996; Das and Pandey, 1999). The irreparable harm caused to the snowtrouts and its habitat need to be compensated through strict implementation of Indian Fisheries Act, 1897 (modified, 1956) along with adoption of the following strategies. Legal measures for protecting and propagating fish germplasm resources cannot be implemented without taking the ground socio-economic realities in the consideration.

In situ Protection and Propagation

In situ protection of fish is useful where genetic diversity exists and where wild forms are present. This can be done through their maintenance within natural or man-made ecosystems in which they occur. Rishikesh, Har-ki-Pauri (Haridwar) and Baijnath (Almorah) are some examples of religiously protected areas in Uttaranchal where fishing is totally banned and protected by the community. Such type of protected areas can be developed with the support of community participation. The successful artificial breeding and larval rearing of the snowtrout, *Schizothorax richardsonii* (Agarwal *et al.*, 2002, 2007; Thapliyal, 2002) has opened up the new avenue of river ranching in depleted areas for stock replenishment and conservation. The stocking-restocking pattern by establishing flow through hatcheries may be one of the strategies for the development of snowtrout fishery in rivers and reservoirs. But the natural breeding grounds of the snowtrout should also be protected and developed all along the riverside and near the headwater of reservoir.

Ex situ Conservation

In this measure, the targeted species are conserved outside their natural habitats. The two main pillars of *ex-situ* conservation programmes are live gene bank and Gamete/Embryo banks. In a live gene Bank, the endangered species are reared in captivity, bred therein and genetically managed avoiding inbreeding depression, domestication and unintended selection (Minckley and Deacon, 1991). In Gamete/Embryo Banks, adequate samples representative of the natural genetic variations of the threatened species are kept in liquid nitrogen (LN_2) at extreme low temperatures (-196° C). Establishment of Sperm Bank by cryopreserved milt assures further availability of genetic materials of desired species for intensive breeding programmes. Cryopreservation of snowtrout milt (Semen) may play an important role in the protection and propagation of species through intensive seed production in the hatcheries.

Cryopreservation of Snowtrout Milt

Cryopreservation of fish milt (sperm) without loss of viability is of considerable value in propagation of snowtrout fishery. It helps in the availability of male gametes round the year for seasonal breeders and in organizing easy transport of germplasm, genetic selection and hybridization programmes. The cryopreservation of snowtrout milt overcome the problem of asynchronization of maturation of males and females by making available cryopreserved milt round the year for fertilizing eggs in hatcheries. This will not only improve the snowtrout seed production efficiency of the hatchery but also greatly minimize the cost of maintenance of male brooders. By raising snowtrout seed in the hatchery, river ranching programme can be initiated for stocking the rivers and streams to replenish the snowtrout in the wild. The research group has developed short term and long term preservation techniques for snowtrout milt. The milt of *Schizothorax richardsonii* and *Schizothoraichthys progastus* can be stored in refrigerator (0-4 °C) upto 5 days with 50 per cent motile sperm in Mounib's and KCl extender respectively (Agarwal, 2005). The protocol for the cryopreservation of snowtrout (*S. richardsonii* and *S. progastus*) milt in Liquid nitrogen (-196 °C) has been developed (Agarwal *et al.*, 2009). The initial success has been achieved in fertilizing the eggs with 375 days old cryopreserved milt and developing the snowtrout seed. Commercialization of the cryopreservation technique for snowtrout milt at the level of field trial is now awaited. This will certainly open up an avenue for the development of cold-water fishery through protection and propagation of snowtrout in cold-water bodies of Indian uplands.

Acknowledgements

The senior author is indebted to Prof. H. R. Singh, former Vice-Chancellor Allahabad University, and Dr. W. S. Lakara, Director, NBFGR Lucknow for constant encouragement. The author is grateful to ICAR New Delhi for financial assistance under the project no. 4(20)2000-ASR-1.

References

Agarwal, N.K., 1996. *Fish Reproduction*. APH Publishing Corporation, New Delhi, pp. 157.

Agarwal, N.K., 2001. *Reproductive Strategies of Snowtrout (Schizothorax sp.) with Reference to the Environment of the River Bhagirathi and Bhilangana of Tehri Garhwal*. Final Technical Report of ICAR Scheme, pp. 1–90.

Agarwal, N.K., 2005. *Developing Cryopreservation Protocols for the Spermatozoa of Snowtrout and Other Endangered Coldwater sp.* Final Report, ICAR Scheme, pp. 105.

Agarwal, N.K., Khanna, D.R., Thapliyal, B.L. and Rawat, U.S., 2005. Resource assessment of hill fisheries in Garhwal Himalayan region of Uttaranchal: A Perspective. In: *Aquatic Biodiversity in India: The Present Scenario*, (Eds.) D.R. Khanna, A.K. Chopra. and G. Prasad. Daya Publishing House, New Delhi, p. 81–97.

Agarwal, N.K., Raghuvanshi, S.K. and Saini, V., 2009. Cryopreservation of snowtrout (*S. richardsonii*) milt as a means for propagation and *ex situ* conservation of species. In: *Fish Genetic Resources*, (Eds.) W.S. Lakara, A.K. Singh and P.C. Mahanta. Narendra Publishing House, Delhi, p. 273–284.

Agarwal, N.K., Singh, W. and Singh, H.R., 1988. Fecundity of the snowtrout, *Schizothorax plagiostomus* (Heckel) from Garhwal Himalayas. *J. Indian Fish. Assoc.*, 18: 537–548.

Agarwal, N.K., Thapliyal, B.L. and Raghuvanshi, S.K., 2007. Induced breeding and artificial fertilization of snowtrout, *Schizothorax richardsonii* through the application of ovaprim. *J. Inland Fish. Soc. India*, 39(1):12–19.

Agarwal, N.K., Thapliyal, B.L. and Rawat, U.S., 2002. Artificial breeding of a snowtrout, *Schizothorax richardsonii* inhabiting the Bhilangana of Garhwal Himalaya. *J. Inland Fish. Soc., India*, 33(1):77–80.

Das, P. and Pandey, A.K., 1999. Endangered fish species: Measures for rehabilitation and conservation. *Fishing Chimes*, 19(6): 31–33.

Dehadrai, P.V., Das, P. and Verma, S.R., 1994. Threatened fishes of India. *Nature Conservators*, Muzaffarnagar, p. 412.

Gjedrem, T., 1981. In: Fish Gene Pools. *Ecol. Bull.* (Stockholm), 34: 33.

Jhingran, A.G., 1991. Impact of environmental stresses on freshwater fisheries resources. *J. Inland Fish. Soc. India*, 23: 20–32.

Kirchhoffer, A. and Hefti, D., 1996. *Conservation of Endangered Freshwater Fish in Europe*. Birkhauser-Verlag, Basall, Switzerland.

Minckley, W.L. and Deacon, J.E., 1991. *Battle Against Extinction: Native Fish Management in the America*. Western University of Arizona Press, Tucson and London.

Pavolov, D.S., 1993. Strategies for preserving rate and endangered fishes. *J. Ichthyol.*, 33(1): 109–127.

Penczak, T., 1996. Natural regeneration of endangered fish populations in the Pilica Drainage Basin after reducing human impact. In: *Conservation of Endangered Freshwater Fish of Europe*, (Eds.) A. Kirchhoffer and D. Hefti. Birkhaŭser-Verlag, Basel, p. 121–133.

Ponniah, A.G., Das, P. and Verma, S.R., 1998. *Fish Genetics and Biodiversity Conservation*. Nature Conservators, Muzaffanagar, p. 474.

Pullin, R.S.V., 1994. Exotic species and genetically modified organisms in aquaculture and enhanced fisheries: ICLARM position. *Naga (ICLARM)*, 17: 19–24.

Shrestha, T. K., 1997. *The Mahseer in the River of Nepal Disrupted by Dams and Ranching Strategies*. R.K. Printers, Kathmandu.

Singh, H.R., Dobriyal, A.K. and Nautiyal, B.P., 1985. Spawning patterns and environmental regulation of spawning in hill-stream fishes. In: *The Endocrine System and the Environment*, (Eds.) B.K. Folleit, S. Ishii and A. Chandola. Japan Sci. Soc. Press Tokyo/Springer-Verlag, Berlin, p. 1–11.

Singh, A.K. and Pandey, A.K., 1995. Genetic constraints in management of endangered fishes. *J. Natcon.*, 7: 99–105.

Singh, H.R. and Agarwal, N.K., 1986. Spawning ecology *of Schizothorax plagiostomus* from Garhwal Himalaya. *Proc. Symp. Fish*, p. 82–85.

Singh, N., Bhatt, K.C. and Agarwal, N. K., 1992. The Alaknanda catchment area (Garhwal Himalaya): A case study of developmental strategies vis-à-vis Ichthyofauna and environmental imbalances. In: *Aqutic Environment*, (Ed.) A. Gautam. Ashish Publishing House, New Delhi.

Singh, W., Agarwal, N.K. and Singh, H.R., 1987. Present status of snowtrout in Garhwal Himalaya. *U.P. J. Zoology*, 7(1): 85–88.

Thapliyal, B.L., 2002. Study on artificial breeding and embryonic development in a snowtrout, *Schizothorax richardsonii* (Gray) inhabiting snowfed river Bhilangana of Garhwal Himalaya. *D. Phil. Thesis*, HNB Garhwal University, Srinagar Garhwal, India.

Water and Wastewater Management in 2 Vols (2011) *Pages* **634–642**
Editors: **D.R. Khanna, R. Bhutiani & Gagan Matta**
Published by: **DAYA PUBLISHING HOUSE, NEW DELHI**

Chapter 94

Fish and Fisheries of Virla Reservoir of West Nimar, District M.P.

S.K. Pathak[1] and K. Mudgal[2]
[1]Department of Zoology, Govt. P.G. College, Mhow
[2]Govt. Girls P.G. College, Moti Tabela, Indore

ABSTRACT

After the construction of reservoir this is the first limnological study on viral reservoir West Nimar (Kargone) M.P., India. 29 species of fish biodiversity were recorded, these belong 6 orders, 10 families and 16 genera. On the basis of economic importance fishes were divided into three major groups *viz.* catfish (*Mystus seenghala, Mystus teengara, Wallago attu* and *Ompok bimaculatus*), major carp (*Catla catla, Labeo rohita* and *Cirrhinus mrigala*) and miscellaneous (*Channa* sps., *Mastacembelus* sps. and *Notopterus notopterus*). The water body were divided into three zones *viz.* upstream zone, reservoir zone and downstream zone for fish catch. The total annual fish catch was computed in this water body 2465.40 kg. and 2576.24 kg. during the 2001-02 and 2002-03 respectively. The minimum catch was observed in August while maximum in May. The percentage composition of major carp showed decline trend and catfishes showed increasing trend during two years study time. Catch per unit effort (CPUE) ranged between 0.015 to 0.191 kg/hrs. The water level fluctuations and fish production showed inverse relationship.

Keywords: Fisheries, Virla reservoir, Economic, Carp.

Introduction

India has third rank in inland fish production in the world after China and Russia. The total fish production is 47.89 lakh tones contributed 20.97 lakh tones from inland and 26.92 lakh tones from marine sector respectively in the year 1994-95 (Pandey and Unyal, 1907-1998). The contribution of inland sector nearly one third of the total fish production with 57 per cent of total domestic consumption supply. The major portion of protein rich food of domestic consumption come from inland fish production. Madhya Pradesh is occupy 2.75 lakhs hectares area with 60 reservoir and got second position in India. The total fish catch is 2619 tones per year from reservoirs of M.P. The state fishery department governs the development of reservoir fisheries since 1962 and has only 40 per cent stocking facility of total production at major reservoirs. The management of reservoir fisheries in India is not well developed when we compared with other countries. Many impending factors *i.e.* regular monitoring of reservoir, excessive growth of aquatic vegetation, introduction of commercial fish seed, use of fishing gears, untrained staff, transportation and marketing facilities are responsible for development of reservoir fisheries in the country. A number of national and foreigner researchers (scientists) have contributed our knowledge in the fisheries aspects. Their review of literature indicates that not much work has been done on the reservoir of M.P. and no attempts has been made on Virla reservoir in this aspect. The first limnological study in this reservoir has been conducted and details of fish biodiversity and potential structure were observed and will be helpful in the management and development of fishery in the reservoir.

Materials and Method

The present study was done for a period of two years from July, 2001 to June, 2003. Four collection centers were selected in the studied water body *viz.* one at upstream site, two at reservoir site and one at downstream site. Besides this, fish collection and marketing survey were done in one week of every month during the study period. The main fishing gears operated in the reservoir were gillnets of mesh was ranging from 23–180 mm bar. Further the fish caught by small meshed gill net, cast net and traps. The fishes soon after collection were cleaned and preserved in 5 per cent formaline. The fishes were identified to genus/species with the help of keys provided by Day (1958), Shrivastav (1968) and Jhingran (1992). Catch per unit effort (CPUE) was calculated based (by) on the mean monthly fluctuation of fish landings.

Results and Discussion

The reservoir receives rainwater through local nalla started from Jalabad of Satpura ranges. The west of the reservoir is covered by the agriculture fields and remaining three fields covered by hill and village of Virla. The reservoir is mainly used for culturing of fishes. The near by villages also used the water for irrigation, bathing, washing and drinking for domestic animals. The dam is constructed by cement concrete and is provided with one small siphon to let out the excess water from the reservoir particularly during rains. The length of dam is 270 m and height is 19.29 m. The first limnological study has been conducted in this water body for fish biodiversity and its potentials. It will be useful to fishery department, for future planning in reservoir. The above details of morphometry and hydrology of reservoir denote the static value. The water volume and water level changed seasonally and depend upon rainfall, water inflow and erosion inputs from catchment area.

Morphometric and Hydrological Characteristics Particulars of the Dam

Water Elevation

Maximum water level-154.40 m

Full reservoir level- 153.20 m

River bed level- 136.91 m

Capacity

Gross storage- 4.47 M.CuM.

Live storage- 4.29 M.CuM.

State-M.P.

District- Khargone

Tehsil- Segon

River- Local nala

Location- Near village Virla

Total length- 270 m

Maximum height – 19.29

Area

Gross storage level- 525 Hectares

Full reservoir level- 445 Hectares

Average level- 495 Hectares

Depth

Maximum at FSL- 16.29

Maximum at DSL- 9.69

Fish Diversity

The fish diversity mainly depends on the type of habitat and its abiotic and biotic factors. The present limnological study on Virla reservoir of west Nimar (Khargone) was done and 29 species of fish diversity were noted which belong to 6 orders, 10 families and 17 genera (Table 94.1). Dubey and Mehra (1959) reported 71 species from the Chambal river; Dubey and Verma (1965) reported 104 species during the study time from M.P., Dubey and Chatterjee (1977), Choudhary (1977) observed 39 species after the impoundment of Gandhi Sagar reservoir. Kartha (1987) and Choubey reported 41 species of fish from Gandhi Sagar reservoir. Singh (1993) reported 84 species from Sardar Sarovar dam of Narmada River. Saxena (1997) reported 42 species from upstream region and 35 species from down stream region in river Satluj. Pathak and Pathak (2000) reported 40 species from various water bodies of west Nimar (Khargone). When Virla reservoir compared with above water bodies it shows less fish diversity. Prior to present study the account of picscifauna in Virla reservoir was not available, that's why it was not possible to determine whether the piscifauna is depleted or not.

Commercial Fisheries

Due to lack of knowledge of fisheries management techniques in Virla reservoir the commercial fishery is unorganized. The history of Virla reservoir represents the natural stock. The commercial fishes of Virla reservoir were dominated against non commercial fishes. In the present study fishes were classified as major carps, cat fishes and a group of miscellaneous fishes. The dominated fisheries include *Catla catla, Labeo rohita, Mystus seenghala, Cirrhinus mrigala, Wallago attu, Channa* sps. *Notopterus notopterus* and *Mastacembelus* sp.

Table 94.1: List of Fishes Recorded in Virla Reservoir during July 2001 to June 2003

Order	*Family*	*Genera*
Cypriniformes	Cyprinidae	*Catla catla* (Ham)
		Cirrhinus mrigala (Ham)
		Cirrhinus rebo (Ham)
		Labeo rohita (Ham)
		Labeo calbasu (Ham)
		Labeo bata (Ham)
		Puntius ticto (Ham)
		Puntius sophore (Ham)
		Rasbora daniconius (Ham)
		Nemacheilus botia (Ham)
		Nemachellus beavani (Ham)
		Nemacheilus aurius (Ham)
	Siluridae	*Ompok bimaculatus* (bloch)
		Wallago attu (Sclin)
	Bagridae	*Mystus seenghala* (Skyes)
		Mystus tengaara (Ham)
		Mystus aor (Ham)
	Schielbeidae	*Eutropiichthys vacha* (Ham)
	Saccobranchidae	*Heteropneustes fossilis* (Bloch)
Clupeiformes	Notopteroidae	*Notopterus notopterus* (pallus)
Belniformes	Belonidae	*Xenentodon cancila* (Ham)
Ophiocephalieoformes	Ophiocephalidae	*Channa marulius* (Ham)
		Channa gachua (Ham)
		Channa striatus
		Channa punctatus
Perciforemes	Cenyropomidae	*Ambasis ranga* (Ham)
		Ambasis nama (Ham)
Mastacembeleformes	Mastacembelidae	*Mastacembelus armatus*
		Mastacembelus pancalus (Ham)

The commercially important species of fishes of Virla reservoir are grouped according to their economic importance into three major categories *viz.* major carp (*Catla catla, Labeo rohita* and *Cirrhinus mrigala*) cat fish *Wallago attu, Mystus teengara, Mystus seenghala, Ompok bimaculatus*) and miscellaneous (*Notopterus notopterus, Channa* species and *Mastacembelus*). Dubey and Mehra (1959) reported 70 species from Chambal river in which 46 species were identified as commercially important fishes. Rao *et al.* (1988 a). Nine commercially important fishes from Gandhi sagar reservoir. Kartha (1987) reported 77 per cent of growing rate of *Catla catla* during the period of 1980-81 to 1985-86 in Gandhi Sagar Reservoir. Gandhi Sagar Reservoir is known as Catla sagar because of increased landing of *Catla catla* in this reservoir.

In the present study of Virla reservoir the percentage contribution of commercially important fishes like *Catla catla, Labeo rohita* and *Cirrhinus mrigala* showed 7.57 per cent, 6.93 and 3.79 per cent of the total catch respectively. There are no storage facilities available for catched netted fishes from the reservoir. Hence, after catching they have to sold out at any rate. The seed of major carps should be introduced for the enhancement of further production of commercially important fishes in the Virla reservoir. Kartha (1987) in Gandhi Sagar and Pandey (1998) made similar observations in Gambhir dam.

Annual Fish Yield

For the fisheries output, hydrological characteristics of the collection centers play an important role to a greater extent. In the present study of Virla reservoir four sampling stations were identified according to their morphometry, water flow and hydrobiological charcteristics. The sampling station I exhibits lotic condition to some extend. This centers contributed 26.85 per cent of total fish catch of Virla reservoir. The sampling station II and III represents lentic habitat and contributed total fish catch 28.33 per cent and 34.27 per cent respectively. The sampling station IV (downstream) exhibits rapid water current very similar to river during rainy season. It contributes 10.55 to the total fish catch per cent. The sampling station wise data of annual catch showed minimum 251.87 kg/year at collection station IV and maximum 873.58 kg/year at collection station III (Table 94.2). In the present study fish catch includes 4.69 and 4.90 kg/ha/yr. landings were estimated in 2001-02 and 2002-03 respectively. The present results indicate that the total fish yield/ha/yr. showed increasing trend. Similarly the total fish yield at Virla reservoir also showed increasing trend *i.e.* 2465.40 kg/yr. in 2001-02 and 2576.24 kg/yr. in 2002-03. Kartha (1987) reported the total fish landings at Gandhi Sagar reservoir 2437.0 tons/yr. and 13420 tons/yr. in 1984-85 and 1985-86 respectively. According to All India Coordinator Reservoir Research Project the annual fish production from important reservoir during 1983-84 were 1.4, 4.1, 41.2, 1.1, 25.3, 4.3 and 2.8 kg/ha/yr. from Ukai, Nagarjunsagar, Getal Sud, Kangabati reservoir respectively (Saxena, 1990). Kumar (1990) reported 78.4 kg/ha/yr. annual landing from Govind Sagar reservoir during 1988-89. Sinha (1990) reported 71.55kg/ha annual landing from Gumti reservoir (Tripura) during 1987-88. Selvaraj and Murugesan (1990) reported 167.23kg/ha/yr. landing from Aliyar reservoir during 1988-89. National commission on agriculture reported a variation from 5-8 kg/ha/yr. on yield of Indian reservoirs in (1976). Jhingran (1983) reported yield of Indian reservoir 16 kg/ha/yr. The per hectare of some important reservoir like Ukai, Rihand, Keethan, Ghambir *etc.* were found in decline trend. Per hectare yield of the small reservoir is not appreciably higher than medium reservoir. For the unsatisfactory results so many problems are responsible and they can be classified as biological, technological, economic, administrative and legislative (Saxena, 1990).

Table 94.2: Annual Fish Yield in Virla Reservoir Zone and Site-wise during July 2001 to June 2003

Name of the Zone	Name of the Site	2001-02		2002-03	
		Yield	Per cent	Yield	Per cent
Up stream	Site No. 1	654.39	26.54	699.73	27.16
Reservoir zone	Site No. 2	705.93	28.64	722.26	28.03
Reservoir zone	Site No. 3	853.21	34.62	873.58	33.92
Down stream	Site No. 4	251.87	10.21	280.67	10.89
Total		2465.40	100	2576.24	10

Species-Wise Percentage Composition

In the present study composition of *Catla catla* was 48.88 per cent and 49.36 per cent to the total fish landings during 2001-02 and 2002-03 respectively. *Labeo rohita* was recoded 20.44 per cent in 2001-02 but in 2002-03 it was decreased upto 19.80 per cent. *Cirrhinus mrigalra* contributed 7.38 per cent during 2001-02 and it incrased upto 7.39 during 2002-03. Among the catfishes *Mystus seenghala* was 8.05 per cent in 2001-02 and this increased upto 8.16 per cent in 2002-03 the presentage contribution was *Mystus teengara* 6.78 per cent and 6.73 per cent in 2001-02 and 2002-03 respectively. *Wallago attu* contributed 2.975 in 2001-02 and 3.00 in 2002-03. *Ompok bimaculatus* contributed 1.82 per cent in 2001-02, which further increased upto 1.90 per cent in 2002-03.

Among the miscellaneous group the *Channa* sps. contributed 0.77 per cent and 1.03 per cent. *Mastacembelus* sps., 1.46 per cent and 1.36 per cent and *Notopterus* sps. 0.62 per cent and 0.61 per cent during 2001-02 and 2002-03 respectively. Other small fishes shared 0.83 per cent in 2001-02 and 0.66 per cent in 2002-03 (Table 94.3).

Table 94.3: Percentage Composition of Fishes at Virla Reservoir: Group-wise during July 2001 to June 2003

Group of Fishes	2001–02		2002-03	
	Catch (kg)	Per cent Composition	Catch (kg)	Per cent Composition
Major carps	1891.04	76.70	1971.85	76.55
Cat fishes	483.70	19.62	509.90	19.79
Miscellaneous	70.26	2.85	7741	3.00
Other small fishes	20.40	0.83	17.08	0.66
Total	2465.40	100	2576.24	100

Kartha (1987) reported maximum contribution of *Catla catla* (58.67 per cent), *Cirrhinus mrigala* (22.18) and *Labeo rohita* (17.14) in Gandhi Sagar reservoir. Singh (1993) reported the maximum Catch of *Tor tor* (23.84 to 26.66 per cent), *Labeo fimbratus* (18.1to 18.8 per cent), *Labeo calbasu* (5.2 to 5.66 per cent), *Cirrhinus mrigala* (2.26 to 3.06 per cent), *Mystus seenghala* (7.7 to 8.6 per cent), *Mystus teengara* (0.26 to 0.4 per cent), *Wallago attu* (7.3 to 8.14 per cent) and Miscellaneous species (5.1 per cent) in Sarover of Narmada River. Pandey (1980) reported maximum contribution of *Mystus seenghala* (49.35 to 49.64 per cent), *Mystus teengara* (19.45 to 20.12), *Wallago attu* (8.80 to 10.35 per cent), *Catla catla* (6.78 to 7.45 per cent), *Labeo rohita* (4.33 to 4.72 per cent), *Cirrhinus mrigala* (3.65 to 3.77 per cent), *Ompok bimaculatus* (1.58 to 1.66 per cent), Channa species (0.99 to 1.185), *Mastacembelus* sps. (0.69 to 1.46 per cent) and *Notopterus notopterus* (1 to 1.21 per cent) and other small fishes (0.77 to 0.98 per cent), in Gambhir dam. In Virla reservoir the major carps were the most dominant followed by catfish and miscellaneous group. The percentage contribution of major carps showed increasing trend except *Labeo rohita*. Similarly, among catfishes except *Mystus teengara*. Other fishes showed decreasing trend. It is therefore, suggested that for further increase in the production of fishes appropriate fish seed stocking is required. Special attention is needed for *Labeo rohita* and *Mystus teengara*.

Water Level Variation and Fish Catch

The impact of water level variation on total fish landings were observed at Virla reservoir (Table 94.4 and Figure 94.1). The value of average water level and total fish landings were recorded 145.59 m. and 2465.40 kg. during 2001-02 respectively and in 2002-03 the value of average water and total fish landing were 140.80 m. and 2576.24 kg. respectively. The impact of water level variation on total fish landings is inversely proportional. Pandey (1998) supported my view.

Table 94.4: Water Level Variation and Fish Catch at Virla Reservoir During the Year 2001-02 and 2002-03

Water Level in Meter	*Year*	
	2001–02	*2002–03*
Maximum	149.30	145.20
Minimum	142.68	136.40
Average	145.99	140.80
Fish catch (kg)	2465.40	2576.24

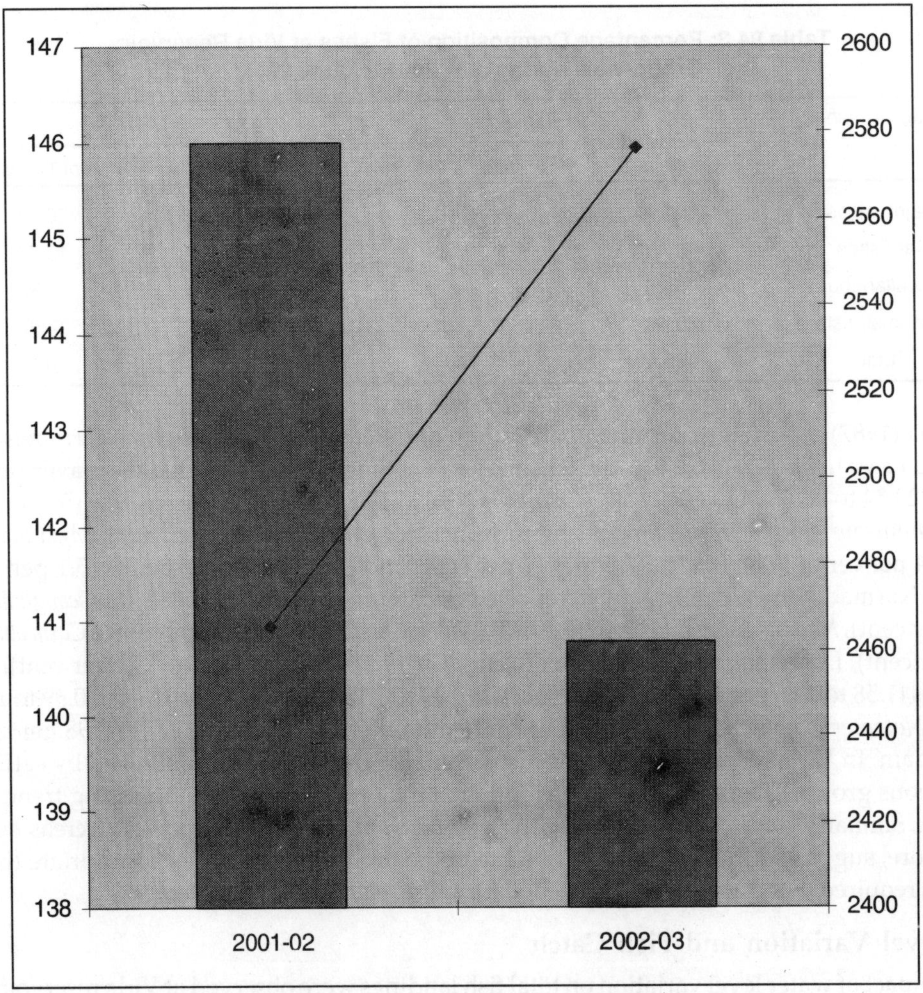

Figure 94.1: Average Water and Fish Landings in Virla Reservoir During July 2001 to June 2003

Catch per Unit Effort (CPUE)

On the basis of monthly mean value of fish catch the catch per unit effort has been calculated. The abiotic biotic factors, volume of water and velocity of current also affected the CPUE. In the present study CPUE was estimated in four study sites of Virla reservoir. The CPUE values ranged 0.015 to 0.191 kg./h the highest value of CPUE in the month of May at site III and lowest in August at site IV (Table 94.5). The value of CPUE ranged 21.57 to 31 kg/hr. at Gandhisagar reservoir (Kartha, 1987). Singh (1993) observed CPUE from 3.30 to 3.41 kg./hr. at Sardar Sarovar dam. Shyam Sundar *et al.* (1995) recorded CPUE in between 0.35 to 639 gms/h at river Guala. Pandey (1998) reported 0.375 to 1.137 kg/hr. at Gambhir dam. The value of CPUE is very less in Virla reservoir when compared with above findings. Probable reason of less CPUE value might be due to non-availability of Stocking facility in this reservoir.

Table 94.5: Catch Per Unit Effort (CPUE) in kg at Four Sites of Virla Reservoir During July 02 to June 03

Month	*Site-1*		*Site-2*		*Site-3*		*Site-4*	
	Production	*CPUE*	*Production*	*CPUE*	*Production*	*CPUE*	*Production*	*CPUE*
July 02	30.10	0.040	28.50	0.038	30.60	0.041	14.35	0.019
Aug.	27.68	0.037	15.90	0.021	27.00	0.036	11.22	0.015
Sept.	32.26	0.044	22.45	0.031	35.60	0.049	15.30	0.021
Oct.	36.41	0.048	30.24	0.040	42.50	0.057	18.40	0.024
Nov.	45.70	0.063	48.62	0.067	64.63	0.089	21.20	0.029
Dec.	60.35	0.081	52.30	0.070	75.25	0.101	24.50	0.032
Jan. 03	67.40	0.090	62.44	0.083	88.20	0.118	26.65	0.035
Feb.	73.00	0.108	82.60	0.122	99.35	0.147	28.15	0.041
March	82.21	0.110	107.35	0.144	108.50	0.145	30.75	0.041
April	86.50	0.120	118.60	0.164	120.40	0.167	35.20	0.048
May	89.95	0.121	134.15	0.180	142.75	0.191	42.85	0.057
June	38.27	0.053	19.11	0.026	38.80	0.053	12.10	0.01

Acknowledgement

The authors are thankful to the Principal, Govt. P.G. College, Khargone for providing laboratory facilities.

References

Choudary, D.K., 1977. Case study of Gandhi Sagar reservoir Fisheries (M.P.). *News Letter, CIFE*, Bombay, July–August, p. 31.

Day, F., 1958. *The Fisheries of India*. William Dawson and Sons Ltd., London, Vol. I, p. 777 and Vol. II, p. 198.

Dubey, G.P. and Mehera, R.K., 1959. Fish and fisheries of Chambal river. *Proc. All India Zool. Congr.*, 1(2): 647–665.

Fisher, A., Hanemman, W.N. and Keelar, A.G., 1991. Integrating fishery and water resource management: A biological model of California Salmon fishery. *J. Environ. Econ. Manage.*, 20(3): 234–260.

Gupta, D.R. and Rao, K.S., 1978. Experimental studies on Tagging as an effective technique in aging and fish growth in Gandhi Sagar reservoir. *Bioresearch*, 2(182): 15–18.

Herpman, D.A., 1990. The value of in stream flow use use to produce a recreational fishery. *Ph.D. Thesis*, Deptt. Agric. and Reser. Econ. Colorado State Univ., Fort Collins.

Herpman, D.A., Sparling, E.W. and Waddle, T.J., 1993. A methodology for quantifying and valuing the impacts of flow changes on a fishery. *Water Resour. Res.*, 29(3): 575–582.

Jayaram, K.C., 1981. *The Freshwater Fishes of India*. Zoological Survey of India, Calcutta, Quoted by Gopinath and Jayakrishnan (1984).

Jhingran, V.G., 1982. *Fish and Fisheries of India*. 2nd edn. Hindustan Publishing Corp. (India) Ltd., Delhi.

Jhingran, V.G., 1992. *Fish and Fisheries of India*, 3rd edn. Hindustan Publishing Corp. (India) Ltd., Delhi.

Johnson, D.M. and Walsh, R.G., 1989. Effect of participant skill on the value of alternative fishery management practices. In *The Transactions of the 54th North American Wildlife and Natural Resources Conference*, WMI, Washington D.C., p. 247–254.

Kartha, K.N., 1987. Studies on experimental trawl fishing in Gandhi Sagar reservoir, Mandsaur (M.P.), India. *Ph.D. Thesis*, Vikram University, Ujjain, India.

Kumar, A., 1990. Management and devolepment of Govind Sagar reservoir: A case study. In: *Reservoir Fisheries in India*, (Eds.) A.G. Jhingran and B.K. Unnithan, p. 13–20.

Nanda, S.N. and Tiwari, T.N., 2001. A survey of fish fauna in the Sambalpur, Hirakund, Burla region, Orissa. *Environment and Pollution*, 8(1): 43–44.

Pandey, Ail, 1998. Hydrobiological studies on Gambhir dam, Ujjain (M.P.) including up and down stream with special reference to fisheries. *Ph.D. Thesis*, Vikram University, Ujjain, M.P.

Pathak, S.K. and Pathak, S.K., 2000. Fish fauna and fish production of tribal district West Nimar (Khargone) of M.P., India. *Environ.Cons. J.*, 1(23): 95–98.

Pathak, S.K. and Mudgal, L.K., 2005. Limnology and bioidiversity of fish fauna in Virla reservoir M.P., India. *Environ. Cons. J.*, 6(1): 41–45.

Rao, K.S., Kartha, K.N., Gupta, D.R. Pandya, S.S. and Iter, H.K., 1988. Studies on morphometry and hydrology of Gandhisagar reservoir with special reference to its fisheries. *Fisheries Techno. Cochin.*, 25(1): 21–28.

Rawat, V.S. and Nautiyal, P., 1995. National history of hillstream cyprinid *Barillius bendelisis*: Food and feeding habits in relation to growth rate. *J. Inland. Fish Soc. India*, 27(2): 6–13.

Saxena, B.S., 1990. Potentialities production and problems of Indian reservoir fisheries. In: *Reservoir Fisheries in India*, (Ed.) A.G. Jhingran, p. 79–84.

Saxsena, P.K., 1997. Impact of pollution on the fisheries of river Satluj. *Environ. Manag. Int.*, 1(1): 69.

Singh, A.K., 1993. Pre impoundment studies on Sardar Sarovar area of Narmada River (Western Zone) with special reference to fisheries. *Ph.D. Thesis*, Vikram University, Ujjain, India.

Srivastava, G.J., 1980. *Fishes of Eastern Uttar Pradesh*. Viswavidyalaya Prakashan, Varanasi, pp. 207.

Water and Wastewater Management in 2 Vols (2011) *Pages* **643–651**
Editors: D.R. Khanna, R. Bhutiani & Gagan Matta
Published by: DAYA PUBLISHING HOUSE, NEW DELHI

Chapter 95

Abiotic Factors of Yashwant Nagar Talab, Mhow, M.P.

Gopal Saxena[1] and M.M. Prakash[2]
[1]*Department of Zoology, Govt. P.G. College, Mhow, M.P.*
[2]*Department of Biotech and Bioinformatics, Govt. Holker Science College, Indore, M.P.*

ABSTRACT

In the present paper the status of water quality of Yashwant Nagar talab was determined. The different parameters which were analyzed during the study are temperature, conductivity, pH, total suspended solid, DO, BOD, silicates, chloride, nitrate, phosphate, sulphate. During course of study minor differences were observed in parameter studied.

Keywords: Abiotic factor, Yashwant Nagar, DO, BOD, Water quality.

Introduction

Aquaculture demands productive status of water but productivity of water body is determined by its abiotic and biotic factors. Thus for the aquaculture point of view both the factors of urban water body have sufficient role in limnological study. A lot of work has been done on abiotic factors of urban water body in comparison to rural water body.

Materials and Method

For the present study a perennial lentic water body- Yashwant Nagar Talab was selected, situated on Mhow– Manpur, AB Road at a distance of about 23 km from Mhow. Geographically it is located at latitude 22.37′ 30" and longitude 75.37′00" at 579.2 m above the mean sea level. Yashwant Nagar Talab is a multipurpose water body, used to irrigate the adjacent agriculture lands to provide drinking water *etc.* to nearby village population. Investigation conducted for the present study on perennial Yashwant Nagar Talab, Mhow (M.P.) from April 2005 to March 2007 at four different study points. To give overall picture we used mean values obtained by following formula for results:

$$\text{Mean} = A+B+C+D/4$$

Where A,B,C and D are the value of study point I, II, III and IV respectively. Water samples were collected monthly in sterilized glass bottles throughout the period of investigation from surface. Collection time was between 8 A.M. to 11 A.M. Collected samples were kept in an ice-box and brought to the laboratory for further analysis. Analyses of various parameters were done according to the standard method of APHA (1989).

Results and Discussion

The mean values of physico-chemical parameters of present water are described in the Tables 95.1 and 95.2, the physico-chemical character of water play an important role in maintaining pond eco-system, which directly have relationship with biological characters of pond.

Temperature

The temperature is the key parameter governing metabolic activity of aquatic organisms. Sheikh (1993) observed surface water temperature ranged 16.5 to 30.50 °C at Hathikheda reservoir. Prakesh (1994) recorded the temperature variation between 17.1 to 39.50 °C at Mehta pond, Jhabua. Shukla (1995) observed water temperature ranged in between 17.20 to 34.20 °C at Gandhi Sagar reservoir. Dave *et al.* (1999) reported surface water temperature range from 16.2 to 30.4 °C (1998-90) and 14.0 to 30.2 °C (1990–91) in Kaila pond, Dhar, Ansari and Prakesh (2000) reported water temperature range from 12°C to 33°C with maximum and minimum value in summer and winter months in Motisagar Tal (U.P.). The range of surface water temperature of Yashwant Sagar, Indore was from 13.0 to 31.0 °C (Shrivastava, 2002). Pathak (2004) observed surface water temperature range in between 19.2 to 34.2°C at Virla reservoir Khargone (M.P.). Garg and Rao (2005) described that water temperature ranged 19.15 to 29.52°C in Harshi reservoir Gwalior (M.P.). In the present investigation surface water temperature ranged from 14.44°C (Dec.) to 32.22°C (May) during 2005-06 and from 14.33°C (Dec.) to 31.33°C (May) during 2006-07. These values are closely similar to findings of Shrivastava (2002) and Pathak (2004) and very much suitable for fish culture.

George *et al.* (1969) recorded higher water temperature of reservoir during summer and early part of rainy season. According to Dhawan (1970), the occurrence of minimum and maximum coincide with winter and summer respectively. In the present study minimum and maximum water temperature was in the winter and summer as reported by earlier scholars. The air temperature of water body ranged from 16.4°C (Jan.) to 33.54°C (May) during 2005-06 and 16.3°C (Jan.) to 33.22°C (May) during 2006-07. These readings suggested that warming and cooling of the suface water of Yashwant Nagar Talab was infuenced considerably by meterological factors such as air temperature and wind as described by Hutchinson (1957).

Conductivity

Conductivity (specific conductance) is the numerical expression of the water's ability to conduct an electric current and depends on the total concentration, mobility and the temperature of the solution of ions. In the present investigation conductivity ranged from 14.0 μmoh/m to 45.0 μmoh/m lowest and highest in August and April during 2005-06. In the second consecutive year (2006-07) of study the minimum value (13.0 μmoh/m) was observed again in the month of August and maximum value

Table 95.1: Mean Value of Physico-chemical Parameters of Yashwant Nagar Talab, Mhow 2005–2006

Sl.No.	Parameter	Unit	Apr	May	Jun	Jul	Aug	Sep	Oct	Nov	Dec	Jan	Feb	Mar
1.	Air temp	°C	29.11	33.54	30.11	25.22	23.54	22.6	20.20	18.16	18.67	16.4	21.1	26.03
2.	Water temp	°C	28.9	32.22	29.09	24.11	22.37	20.2	18.1	15.33	14.44	15.4	20.1	25.77
3.	Conductivity	µmoh/m	45	41	21	19	14	18	24	28	32	35	38	42
4.	pH		8.5	8.2	8.1	8.0	7.8	7.4	7.3	7.2	7.8	8.1	8.2	8.3
5.	TDS	mg/l	345	395	449	585	605	580	525	435	400	305	285	325
6.	SS	mg/l	34	34	38	43	50	54	50	48	45	40	38	36
7.	CO_2	mg/l	0.0	0.0	0.0	0.0	2.61	2.0	2.51	0.0	0.0	0.0	0.0	0.0
8.	Silicate	mg/l	2.18	2.2	2.28	1.22	1.21	1.03	106	1.21	1.26	1.23	1.2	1.26
9.	DO	mg/l	6.9	6.4	3.6	3.4	3.1	3.0	7.6	7.7	76	7.7	7.9	6.4
10.	BOD	mg/l	3.3	3.5	4.1	3.9	3.1	2.8	2.5	2.6	2.8	3.2	3	3.1
11.	COD	mg/l	30	30	30	29	28	26	25	22	24	26	28	26
12.	Chloride	mg/l	33	33	29'	26	24	22	24	26	28	30	28	32
13.	Sulphate	mg/l	4.2	4.7	5.1	5.6	5.7	6.1	6.3	6.9	6.9	5.1	4.6	4.5
14.	Nitrate	mg/l	0.18	0.20	0.34	0.26	0.36	032	0.34	0.32	0.26	0.2	0.13	0.11
15.	Phosphate	mg/l	0.34	0.62	0.36	0.38	0.36	0.3	0.31	0.39	0.3	0.46	0.42	0.45

Table 95.2: Mean Value of Physico-chemical Parameters of Yashwant Nagar Talab, Mhow 2006–2007

Sl.No.	Parameter	Unit	Apr	May	Jun	Jul	Aug	Sep	Oct	Nov	Dec	Jan	Feb	Mar
1.	Air temp.	°C	28.1	33.07	30.22	24.11	22.7	20.3	18.4	18.43	16.88	16.3	16.4	20.11
2.	Water temp.	°C	27.09	31.33	27.23	21.00	19.76	19.1	17.8	15.98	14.33	15.6	14.3	19.09
3.	Conductivity	µmoh/m	42	39	17	15	13	15	22	26	33	34	36	39
4.	pH		8.4	8.4	8.6	8.4	8	7.8	7.6	7.2	7.6	7.9	7.8	8
5.	TDS	mg/l	335	389	439	580	598	560	524	455	390	398	290	320
6.	TSS	mg/l	31	32	36	40	45	48	46	44	43	38	36	32
7.	CO_2	mg/l	0.0	0.0	0.0	0.0	2.32	1.86	2.33	0.0	0.0	0.0	0.0	0.0
8.	Silicate	mg/l	2.2	22	1.88	1.67	1.08	1.1	1.15	1.18	1.03	1.08	1.04	1.98
9.	DO	mg/l	7.8	74	4.4	4.2	4	3.6	6.4	6.3	6.2	6.8	6.6	6.9
10.	BOD	mg/l	2.1	2.4	3.3	2.5	2	1.9	1.6	1.9	2	2.1	2.2	2.3
11.	COD	mg/l	26.3	27.5	26.5	25.7	26.6	24.5	23.5	21.88	22.3	24.8	25.1	24.6
12.	Chloride	mg/l	29.2	26.6	25.3	24.4	23.5	20.7	21.4	23.8	24.98	26.9	25.3	29.54
13.	Sulphate	mg/l	3.8	3.6	4.44	4.5	477	5.08	5.1	5.66	4.9	4.6	3.8	3.7
14.	Nitrate	mg/l	0.16	0.98	0.54	0.44	0.33	0.36	0.34	0.32	0.23	0.26	0.21	0.18
15.	Phosphate	mg/l	0.24	0.56	0.26	0.28	0.27	0.28	0.27	0.29	0.28	0.35	0.36	0.28

(42.0 µmoh/m) was noted in month of April. The highest values of conductivity in summer is mostly due to the increased concentration of salts because of evaporation, where as lowest values of conductivity in rainy months due to dilution.

pH

In the present investigation pH range was founding between 7.2 to 8.5 during 2005-06 and 7.2 to 8.6 during 2006-07. These were matched with findings of Huntchison (1957) who reported pH range between 7.5 and 9.0. In monsoon months, Hydrogen-ion concentration of water at Yashwant Nagar Talab was low may be due to mixing of rainwater. Ellis (1937) pointed out that pH values between 6.7 and 8.4 are suitable for survival and growth of fish. Sreenivasan (1969) stated that the range from 5.0 to 6.6 and 9.1 to 11.0 give low productivity results. Banerjee (1967), described pH range 6.5 to 7.5 as most favorable for production in ponds and 7.5 to 8.5 just favorable for an average production. In the present investigation, water of the Yashwant Nagar Talab was noted on an alkaline range mean value varying from 7.3 to 8.68, which is considered as productive more over the water of this talaab is also suitable for potable purpose as pH well with in permissible limits as described by Kadam *ct al.* (2008).

Total Suspended Solids

Total solid is the term applied to the materials residue left in the vessel after evaporation of the sample and its subsequent drying in an oven at a temperature of 103-105 °C. Total solids (TS) include Total Suspended Solids (TSS) and Total dissolved solids (TDS). In the present investigation TS (during 2005-06) ranged from 305 mg/l to 605 mg/l with lowest and highest value in the month of February Regulates biological process among the aquatic communities and can form many compounds (Agarkar *et al.*, 1994).

Free CO_2

Unni (1992) emphasized that the rate of changes in the free carbon-dioxide concentration is considerable due to composition of organic matter at the bottom. In the present investigation carbon dioxide during 2005-06 ranged between 2.00 mg/l to 2.61 mg/l with the lowest in September and highest in August. During 2006-07 Carbon dioxide range was in between 1.86 mg/l to 2.33 mg/l with lowest September and maximum in October. Except Aug to Oct in the rest of the months CO_2 was absent. Kumar *et al.* (2008) also reported absence of CO_2 in his observations. The presence and absence of the free carbon-dioxide in the surface water is mostly governed by the utilization by algae during photosynthesis at Yashwant Nagar and also through its diffusion from air (Sreenivasan, 1974).

Dissolved Oxygen

Dissolved Oxygen is an important limnological indicator of water quality and organic production in the lake, Wetzel *et al.* (2006). Rawson (1939) stated variation in dissolved is related with utilization of Oxygen during chemical oxidation and respiration of animals, plants and bacteria present in the water. In the present investigation dissolved oxygen ranged between 4.3 to 7.9 mg/l (2005) and 4.7 to 7.8 mg/l (2006-07) with maximum value in April and minimum in September. These findings corroborate with the observations made by Agarwal (1978), Nair *et al.* (1988), Prakash (1994), Kumar and Singh (2002), Shrivastava (2002). However, Hannon *et al.* (1979), Bagde and Verma (1985), Kaushik (1992), Agarkar *et al.* (1994), Pathak (2004), Kadam *et al.* (2008).

Silicates

Silicates in 2005-06 ranged from 1.03 mg/l to 2.28 mg/l. Minimum value were 1.03 mg/l in the month of September. Maximum value 2.28 mg/l was in the month of June. While in 2006-07 it ranged from 1.04 mg/l to 2.20 mg/l. minimum 1.04 mg/l in the month of February and maximum (2.20 mg/l) in the month of April and May.

Biochemical Oxygen Demand

Biological Oxygen demand and chemical oxygen demand are important parameters to estimate the pollution level and water quality of a particular water body. In the present investigation BOD value ranges between 2.6 to 4.1 mg/l with maximum value was in June and minimum in October. Similar trends were obtained by Belsare (1984), Badge and Verma (1985), Pant *et al.* (1985) and Khanna and Chugh (2004). In the present investigation higher value of BOD was observed in summer month may be due to high temperature, rich of eutrophication and decreased water level as described by Kumber *et al.* (2008).

COD

Chemical oxygen demand is a measure of oxygen required for complete oxidation organic matter by a strong oxidant. In the present investigation COD in 2005-06 ranged from 22 mg/l to 30 mg/l. minimum (22 mg/l) value was in the month of November and maximum (30 mg/l) value was in the month of April, May and June. COD in 2006-07 ranged from 21.88 mg/l to 27.5 mg/l with minimum (21.88 mg/l) in the month of November and maximum (27.5 mg/l) in the month of May. Higher value of COD in summer was also reported by Manimepalai *et al.* (2008).

Chloride

The chloride in the present study varied in between 22.0 to 33.0 mg/l during 2005-06 and 20.66 to 29.54 mg/l during 2006-07. The minimum value was recorded in the month of September and maximum in April and May. Gonzalves and Joshi (1946); Singh (1960); Ganpathi (1962); Zafar (1964); Singh (1986); Nair *et al.* (1998); Dave *et al.* (1999); Shrivastava (2002) and Pathak (2004) had also found higher chloride in summer due to increased rate of evaporation of water.

Nitrate

The rate of Nitrate in present investigation between 0.11 to 0.36 mg/l during 2005-06 and 0.16 to 0.58 mg/l during 2006-07. Shrivatava (2002) observed nitrate range between 0.13 to 0.73 mg/l at Yashwant Sagar, Indore. Garg *et al.* (2005) described its range from 0.010 to 0.028 mg/l at Harshı reservoir, Gwalior. In the present study maximum nitrate was observed in summer (May) and minimum in December and April.

Phosphate

In aquatic ecosystem phosphorus occurs both in organic and inorganic forms.The main source was detergent and sewage containing high concentration of phosphorus. In the present investigation Phosphorus content varied from 0.03 to 0.62 mg/l (2005-06) and 0.24 to 0.56 mg/l (2006-07). Maximum value was recorded in May and minimum in September may be related with concentration and dilution of water body.

Sulphate

On the basis of the observations made on physico chemical features of water body, it is regarded as of great economic, ecological and biological importance. The present water body would prove ideal, specially for the purpose of fish culture.

Acknowledgements

Authors are thankful to Dr. Harbans Singh, Principal, Govt. Holker Science College, Indore and Dr. Ashok Bajpai, Principal Govt. P.G. College, Mhow for providing necessary laboratory facilities.

References

Agarkar, M.S., Goswami, H.K., Kaushik, S., Mishra, S.M., Bajpai, A.K. and Sharma, U.S., 1994. Biology, conservation and management of Bhoj Wetland. I. Upper lake ecosystem in Bhopal. *Bionature*, 14(2): 1–119.

Agarwal, S.S., 1978. Hydrobiological survey of Janaktal tank, Gwalior (M.P.) India. *Proc. All India Semin. Ichthyol.*, p. 20–26.

Alikunhi, K.H., 1957. Fish culture in India. *Farm Bull. Indian Coun. Agri. Res.*, 20: 1–144.

Ansari, K.K. and Prakash, S., 2000. Seasonal variations in rotifers population and some physico–chemical factors of Motisagar Tal of Tarai region, Balrampur, U.P. *Flora and Fauna*, 6(1): 19–20.

APHA, AWWA and WECF, 1989. *Standard Method for the Examination of Water and Wastewater*, 18th edn. (1992). APHA, Inc. New York.

Belsare, D.K., 1994. Inventory and status of Vanishing wetland wildlife of Southeast Asia and a operational management plan for their conservation in global wetlands: Old world and new (Eds.) William J. Mitsch. Elsevier, Amsterdam, Lausanne, New York, Oxford, Shannon, Tokyo.

Bagde, V.S. and Verma, A.K., 1985. Physico-chemical characteristics of water JNU lake at New Delhi. *Indian J. Ecol.*, 12(1): 151–156.

Dave, R.K., 1994. Hydrobiological studies of a perennial pond at Dhar (M.P.). *Ph.D. Thesis*, D.A.V.V. Indore.

Dave, R.K., Prakesh, M.M. and Dhakad, N.K., 1999. Seasonal trend in abiotic factors of the lentic habitate of Kalika pond, Dhar. In: *Limnological Research of India*, pp. 157–171.

Dhawan, R.M., 1970. Plankton and hydrobiological factors at Kandla in the Gulf of Kutch during 1960–63. *Indian J. Fish*, 17: 122–131.

Ganpati, V.S., 1962. A five year investigation of the Almati reservoir, Chennai. Parts 1–4. *Environ. Hlth.*, 6: 1–39.

Garg, R.K., Saksena, D.N. and Rao, R.J., 2005. Physico-chemical characteristics and trophic status of Harshi reservoir Gwalior (M.P.). In: *Abstract Booklet of Nat. Sem. on Recen. Advances in Aquatic Biology*, March 15–16, pp. 12.

Gonzalves, E.A. and Joshi, D.B., 1946. Freshwater algae near Bombay. I. The seasonal Succession of the algal in the tank of Bandra. *J. Bombay Nat. Hist. Soc.*, 46: 154–175.

Hannan, H.E., 1979. Chemical modification in the reservoir regulated streams. In: *The Ecology of Neglected Streams*, (Eds.) J.W. Ward and J.A. Standford. Plenum Corporation Publication, p. 75–94.

Hutchinson, G.E., 1957. *A Treastise on Limnology*. John Wiley and Sons Inc., New York, London and Sydney.

Kadam, M.S., Narware, S.S. and Ambore, N.E., 2008. Quality of lentic water of Bhategaon Pond in Parbhini Dist., Maharashtra. *Life Science Bulletin*, 5(1): 27–29.

Kaushik, S., 1992. Ecological studies of some water bodies in Gwalior, (M.P.). *Ph.D. Thesis*, Jiwaji University, Gwalior.

Khan, N.A. and Zutshi, D.P., 1980. Primary productivity and trophic status of a Kashmir Himalaya Lake. *Hydrobiologia*, 68(1): 3–8.

Khanna, D.R., Gautam, A., Chugh, T. and Sarkar, P., 2000. Impact of abiotic factors on the phytoplankton population of a pond. *Env. Cons. J.*, 1(1): 41–46.

Kumar, A., 1996. The limnological profile of a tropical fish farming pond at Dumka (Santhal Pargana) Bihar. *J. Ecobiol.*, 8: 177–122.

Kumar, A. and Bohra, C.J., 2002. Faunistic composition, communities structure, population dynamics and ethology of the birds of Udhawa lake (Santhal Pargana) and the strategies for the conservation in new millennium. In: *Ecology and Conservation of Lakes, Reservoir and Rivers*, Vol. II, (Ed.) A. Kumar. ABD Publisher, Jaipur, pp. 411–434.

Kumar, A. and Singh, A.K., 2002. Ecological conservation and management of river Mayurakshi in Santhal Pargana (Jharkhand) with special reference to effect of sewage pollution on abiotic and biotic potential. In: *Ecology and Conservation of Lakes, Reservoir and Rivers*, (Ed.) A. Kumar. ABD Publisher, Jaipur, pp. 1–43.

Kumar, R.S., 2002. Spatial and temporal diversity of zoobenthocoenosis in a freshwater ecosystem of Pathanamthitta in Kerela. In: *Ecology and Conservation of Lakes, Reservoir and Rivers*, (Ed.) A. Kumar. ABD Publisher, Jaipur, pp. 329–343.

Kumbar, S.M., Ghardage, A.B. and Patil, S.S., 2008. Preliminary studies on physico-chemical parameters of pond water of Palus, Dist. Sangli (MS). *Life Science Bulletin*, 5(1): 59–61.

Manimegalai, M., Umarnath, S. and Binukumari, S., 2008. Limnological studies of Ooty lake in Tamil Nadu. *Indian J. Environ. and Ecoplan.*, 15(1–2): 427–433.

Nair, M.S.R.; Seth, O.N. and Mishra, K.D., 1998. Limnological studies on a village pond Imalia (Vidisha). I. Physico-chemical characteristics. In: *Proc. Nat. Symp. Past, Present and Future of Bhopal Lakes*, pp. 195–201.

Pathak, S.K., 2004. Limnological status of Virla reservoir, District West Nimar (Khargone), M.P., India *Ph.D. Thesis*, D.A.V.V., Indore (M.P.).

Prakash, M.M., 1994. Seasonal variation in the physico-chemical characteristics of a perennial pond Jhabua (M.P.). *Aqu. Biol. Fish*, 1(1): 43–47.

Rajput, A. and Bhadoria, A.K.S., 1996. The study of seasonal parameters of water in Hanumantal Lake of Jabalpur. In: *Assessment of Water Pollution*, (Ed.) S.P. Mishra. APH Publishing Corporation, New Delhi, pp. 342–348.

Rawson, D.S., 1930. *The Bottom Fauna of Lake Simcoe and its Role in the Ecology of the Lake*. Univ. Torono Studies Publ. Ont. Fish. Res. Lab., 40: 1–180.

Sharma, D.K., Sharma, M. and Jain, R., 2002. Physico-chemical parameters of Halali reservoir (M.P.) in relation to aquatic organism. In: *Abstract Booklet on Nat. Sem. on Env.Poll. and Fish Manag.* Vikram University, Ujjain, Aug. 8–10, p. 23.

Shastree, N.K., Islam, M.S., Pathak, S. and Afshan, M., 1991. Studies on the physico-chemical dimensions of the lentic hydrosphere of Ravindra Sarovar (Gaya). In: *Current Trends in Limnology.* Narendra Pub. House, Delhi, 1: 133–152.

Sheikh, R., 1993. Studies of physico-chemical characteristics of Bilawali talab. Indore. In: *Abst. in 14th Ann. Sers. Acad. Environ. Biol.*, Indore, pp. 110.

Shrivastava, S., Shukla, A.N. and Rao, K.S., 2003. Biodiversity of Undasa wetland Ujjain (India) with special reference to its conservation. *J. Exp. Zool. India*, 116(1): 125–135.

Shrivastava, C.S., 2002. Ecology and dynamics of plankton in lotic habitat. *Ph.D. Thesis,* DAW, Indore (M.P.).

Shukla, A.N., 1995. Limnological studies on Gandhi Sagar reservoir with special reference to Macro-zoobenthos. *Ph.D. Thesis,* Vikram University, Ujjain, India.

Singh, R.K., 1986. Fluctuations in the composition of zooplankton population in relation to hydrobiological conditions of a reservoir. *J. Hydrobiol.*, 2(3): 37–42.

Sing, S.B. and Shani, R., 1978. Seasonal fluctuation of Zooplanton population in relation to certain physico-chemical factors of a Jalwana pond of Gorakhpur. *Geobios.*, 5: 228–230.

Sreenivasan, A., 1974. Limnological factors of trophical impoundment Bhavani Sagar reservoir, Tamil Nadu, India. *Int. Rev. Ges. Hydrobiol.*, 59: 327–342.

Unni, K.S., 1972. An ecological study of Macrophytic vegetation of Doodnadhani lake, Raipur. III. Chemical factors. *Hydrobiol.*, 39.

Welch., P.S., 1952. *Limnology*, 2nd edn. McGraw Hill Book Co. New York.

Wetzel, R.G., 1975. *Limnological Analysis*. Saunders, W.B. Philadelphia.

Wetzel, R.G., 1983. *Limnological Analysis*. W.B. Saunders Company, pp. 743.

Zafar, A.R., 1964. On the ecology of algae in certain fish ponds of Hyderabad, India. I. Physico-chemical complexes. *Hydrobiol.*, 23: 179–195.

Water and Wastewater Management in 2 Vols (2011) *Pages* **652–657**
Editors: **D.R. Khanna, R. Bhutiani & Gagan Matta**
Published by: **DAYA PUBLISHING HOUSE, NEW DELHI**

Chapter 96

Effect of Textile Effluent on Chemical Characteristics of Wunna River of Wardha District, Maharashtra

S.S. Gandhewar[1], K. Vasudevan[2] and S.B. Zade[3]
[1]*Department of Fisheries, Jankidevi Bajaj College of Science, Wardha*
[2]*Department of Zoology, DDE, Annamalai University, Annamalai Nagar, Tamil Nadu*
[3]*P.G. Department of Zoology, M.B. Patel College, Sakoli*

ABSTRACT

The chapter describes the chemical characteristics of Wunna River at Hinganghat, Wardha district of Maharashtra State. Various chemical characteristics such as BOD, COD, total alkalinity, total hardness, total dissolved solids, chloride, sodium, nitrate, chromium and iron were studied at 3 different station of Wunna River *viz.* upstream region (S_1), effluent mixed region (S_2) and 700 m away from effluent mixed region (S_3). Marginal variations in chemical characteristics were noticed at station 1 and 3. However, higher values were recorded at station 2 *i.e.*, effluent mixed region. Data analysis indicated heavy pollution of ecosystem of the river at the site of release of textile industry wastewater *i.e.* S_2.

Keywords: *Analysis, BOD, COD, Chemical characteristics, Ecosystem, Textile industry, Wastewater, Wunna River.*

Introduction

Water is one of the most precious gifts of nature. It occupies over 71 per cent of the earth surface (Miller, 1991) of which about 97 per cent is saline in nature while 2.14 per cent water is trapped in the giant glaciers and polar ice caps. This means that not even 1 per cent quantity of water is available for drinking, agriculture, domestic and industrial consumption.

There are 14 major Rivers in India that share 83 per cent of the total drainage basin and contribute 85 per cent of the total surface flow (Chaudhari, 1982). Water used by the consumers must be free from microbial contamination, toxic substance and excessive amount of minerals and organic matter (Godi *et al.*, 2003). Anthropogenic activities like rapid industrialization, urbanization and improper waste management techniques (Rao and Rao, 1995; Todd, 1995, Ranga Raj *et al.*, 1996) lead to heavy pollution of water.

The effluents of one of the textile industries involved in fabric dyeing are being released in Wunna River. Hence, the present investigation was undertaken to assess the changes in chemical characteristics of the River due to release of industrial effluents.

Materials and Method

Wunna River

It originates from Pilkapar rows of Mahargarh valley in Tahsil Katol, District Nagpur of Maharashtra State. The latitude of the river is 20°32′58″, the longitude is 78°49′00″, and altitude is 214.2 metre. It is the main tributary of river Wardha that joins the Pranhita River which ultimately flows into the Godavari River. People residing in the vicinity of Wunna River heavily depend on it for drinking and other domestic purposes.

Study Stations

Water samples from three stations of the Wunna River were collected monthly from Station-1 (upstream region), Station-2 (effluent mixed region) and Station-3 (700 m away from effluent mixed region) during June 2007 to March 2008 between 10 A.M. to 11 A.M. The samples were collected 2 meter away from the bank and 0.3 meter below surface to prevent the surface micro layer.

Sample Collection and Analysis

The samples were collected in pre-cleaned plastic bottles. The samples were brought to the laboratory and stored at 4 °C till the analytical work was carried out. The chemical analysis of samples were undertaken as per American Public Health Association (APHA, 1989) and Bureau of Indian Standards (IS-3025).

Results and Discussion

The chemical analysis of samples of three different locations distribution of these parameters for the three stations of the river was analyzed using descriptive statistics and the results are given in the Tables 96.1 and 96.2.

Alkalinity

The estimated levels of alkalinity at different stations along the course of the river were presented in Tables 96.1–96.2. At station 1, maximum value of alkalinity recorded was 303 mg/l, while minimum value was found to be 105 mg/l. For station 2, the maximum value was 696.6 mg/l and minimum value was found to be 309 mg/l. For station 3, the maximum value was 397 mg/l and minimum value was found to be 132 mg/l.

Hardness

At station 1, maximum value of hardness was found to be 332.8 mg/l, while minimum value was 109 mg/l. For station 2, the maximum value was found to be 450 mg/l. For station 3, the maximum value was found to be 327 mg/l and minimum value was 128 mg/l (Tables 96.1–96.2).

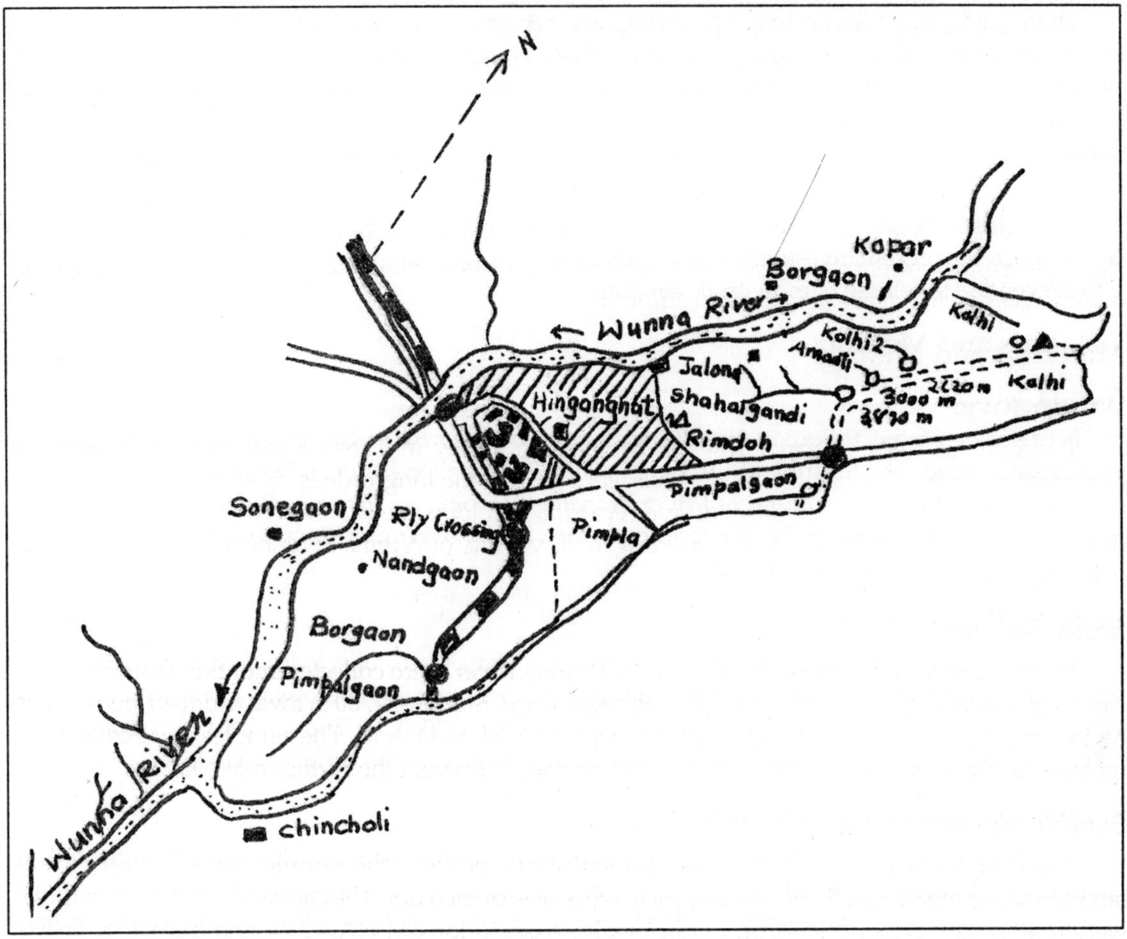

Figure 97.1: Location Map of River Wunna at Hinganghat

BOD

In the present study, the estimated levels of BOD at station 1, maximum value of BOD recorded was 4.9 mg/l, while minimum value observed was 2.0 mg/l. For station 2, maximum value was 112 mg/l in October 2007 which indicated heavy organic load. While minimum value was 69 mg/l. For station 3, maximum value of BOD recorded was 19 mg/l and minimum value was 13 mg/l (Tables 96.1–96.2).

Chemical Oxygen Demand (COD)

The estimated levels of COD at station 1 had ranged from 23-54mg/l (Table 96.1). Station 2 was recorded with minimum level of 240mg/l and the maximum levels of COD recorded were 289 mg/l. At station 3 recorded levels ranged from 55-119 mg/l (Table 96.2). Variations in the estimated levels of COD at all the stations studied during the period of investigation, station 2 shows estimated COD levels more than 10 folds increased than the reference site (Station 1) due to the influx of industrial wastewater (Tables 96.1–96.2).

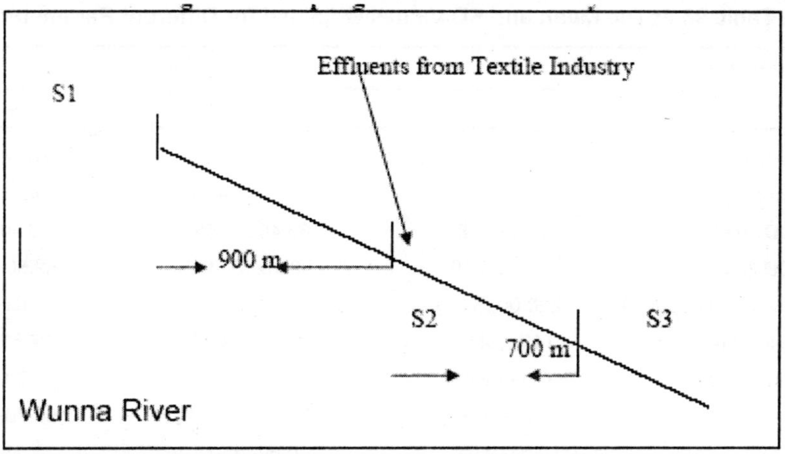

Figure 96.2: Sampling Locations in the Study Area

Total Dissolved Solids

At station 1 TDS levels ranged from 180-290 mg/l during the period of study; the minimum levels were recorded during the November 2007 and March 2008; the maximum level during the month of January 2008 (Table 96.1). Station 2 were ranged from 1600-2050 mg/l (Table 96.2). And 410-650 mg/l respectively at station 3 (Table 96.2) Among the stations studied Station 2 were recorded with highest levels of TDS during the period (June 07-March 08) of study.

Table 96.1: The Ranges of Values Obtained for Different Parameters

Parameter	*WHO/Indian Standards*	*Ranges of Values*		
		Station 1	*Station 2*	*Station 3*
Alkalinity (mg/l)	200	105-303	399-696.6	132-397
Hardness (mg/l)	200-300	109-332.8	266.24-450	128-327
BOD (mg/l)	1-2	2-4.9	69-112	13-19
COD (mg/l)	0.5	23-54	240-289	55-119
Total Dissolved Solids (mg/l)	500	180-290	1600-2050	410-650
Chloride (mg/l)	200-300	11.1-19.7	750.9-905.1	30-48
Sodium (mg/l)	–	11-19	611-938	29-58
Nitrate (mg/l)	45	1.1-3.1	550-620	40-80
Chromium (mg/l)	0.05	0.02-0.05	0.08-0.10	0.04-0.07
Iron (mg/l)	0.3	0.1-0.2	0.15-0.29	0.12-0.27

Chloride

At station 1 recorded levels of chloride ranging from 11.1 to 19.7 mg/l (Table 96.1) For station 1, maximum value was 19.7 mg/l in October 2007 and minimum 11.1 mg/l in July 2007. For station 2, maximum value observed was 905.1 mg/l in January 2008 and minimum was 750.9 mg/l in July 2007. For station 3, maximum value was 30 mg/l recorded in January and March 2008 and minimum was 30 mg/l recorded in August 2007 (Tables 96.1–96.2).

Table 96.2: The Mean and SD Values Obtained for Different Parameters

Parameter	*Mean and SD Values*		
	Station 1	*Station 2*	*Station 3*
Alkalinity (mg/l)	213.45±74.39	564.06±93.27	270.88±93.34
Hardness (mg/l)	140.88±67.83	350.02±54.27	164.10±58.78
BOD (mg/l)	3.53±0.79	83.40±12.55	16.40±2.01
COD (mg/l)	32.80±9.08	265±13.39	69.50±18.76
Total Dissolved Solids (mg/l)	220.00±3.97	1779.0±13.67	516.0±7.66
Chloride (mg/l)	14.59±2.36	814.34±44.13	38.59±5.52
Sodium (mg/l)	15.90±2.60	776.50±106.27	43.90±10.39
Nitrate (mg/l)	2.00±0.56	588.10±29.74	56.70±12.75
Chromium (mg/l)	0.04±0.01	0.09±0.01	0.06±0.01
Iron (mg/l)	0.14±0.03	0.22±0.05	0.19±0.05

Sodium

At station 1 recorded maximum value of 19 mg/l in August 2007 and February 2008 and minimum was 11 mg/l in November 2007. For station 2, maximum value was 938 mg/l in October 2007 and minimum 611 mg/l in February 2008. At station 3, maximum value was 58 mg/l in December 2007 and minimum 29 mg/l in July 2007 (Tables 96.1–96.2).

Nitrates

Tables 96.1–96.2 provided the data on variations in the nitrate levels at different stations along the course of Wunna river. At station 1, maximum value was found to be 301 mg/l in July 2007 and minimum was 1.1 mg/l in February 2008. For station 2, maximum was 620 mg/l in December 2007 and minimum 540 mg/l in November 2007. For station 3, maximum value recorded was 80 mg/l in September 2007 and minimum value was 40 mg/L in the month of March 2008.

Chromium

At station 1, maximum value recorded was 0.05 mg/l in the month of October 2007 and February 2008 and minimum was 0.02 mg/l recorded in June 2007. For station 2, minimum value was 0.1 mg/l in the month of August 2007, while minimum value was 0.08 mg/l in December 2007. For station 3, maximum value was 0.07 mg/l in July 2007, while minimum value was 0.04 mg/l in June 2007 (Tables 96.1–96.2).

Iron

The estimated levels of iron content at different stations during the period of study were presented in the Tables 96.1–96.2, maximum value observed was 0.20 mg/l in March 2008, while minimum value was 0.10 mg/l in August 2007. Station 2, maximum value was 0.29 mg/l in March 2008 while minimum value was 0.15 mg/l in January and February 2008. For station 3, maximum value was 0.27 mg/l in January 2008 while minimum value was 0.12 mg/l in August 2007.

Conclusion

From the exceptionally higher values of BOD, COD, TDS, Hardness, Alkalinity, Sodium, Chloride and Nitrate at station 2 *i.e.*, effluent mixed region, it may be concluded that the water of station 2 is unsuitable for drinking and other domestic purposes.

Acknowledgements

The authors are thankful to Dr. Om Mahodaya, Principal, J.B. College of Science, Wardha and Prof. R. D. Nehete, Dr. G. N. Paliwal and P.V.Tekade for their help during this research work.

References

APHA, 1989. *Standard Methods for the Examination of Water and Wastewater*, 17th edn. APHA, New York, USA.

Bureau of Indian Standards–IS: 3025. *Methods of Sampling and Test (Physical and Chemical) for Water and Wastewater*.

Chaudhari, N., 1982. *Water and Air Quality Control: The Indian Context for the Prevention and Control of Water Pollution*, New Delhi, India.

Engleman, R. and Roy, P.,1993. Cited in: *Sustaining Water, Population and the Future of Renewable Water Supplies*, Washington D.C., Population and Environment Programme, Population Action International.

Godi, S.D., Barbudde, S.B., Hazel, D. and Dolley, C., 2003. Physico-chemical characteristics of drinking water at Velsao (Goa). *J. Ecotoxicol. Env. Monitor*, 13(3): 203–209.

Water and Wastewater Management in 2 Vols (2011) *Pages* ***658–662***
Editors: **D.R. Khanna, R. Bhutiani & Gagan Matta**
Published by: **DAYA PUBLISHING HOUSE, NEW DELHI**

Chapter 97

Physico-chemical Analysis of River Panvdhoi at Saharanpur, U.P.

D.R. Khanna[1], J. Ashraf[2], Beena Chauhan[1] and R. Bhutiani[1]
[1]Department of Zoology and Environmental Sciences, Gurukul Kangri University, Haridwar
[2]Department of Zoology, M.S. College, Saharanpur

ABSTRACT

The seasonal physico-chemical and heavy metals concentration of Panvdhoi river Saharanpur has been analysed. Various parameters *viz.* temperature, conductivity, turbidity, total solids, total dissolved solids, total suspended solids, pH, BOD, COD, DO, free CO_2, alkalinity, hardness and chloride were observed. In Heavy metals lead, copper, zinc and mercury were observed. Less concentration of these parameters except DO were obtained at Sampling station A, B and C while greater amount were observed on sampling station D and E.

Keywords: Rher, Water, BOD, COD, DO, Alkalinity.

Introduction

Water is most precious commodity of life. It is not only the basic need for sustaining human life but also vital to all the segments of economic development. Water is being adversely affected, qualitatively and quantitatively by all kinds of human activities on land, in air and/or in water. According to Bernar (1951) domestic sewage consists of about 99.9 per cent water coming from washing, rinsing, flushing and other activities. The major part of pollution takes place by industrialization because in the industrial sewage there are large number of chemicals and heavy metals.

District Saharanpur is situated in the north of Uttar Pradesh. Many rivers flows through the Saharanpur *viz.* Yamuna, Hindon, Damola, Maskara, Krishna, Kalinadi, Panvdhoi *etc.*

River Panvdhoi flows through Saharanpur district, it is a stream fed river and a tributary of Hindon. This river originates near Shanklapuri Shiv Mandir of Panwarka, then it goes to Saharanpur. It is about 16 km. in length and then in Saharanpur it mixes with river Dhamola. Due to industrialization in district Saharanpur main drain carries away some effluent from the factories and also from the residential colonies. This drain carries domestic sewage, which is poured into river Panvdhoi. Therefore, it can be well considered that it carries a variety of pollutants of equally different in physico-chemical nature. These effluents changes the physico-chemical characteristics to such an extent that they sometimes lead to high fish mortality.

The location of Saharanpur on Globe is on latitude 29°58′ North and longitude 77°33′ East while the height from sea level is 270.50 meter. Present study was carried out on stretch of 16 Km from Shanklapuri Shiv Mandir to near Dhamola which is having a width of 1–6 meter and depth of 0.5 to 1.5 meter.

River Panvdhoi is a small tributary having a length of 16 kms. It originates from Panwarka and in Saharanpur conflicts into Dhamola. The place where it originates is called Sarkadi Ki Puliya. In this point the width of the river is 1–4 meter and depth is 1.5 meter.

Due to new developing urban areas however, so far no study has been made to find out the seasonal variation in the ecology of the river Panvdhoi at Saharanpur. Hence the study was conducted to fill up this Lacuna.

Materials and Method

Sampling Stations of the Study

1. Shanklapuri Shiv Mandir
2. Makhraj Ka Pul
3. Lal Das Ka Baada
4. Jogyan Pul
5. Near Dhamola

For ecological study of river Panvdhoi the water samples were collected fortnightly from different sampling stations for a period of two years.

Various physico-chemical parameters and Heavy metals were analyzed following the standards methods of APHA (1998), Trivedi and Goel (1986), Mathur (1982).

Results and Discussion

The results (average annual values) of the present study are given in Tables 97.1–97.2. In the present study a difference in the fluctuation of water temperature was observed 15.6°C±4.5°C to 25.60°C±4.0°C. The water temperature showed an upward trend from winter season to summer season followed by a downward trend from monsoon season onwards. A more or less similar trend has been observed in the river Yamuna by Chakrabarty *et al.* (1959) and in the Kallayi by John (1976). Same trend of temperature was observed by Khanna (1993) in river Ganga at Haridwar.

The annual average value of dissolved oxygen ranged between 10.047 mg/l±2.09 mg/l to 1.20 mg/l±1.07 mg/l, Chopra *et al.* (1990), Gopal and Shah (1993), Joshi *et al.* (1993) and Sharma (1999) also got the same results and have opined that low temperature in winter increases the oxygen retaining capacity of water and solubility of oxygen in water. A negative relationship has been observed between

Table 97.1: Average Annual Value of Physico-chemical Parameters of River Panvdhoi

Physico-chemical Parameters	Sampling Station 'A'		Sampling Station 'B'		Sampling Station 'C'		Sampling Station 'D'		Sampling Station 'E'		Average
	2002–2003	2003–2004	2002–2003	2003–2004	2002–2003	2003–2004	2002–2003	2003–2004	2002–2003	2003–2004	2002–2004
Temperature (ºC)	15.6±4.50	16.5±6.30	17.6±4.20	19.3±4.30	20.2±5.20	20.4±5.70	23.4±5.70	23.1±4.40	25.0±4.40	25.6±4.00	20.7±3.50
Conductivity (µmhos)	191.44±127.32	206.70±137.55	191.44±121.15	206.84±137.63	165.56±117.35	166.33±116.22	148.96±110.11	137.11±102.53	141.07±104.39	149.20±105.64	170.62±26.56
Turbidity	57.78±54.54	66.81±70.78	49.58±40.44	44.89±27.63	243.33±30.55	242.20±39.75	533.93±126.08	524.18±85.38	908.46±101.05	901.07±198.88	351.42±341.56
Total Solids (mg/L)	179.53±168.53	210.96±25.71	181.92±169.96	213.17±219.34	186.60±23.33	187.44±28.89	1627.40±647.20	1598.38±629.34	1126.34±651.22	1825.66±748.50	793.80±777.64
TDS (mg/L)	121.67±113.62	141.74±149.86	123.03±114.19	142.78±149.85	105.24±13.10	105.58±13.30	720.45±313.86	787.93±326.17	766.66±351.18	500.00±379.86	351.51±304.54
TSS (mg/L)	57.85±54.94	69.22±70.70	58.88±55.79	70.39±69.73	81.53±10.13	81.85±15.52	906.93±347.60	810.45±343.52	960.26±307.78	965.83±378.07	406.31±14.53
pH	7.32±0.25	7.20±0.18	7.34±0.25	7.24±0.16	7.50±0.09	7.51±0.07	8.42±0.42	8.51±0.40	8.65±0.37	8.65±0.40	1.63±0.63
BOD (mg/L)	2.26±0.82	2.29±0.53	2.54±0.85	2.57±0.74	8.42±2.74	8.80±3.12	281.66±66.58	293.33±73.71	468.33±80.98	475.00±78.10	154.52±203.23
COD (mg/L)	5.69±1.45	4.81±1.59	5.62±0.69	4.95±1.69	28.11±6.56	25.90±6.37	768.33±173.22	876.33±174.73	1075.00±220.17	1156.66±188.76	395.34±504.84
DO (mg/L)	9.55±2 18	9.56±1.99	9.61±1.41	10.47±2.09	7.56±0.83	7.57±1.62	4.66±0.58	4.44±0.69	1.68±1.47	1.20±1.07	6.63±3.41
Free CO_2 (mg/L)	2.02±1.95	1.55±1.51	2.56±1.52	2.22±1.55	2.69±1.15	2.59±1.17	3.93±1.36	4.29±1.39	4.74±1.56	4.89±1.34	2.11±1.22
Alkalinity (mg/L)	50.02±15.76	51.93±15.74	50.89±15.94	52.70±15.82	261.17±11.99	244.25±39.71	711.08±3.57	701.05±92.54	181.66±62.11	761.24±112.55	366.70±330.46
Total Hardness (mg/L)	95.66±11.71	97.57±14.10	96.44±12.19	100.66±14.01	221.94±6.96	217.58±7.46	275.83±16.75	275.26±51.77	279.74±17.27	280.31±17.12	194.09±86.04
Chloride (mg/L)	5.51±5.29	5.31±5.26	5.85±5.37	5.74±5.42	34.45±4.33	34.96±6.30	57.16±3.59	44.51±8.15	59.67±3.16	60.51±2.09	31.36±23.95

Table 97.2: Average Annual Value of Heavy Metals of River Panvdhoi

Parameters	Sampling Station 'A'		Sampling Station 'B'		Sampling Station 'C'		Sampling Station 'D'		Sampling Station 'E'		Average
	2002–2003	2003–2004	2002–2003	2003–2004	2002–2003	2003–2004	2002–2003	2003–2004	2002–2003	2003–2004	2002–2004
Lead (Pb) (mg/L)	0.0702±0.0076	0.0741±0.0090	0.0711±0.0073	0.0719±0.0087	1.6170±0.3797	1.9533±0.3776	3.3084±0.0776	3.3320±0.1271	4.8452±0.6109	4.9075±0.6582	2.0882±1.9130
Copper (Cu) (mg/L)	0.0550±0.0291	0.0533±0.0324	0.0626±0.0263	0.0602±0.0332	0.0835±0.0170	0.0811±0.0236	0.8308±0.1756	0.8351±0.1757	1.5945±0.3729	1.5596±0.3328	0.5710±0.6171
Zinc (Zn) (mg/L)	3.1651±0.0528	3.2098±0.0331	3.2109±0.259	3.2281±0.0304	3.6755±0.2067	3.6913±0.3360	4.2443±0.0732	4.2517±0.1139	4.8695±0.1763	4.871±0.1750	3.6417±0.6772
Mercury (Hg) (mg/L)	0.0000±00000	0.0000±0.0000	0.0000±0.0000	0.0000±0.0000	0.0000±0.0000	0.0000±0.0000	0.0001±0.0000	0.0003±0.0002	0.0005±0.0006	0.0006±0.0002	0.0001±0.0002

BOD and DO content. A similar pattern has been reported by Khanna (1993) in river Ganga. The annual average value of biochemical oxygen demand ranged between 2.26 mg/l to 78.10 mg/l.

The annual average value of COD ranged between 4.81 mg/l±1.59 mg/l to 1156.66 mg/l±188.76 mg/l. Similar trend of COD have shown by Khanna and Singh (2000) in river Suswa at Raiwala.

Annual Average value of free CO_2 varied between 1.55 mg/l±1.61 mg/l to 4.89 mg/l±1.34 mg/l. Pahwa and Mehrotra (1966) and Ray *et al.* (1966) have reported that the Ganga river contains maximum free CO_2 in rainy Season at Allahabad. Annual average value of total dissolved solids varied between 105.24 mg/l±13.10 mg/l to 787.93 mg/l±326.17 mg/l.

The annual average values of pH varied between 7.20±0.18 to 8.65±0.40. High value obtained during rainy season may be due to rainy water, runoff of sewage drains *etc.* It was recorded that pH remains slightly alkaline in nature throughout the study. Das (1961) reported that pH of water has an important behaviour on plankton and microbial production.

The annual average value of alkalinity varied between 50.02 mg/l±15.76 mg/l to 781.66 mg/l±62.11 mg/l. Similar observations was also obtained by Holden and Green (1960), Tailing and Rzoska (1967) and Abidin (1948a).

The annual average value of total hardness ranged from 95.66 mg/l±11.71 mg/l to 280.31 mg/l±17.12 mg/l. Mishra and Joshi (2003) observed hardness in river Ganga at Haridwar and found more or less similar trend in their study. The annual average value of Chloride varied between 5.31 mg/l±5.26 mg/l to 60.51 mg/l to 2.09 mg/l. Chlorides are present in sewage, sewage effluents and farm drainage. Significant levels of chloride content were shown by many rivers like Yamuna, Sengar *et al.* (1985) and Tungbhadhara, Reddy and Venkateshwarlu (1987).

In the present study heavy metals, lead, copper, zinc and mercury were taken for observation. The annual average value of heavy metals are given in Table 97.2. The annual average of Lead ranged between 0.0702 mg/l±0.02291 to 4.9075 mg/l±0.6582. The minimum value was found from sampling station A in 2002–2003 and maximum from sampling station E in 2003–2004. The annual average value of copper varied between 0.053 mg/l±0.0324 to 1.5945 mg/l±0.3729 in which average value of copper was obtained at sampling station A in 2003–2004 and maximum average value at sampling station E in 2002–2003. The annual average value of Zinc ranged between 3.1651 mg/l±0.0528 to 4.871 mg/l±0.1750 in which minimum average value of Zinc was obtained from sampling station A in the year 2002–2003 and maximum was recorded at sampling station E in the year 2002–2003. The annual average value of Mercury varied between 0.0000 mg/l±0.0000 to 0.0006 mg/l±0.0002. The minimum was found as nil in both the years (2002–2004) at sampling stations A, B, C and maximum at sampling station E in 2003–2004 (Table 97.2). The concentration of these metals at sampling stations C, D and E give a highly misleading picture of the degree of metal pollution. Heavy metals get contaminated into aquatic systems as result of various natural activities (Weathering of soils and rocks from volcanic eruptions) and from a variety of human activities involving the mining, processing or use of metals or substances. Some metals, such as copper and zinc are essential micronutrients, while others such as mercury and lead are not required even in small amount by any organisms. Virtually all metals, including the essential metal micro-nutrients, are toxic if exposure levels are sufficient high. Andren (1974) reported that about 67 per cent of the mercury in the Mississippi river water is associated with suspended sediments.

References

Abdin, G., 1948a. Physical and chemical investigation relating to algal growth in the river Nile. *Cairo Bull. Inst. Egypt*, 29: 20–24.

Andren, A.W., 1974. *Ph.D. Thesis*, Florida State University Oceanography Department, Tallahaseefla.

APHA, 1998. *Standard Methods for the Examination of Water and Wastewater*. American Public Health Association, 1015 Fifteen Street, NW Washington, 15: 1–1134.

Bemer, L.M., 1951. Limnology of the lower Missouri River, *Ecology*, 32: 1–12.

Chakrabarty, R.D., Ray, P. and Singh, S.B., 1959. A quantitative survey of plankton and physiological conditions of the river Yamuna at Allahabad in 1954–1955. *Indian J. Fish.*, 6(1): 186–203.

Chopra, A.K., Madhwal, B.P. and Singh, H.R., 1990. Abiotic variable and primary productivity of river Yamuna at Naugaon, Uttarkashi, Garhwal. *Indian J. Ecol.*, 17(1): 61–64.

Das, S.M., 1961. Hydrogen ion concentration, plankton and fish in freshwater Eutrophic lakes of India. *Nature*, 191(4787): 511–512.

Gopal, B. and Shah, M., 1993. Conservation and management of rivers in India: Case study of the Yamuna. *Environmental Conservation*, 20(2): 243–253.

Holden, J.M. and Green, J., 1960. Hydrobiology and plankton of the river Sokoto. *J. Anim. Ecol.*, 29(1): 65–84.

John, V., 1976. Hydrobiological studies on the river Kallayi in Kerala. *Indian J. Fish.*, 23: 72–85.

Joshi, B.D., Pathak, J.K., Singh, Y.N., Bisht, R.C.S. and Joshi, P.C., 1993. On the physico-chemical characteristics of river Bhagirathi in uplands of Garhwal Himalaya. *Him. J. Env. Zool.*, 7: 64–75.

Khanna, D.R., 1993. *Ecology and Pollution of Ganga River*. Ashish Publishing House, New Delhi, p. 1–241.

Khanna, D.R. and Singh, R.K., 2000. Seasonal fluctuations in the plankton of Suswa river at Raiwala Dehradun. *Env. Cons. J.*, (2 and 3): 89–92.

Mathur, R.P., 1982. *Water and Wastewater Testing*. Neem Chand and Bros. Publishers, Roorkee, p. 1–54.

Mishra, S. and Joshi, B.D., 2003. Assessment of water quality with few selected parameters of River Ganga at Haridwar. *Him. J. Env. Zool.*, 17(2): 113–122.

Pahwa, D.V. and Mehrotra, S.N., 1966. Observation on fluctuation in abundance of plankton in relation to certain hydrological conditions of river Ganga. *Proc. Nat. Acad. Sci.*, 36B(2): 157–189.

Ray, P., Singh, S.B. and Sehgal, K.L., 1966. A study of some aspects of the river Ganga and Yamuna at Allahabad (U.P.) in 1958–59. *Proc. Nat. Acad. Sci., India*, 36B(3): 235–272.

Reddy, P.M. and Venkateshwarlu, V., 1987. Assessment of water quality and pollution in river Tungbhadra near Kurnool (A.P.). *J. Environ. Biol.*, 8(2): 109–119.

Senger, R.M.S., Sharma, K.D. and Pathak, P.D., 1985. Studies on distribution of algal flora in polluted and non-polluted regions in Yamuna River at Agra (U.P.). *J. Indian Bot. Soc.*, 64: 365–376.

Sharma, A., 1999. Limnological studies of Ban Ganga and distributional pattern of stream bottom fauna. *Ph.D. Thesis*, University of Jammu.

Tailing, J.E. and Rzoska, J., 1967. The development of plankton in relation to hydrobiological regime in Blue Nile. *J. Ecol.*, 55: 636–662.

Trivedi, R.K. and Goel, P.K., 1984. *Chemical and Biological Methods for Water Pollution Studies*. Environmental Publications, Karad, p. 1–251.

Water and Wastewater Management in 2 Vols (2011) *Pages* **663–684**
Editors: **D.R. Khanna, R. Bhutiani & Gagan Matta**
Published by: **DAYA PUBLISHING HOUSE, NEW DELHI**

Chapter 98

Environmental Problems of Using Water in India: Need for Generation of High Value Added Products

S.N. Kaul[1], Tapas Nandy[1] and L. Szyprkowicz[2]
[1]*National Environmental Engineering Research Institute, Nehru Marg, Nagpur – 440 020, India*
[2]*Department of Environmental Science, Venice University, Venice, Italy*

ABSTRACT

The present chapter is an attempt to present the status of environmental problems using water with special reference to India. The paper is also highlighting the need to generate high value added products from the wastewater generated from various activities.

Keywords: Environment, Water use, Energy, Value added products.

Introduction

Water is a prime natural resource and is a basic human need. The underlying assumption in traditional water resources planning process still continues to be: "Fresh water is a gift of God which would continue to be available in a prepetuity and in abundance". This is not valid as both quantity and quality of water poses series problems.

Wastewater is a misplaced resource and can be reused to advantage through proper management. A differentiation, however, needs to be made between water reuse and recycling. Reuse implies utilization of water that has been previously used for another purpose, and recycle implies reuse of water more than once for the same purpose. India can take advantage of using bright sunshine, high temperatures and possibility of water usage for irrigating agriculture land for wastewater management. The approaches in wastewater management should essentially entail: waste volume reduction, waste strength reduction; and by-product recovery.

It is estimated that out of the total precipitation of around 400×10^4 million cubic meter per year in India the surface water resources cover as 176.8×10^4 million cubic meter. Out of this only 50 per cent can be put to beneficial use because of topographical and other constraints. Utilization of water resources in India is classified depending on nature of use into three main categories; domestic, agriculture and industry. An estimated total water use in the country is 4074 million meter cube per day for the year 2000–2001, the figures of distribution among various uses are 2.1 per cent, 88 per cent and 2.4 per cent respectively.

Water and Wastewater Status in India

Domestic Category

It is estimated that 31 per cent of rural population and 77 per cent of urban population in India has access to potable water supply. Surface water is the dominant source of organized urban water supply. In rural area groundwater is the basic source. Municipalities and State Water Supply and Sewerage Boards own and operate the urban water supply schemes, however in rural areas there is no such organized water supply. Table 98.1 gives the status of water supply, wastewater collection and treatment in Metropolitan cities of India.

Table 98.1: Status of Water Supply, Wastewater Collection and Treatment in Metropolitan Cities

Name of City	*Water Supply Population Covered (Per cent)*	*Sewerage System Population Covered (Per cent)*	*Treatment Plant*		*Mode of Disposal*
			Level	*Capacity (MLD)*	
Bombay	99	80	Primary	82.0	Creek, Sea
Delhi	96	75	Primary and Secondary	745.0	River Yamuna/ Agricultural land
Kolkata	95	–	Primary and Secondary	–	River Kulti and Hoogly
Hyderabad	100	75	Primary and Secondary	140.0	Land
Ahmedabad	90	75	Primary and Secondary	382.0	River Sabarmati
Bangaluru	100	85	Primary	286.0	Vrushabhavathi Valley, Bellandur Tank
Kanpur	75	60	Secondary	160.0	River Ganga Agricultural land
Lucknow	100	–	–	–	–
Chennai	85	77	Secondary	–	Sea, Irrigation
Nagpur	75	66	Primary (Not in operation)	45.5	Local Nallaha, Agriculture
Pune	78	53	Primary	90.0	River and Agricultural land
Jaipur	80	–	–	–	Agricultural land
Total	93	63		1930.4	

Source: Status of Water Supply and Wastewater Collection, Treatment and Disposal in Class I Cities, CPCB, CUPS/30/ 1989–1990.

The situation in sewerage collection treatment and disposal is discouraging as only 63 per cent of the metropolitan population is provided with sewage collection and treatment system. The data obtained from remaining municipalities shows less than five per cent of the total wastewater generated in Indian urban areas is collected and only about 2 per cent is receiving treatment leaving remaining 98 per cent discharged into the environment untreated.

This indiscriminate discharge of large quantity of untreated sewage into the surface water has deteriorated all the rivers in the country. India has 14 major rivers whose basin area is 83 per cent of the total land area of the country, they contribute 85 per cent of the surface water flow of the country and 80 per cent of the total population is living in the basins of these rivers.

The preservation and restoration of river water quality in India started with the launch of much published Ganga Action Plan (GAP) in 1984. One of the important part of GAP was collection and treatment of domestic wastewater discharged from urban centers located on the bank of Ganga. GAP phase-I has created an infrastructure for diverting 730 millions litre per day (MLD) and treating 540 millions litres per day of municipal sewage. The second phase of GAP aims in collection and treatment of domestic wastewater from 95 towns. The Central Government has also initiated the National River Conservation Plan (NRCP) involving pollution abatement works in polluted stretches of 18 rivers covering 46 towns.

Often construction of conventional sewage treatment plant requires huge capital cost. The Indian municipalities and public health departments are short of funds, thereby leaving most of the domestic wastewater untreated. The viable option is to adopt cost effective/low cost sewage treatment scheme. Sewage farming is one such option, which utilizes partially treated sewage for irrigation to grow crops and vegetables.

Land Application of Sewage for Irrigation: A Case Study of Chandigarh

70 MLD of community wastewater having 210 mg/L of BOD, 423 mg/L of TSS and 32 mg/L of oil and grease were generated from urban area of Chandigarh. About 98 per cent of the entire volume was collected at the sewage treatment plant (STP) and rest was extracted from the sewer trunk for land irrigation before it reaches the treatment plant. This sewage was partially treated in STP to reduce the pollutants levels through 190 mg/L of BOD, 405 mg/L of TSS and 29 mg/L of oil and grease. The partially treated wastewater is applied on land for cultivation of paddy crop. The sewage farm has 1250 acres of land adjacent to STP and is being utilised for past 10–15 years.

However, this sewage farm is not operated properly. There is heavy application of sewage water for cultivation of paddy crop, which has raised the water table of sewage farm upto the ground level. The quality of the grain and yield of paddy is not satisfactory. The accumulation of heavy metals in sewage irrigated soils was high compared to the soils receiving tube well irrigation. The groundwater near the sewage farm was showing high bacteriological contamination.

A study was conducted to improve the quality of the sewage farm. This study has recommended following important modifications and suggestions.

1. Land preparation and drainage to avoid wastewater logging.
2. Arrangement for periodic desludging.
3. Provision for subsurface drains for water table control.
4. Pretreatment of wastewater before application.
5. Optimum loading of wastewater and scientific irrigation schedule for crop.

6. Optimum fertilizer dosing.
7. Soil management practices.
8. Maintenance of soil pH level and organic matter.
9. Crop rotation.

Industrial Category

Industrial growth in India took place after the liberalization of Indian economy in 1991. To focus on industrial pollution the Government of India has identified 19 critically polluted areas in the country and 17 industrial sub sectors which are polluting namely–cement, thermal power plant, distilleries, sugar, fertilizers, integrated iron and steel, oil refineries, pulp and paper, petrochemicals, pesticides, tanneries, basic drugs and pharmaceuticals, dye and dye intermediaries, caustic soda, zinc, copper and aluminium smelters. The list of critically polluted areas is given in Table 98.2. Out of a total of 1551 units identified under these categories by Central Pollution Control Board (CPCB), 1259 units have installed adequate facilities for pollution control, 112 units have been closed down and the remaining are installing the pollution control schemes.

Table 98.2: Critically Polluted Areas with State

Sl.No.	*State*	*Area*
1.	Punjab	Mandi Gobindgarh
2.	Rajasthan	Pail
3.	Delhi	Najafgarh drain
4.	Gujarat	Vapi
5.	U.P.	Singhrauli
6.	Bihar	Dhanbad
7.	West Bengal	Durgapur
8.	West Bengal	Howrah
9.	Orissa	Talcher
10.	Tamil Nadu	North Arcot
11.	Tamil Nadu	Manali
12.	M.P.	Korba
13	Assam	Digboi
14.	Maharashtra	Chembur
15.	A.P.	Vishakhapatnam
16.	Karnataka	Bhadravati
17.	Kerala	Greater Cochin area
18.	Himachal Pradesh	Kala-Amb
19.	Himachal Pradesh	Parwanoo

Since November 1991, World Bank with MOEF is providing finance for pollution control projects to individual industrial units for pollution prevention and control. The funds are disbursed by Industrial Development Bank of India (IDBI) and the Industrial Credit and Investment Corporation of India

Limited (ICICI). From this credit line of the world bank several units have taken loans amounting to more than 90 millions US dollar. In view of such positive response from industry, Government of India has recently negotiated second line under Industrial Pollution Prevention Project where similar funds would again be available. In the case of new units, the cost of pollution control is internalized with the entire project cost and is financed within the overall financing package of the entire project. In general, this cost is less than 5 per cent of the total capital cost of the project except for specific industry sub sectors where the cost may be as high as 10 per cent.

Effective enforcement of environment legislation, increased public awareness and change in industry's perception of its social responsibility have resulted in significant increase in compliance to the various provisions of environmental legislation in the large and medium scale industries.

Status of the wastewater treatment and reuse for selected industrial sector in the country is briefly given below.

Pulp and Paper Industry

In India there are more than 379 paper mills with an annual installed capacity of around 37.78 lakh tonnes. A significant aspect of Indian paper industry is strong presence of small paper mills which contribute 50 per cent to the production of paper and paper board in the country.

The water requirement to the paper mills in India is in the range of 250–440 m^3/tonne of product and the requirement on the quality of paper made and extent of recycle. The water requirement is high compared to 60120 m^3/tonne water requirement achieved in several developed countries. Combined wastewater generation from large mills range between 167–281 m^3/tonne of product. No significant variation in wastewater generation is observed. It is found that lignin bearing colored wastewater accounts about 30 per cent of the total effluent discharged.

A study on wastewater recycle in the paper industry has revealed that wastewater recycle varies between 8 and 48 per cent with an average recycle of 21.4 per cent, which facilitates reduction in fresh water requirement to the extent of 7 to 44 per cent. A wastewater recycle pattern in a typical Indian paper mill is given Table 98.3.

Table 98.3: Wastewater Recycle Pattern in a Typical Indian Paper Mill

Type of Wastewater	*Source*	*Recycle Process*	*Wastewater Recycled (Per cent)*
Docker wash	Unbleached pulp filter	Screening and dilution of unbleached pulp at any place and chipping when required	56–60
Alkali backwater	Alkali extraction stage filter	Diluting chlorination tower contents from 3 per cent to 1 per cent and alkali tower contents	90–95
Hypo backwater	Hydro extraction stage	Some system, alkali extraction sage and rage plant	80–90
Final washer backwater	Bleech plant	Some system and rag plant	80–85
Paper machine	Paper machine	Dilution high density pulp before feeding to paper machine and for hydropulping	50–55

Source: Comprehensive Industry Document for Large Pulp and Paper Industry, Central Pollution Control Board, Delhi COINDS/3611991.

The pulp and paper industry is adopting recycling of wastewater either without treatment or after suitable treatment. It is also observed that there is growing recycle awareness in the industry.

Recovery of Caustic Soda from Black Liquor of Small Scale Pulp and Paper Industries in Common Chemical Recovery Unit

An individual chemical recovery unit is not economically viable for small scale paper mill, hence a common chemical recovery unit for a number of paper mills is a recommended option where black liquor of individual mills can be collected and processed in a central recovery plant.

The common chemical recovery unit can be installed for a group of six to eight units located in the vicinity. Initially, a bench scale plant will be constructed to determine the design parameters. A bench scale study will essentially involve characterization of black liquors from different units to establish the suitability of their use as a feed for recovery and suggest methods to improve the liquor characteristics (to obtain higher solid content black liquor). The black liquor received from mills will be concentrated in multiple evaporation section. The semi concentrated black liquor then will be taken to recovery boiler and after increasing its concentration in cascade evaporator to convert into green liquor. This liquor will be reacted with lime and converted into white liquor for reusing as a cooking chemical in the paper mills.

Tanneries

There are about 2000 tanneries in India. Approximately 75 per cent of these tanneries are in the cottage and small scale sector, 20 per cent in the medium sector and remaining 5 per cent in the large sector. The total wastewater discharged from Indian tanneries is estimated to be about 50 million liters per day. The tanneries in India are concentrated in clusters in the states of Tamil Nadu, West Bengal, Uttar Pradesh, Karnataka, Punjab and Maharashtra. Tanning industry is water-consuming industry, approximately 30–40 liters of water is used for processing 1 kg of raw hide into finished leather. Most of the tanneries in India are located near the river banks or near the natural water bodies to draw surface water. Some tanneries also use groundwater from the captive open wells. The quantity of water usage and nature of wastewater discharge in tanneries depend on type of tanning process, capacity of tannery and its production pattern.

The tanneries in India are grouped into three major categories from wastewater management point of view.

1. Tanneries having adequate land and managerial capacity to built their own effluent treatment plant. There are about 100 large scale tanneries in the country. These tanneries have put up effluent treatment units.
2. Tanneries that are closed to each other and are-amenable to have a Common Effluent Treatment Plant (CETP). Common Effluent Treatment Plants are put up for these tanneries.
3. Scattered tanneries not having adequate land which can neither put up individual effluent treatment plant nor be included in a CETP. The only option for these tanneries is to shift to new location so that they can join the CETP scheme.

In India more than 80 per cent tanneries adopt chrome tanning process, in this process 60 per cent of the applied chromium is taken by the leather and remaining is discharge in wastewater. Research organizations like NEERI CLRI and IIT, Kanpur have developed technologies for reduction in chromium salt utilization for chrome tanning and for recovery of chromium from the wastewater.

Recovery and Reuse of Chromium Salt from Tannery Wastewater

Lot of research work has been conducted in the past to evolve chemical methods of treatment to recover chromium from tannery wastewater. Some of the important methods are as under:

1. Chrome recovery system based on dewatering of chromium sludge followed by incineration to produce ash which is leached with sulphuric acid to produce reuseable chrome liquor.
2. In the second method the chrome liquor after filteration is treated with sodium carbonate to produce chromium sludge which is dewatered in rotary vacuum filter. The filter cakes are dissolved in sulphuric acid to give chrome tanning solution.
3. Another method of chromium recovery is the use of Magnesium Oxide (MgO) which has low reactivity and solubility causing chromium to settle in a very compacted way. The separation from the liquor is obtained by decanting the supernatant. The sludge can be dissolved in sulphuric acid to obtain the reuseable liquor.

Most of the work carried on recovery of chromium is confined to laboratory only. Not much work has been done on full scale application. An in-depth study will be conducted on all the above technologies to estimate the process parameters for design of full scale plant. A cost benefit analysis will be carried to assess economic viability of the process. The study will also include the effect of recovered chromium on quality of leather produced and possibility of reuse the recovered chromium for number of cycles without affecting the quality.

Textile Industries

Textile industries located at Small Scale Industrial (SSI) centers like Pali and Balotra in Rajasthan are known to be source of high wastewater pollution loads. A total of 767 units located in various areas of Pali are collectively contributing a large quantity of wastewater to the region.

NEERI has studied the pollution from these industries and prepared design packages for control of water pollution on a collective basis. Criteria as defined in Table 98.4, and the basis for treatment presented in Table 98.5, are utilized in preparing various options for such collective treatment. These options are presented in Table 98.6. The capital costs, operation and maintenance costs are specific costs and specific costs were studied in detail to arrive at economically feasible option. One CETP at Pail has been commissioned successfully. NEERI has designed nine CETPs for different types of industrial clusters. The details are presented in Table 98.7.

Recycle and Reuse of Small Scale Textile and Dying Industry Wastewater for Industrial Application

Textile processing comprises one or more the following processes desizing/kierning, bleaching, mercerizing, dying and printing. Additionally, carbonization is included for synthetic alkalinity, heavy metals, suspended solids colloidal material and colour followed by aerobic biological treatment to remove biodegradable organic matter.

Initially, bench scale treatability studies will be conducted to identify the suitable treatment unit to estimate the kinetic coefficients. Colour is an important parameter for dying industry. Dying industry wastewater will be subjected to ASP added with activated carbon. This treated effluent from modified ASP will be further treated/polished in sand filters. This treated wastewater can be reused in the process.

Table 98.4: Criteria for the Design of CETP's as Developed by NEERI in Consultation with MOEF

- ☆ Inventry of industry
- ☆ Flow and characteristics of wastewater
- ☆ Classification of wastewater based on biodegrability (A, B, C and D as per Table 98.6)
- ☆ Design of conveyance system
  - ☆ Free from problems
  - ☆ Optimized scheme
- ☆ Treatibility study
  - ☆ Bench scale
  - ☆ Pilot scale
- ☆ Stagnation of wastewater
- ☆ Pretreatment of wastewater
- ☆ Assessment and available technologies for design of CETP to develop treatment packages and optimization based on spatial distribution of CETP
- ☆ Ranking of technology options
- ☆ Reuse/recycling and resource recovery
- ☆ Disposal of treated effluents
- ☆ Cost estimation based on optimized CETP
- ☆ Cost benefit analysis
- ☆ Scheme for sharing financial burden
  - ☆ Annualized capital cost.
  - ☆ Operation and maintenance for different wastewater based on flow and mass
- ☆ Possibilities of using cleaner technologies
  - ☆ Process
  - ☆ Water consumption
  - ☆ Raw material
  - ☆ Energy requirement
  - ☆ Consideration of waste from one industry to be used as raw material for another industry

Enegy from Waste

Realizing the potential and importance of treatment of wastes and resultant recovery of energy from these wastes, the Government of India has launched in June, 1995 a National Programme on Energy recovery from Urban, Municipal and Industrial wastes with a view to promote the adoption of proper technologies as a means of improving waste management practices in the country with the goals and objectives of (*i*) Creation of conducive conditions and environment with fiscal and financial incentives to help, promote, develop, demonstrate and disseminate and utilization of wastes for recovery of energy, (*ii*) improving the waste management practices through the adoption of technologies for conversion of wastes into energy and (*iii*) promoting the setting-up of projects utilizing wastes from urban, municipal and industrial sectors.

Table 98.5: Basis for Wastewater Categorisation

WWC	*Pollutant*	*Treatment*	*Explanatory Notes*	*Pretreatment*	*Acceptability in CETP*
A	High SS	Highly settleable	Non-toxic solids highly settleable	NP	Acceptable
A	High SS	Colloidal needs coagulant	Non-toxic solids settleable after coagulation	NP	Acceptable
A	Organics	Highly biodegradable	Organic solids in solution and/or suspension, highly biodegradable	NP	Acceptable
A	Organics	Slowly biodegradable	Organic solids in solution/or suspension, slowly biodegradable, example: benzene	NP	Acceptable
B	Inorganics	Acid	Mineral acids produced or used in excess	N and SEP	Acceptable on neutralization
B	Inorganics	Alkali	Alkali produced or used in excess	CP and SEP	Acceptable on neutralization
C	High TDS	Membrane separation	Highly soluble solids need membrane separation	NIP	Require expensive collective treatment using membrane separation
D	Organics	Refractory	Naphthalene, anthracene used as raw material and/or product	NIP	Require expensive collective treatment of ozonation and then to CETP
D	Organics	Toxic	Metal complex organo-chlorine pesticides carbonates	NIP	Require expensive collective treatment or incineration

NP: No pretreatment; NIP: No individual pretreatment; N and SEP: Neutralisation and solids separation; CP and SEP: Chemical precipitation and solid separation; WWC: Wastewater category.

Table 98.6: Various Treatment Options for CETP

Option	*Treatment Alternative*
1.	Primary Settling+Aerobic Biological Process (Extended Aeration)+Sand filter+Granular Activated Column+Disinfection
2.	Primary Settling+Anaerobic Biological Process (AFFB, DP 1 day)+Cacade Aeration+Sand filter+Granular Activated Column+Disinfection
3.	Primary Settling+Anaerobic Biological Process (AFFB, DP 0.25 day)+Aerobic Biological Process (Extended Aeration)+Sand filter+Granular Activated Column+Disinfection
4.	Physico-chemical process+Aerobic Biological Process (Extended Aeration)+Sand filter+Granular Activated Column+Disinfection
5.	Physico-chemical process+Sand filter+Granular Activated Column+Disinfection
6.	Physico-chemical process+Anaerobic Biological Process (Extended Aeration)+Disinfection
7.	Physico-chemical process+Anaerobic Biological Process (AFFB, DP 0.5 day)+Cacade Aeration+Disinfection
8.	Primary Settling+Aerobic Biological Process (Extended Aeration)+Disinfection

AFFB: Anaerobic Fixed Film Bed Reactor.

Table 98.7: List of CETP's Designed by NEERI for Control of Water Pollution/Resource/Recovery/ Recycle in Cluster of Small Scale Industries

Sl.No.	State/Town	Industries in Cluster		Flow	Annualised Cost	Annual (Rs.	Salient Features
		Type	No.		(Rs. Lakhs)	Lakhs)	
1.	Gujarat, Vapi	Chemical Dying etc.	750	16.0	271	114	Wastewater recycling for industrial area
2.	Himachal Pradesh, Parwanoo	Electroplating, Engg. *etc.*	76	1.0	29	38	Conforming to standards
3.	Himachal Pradesh, Barotiwala	Chemical, pulp and paper, *etc.*	34	8.4	313	413	Wastewater recycling for industrial area
4.	Himachal Pradesh, Kala-Amb	Paper, textile, etc.	22	16.0	521	715	Wastewater recycling for industrial area
5.	Himachal Pradesh, Mehatapur	Textile, Engg. etc.	111	0.37	15	27	Effluent conforming to standards
6.	Punjab, Amritsar	Textile	50	50.0	297	–	Effluent conforming to standards
7.	Rajasthan, Balotra	Textile	367	7.7	635	191	Wastewater recycling for industrial area
8.	Rajasthan, Pali	Textile	767	36.0	1900	1070	Wastewater recycling for industrial area
9.	Tamil Nadu, Salem	Sago	641	12.0	291	108	Biogas recovery and aquaculture

Under this scheme 2.75 MW power generating plant using rice husk as feed material has been installed in Andhra Pradesh by M/s Gowthami Solvents Limited, Tanuku. 1 MW power generating plant based on biogas, generated from spent wash at K.M. Sugar Mills Distillery at Faizabad is under installation and biogas generating plants in over 50 distilleries in the country utilizing distillery spent wash have already been installed, a 4000 cubic meter per day biogas generating plant utilizing liquid wastes of slaughter house at Hyderabad has been installed by M/s Al Kabir Exports Limited.

Biogas Generation from Industrial Wastewater

The proposed research work on anaerobic system will focus on process modifications. This work will involve estimation of factors like SRT, type of reactor, and loading rates to increase the process efficiencies towards optimal performance level. This work will also involve careful studies of factors like pH, temperature to maximise the growth rate of microorganisms involved in biological degradation of organic compounds. Determination of kinetic constants for different reactor types and industrial wastewaters. Scale up formulation for different types of reactor systems will be developed as scanty information is available on these aspects.

Intensive experimentation in above areas will increase the level of technology in the treatment of industrial wastes by anaerobic process to a much higher plateau than is realized today.

Agricultural Category

India is an agricultural country. Majority of the Indian population live in rural areas and agriculture is their main occupation, According to an estimate 88 per cent of the Indian water usage is consumed by agriculture sector. As per the livestock census of 1992 India has 204.5 million cattle 83.5 million buffaloes, 50.8 million sheeps, 115.3 million goats, 12.8 million pigs, and 307 million poultry birds. This is one of the largest livestock population in the world. The rate of the chemical fertilizers consumption has increased rapidly in our country as the emphasis is on increasing food grain production. The consumption of chemical fertilizers is given in Table 98.8. The compound growth rate in agricultural production during 1950 to 1996 is 2.67 per cent per annum as against the compound rate of nitrogeneous consumption, which is 11 per cent.

Table 98.8: Consumption of Nitrogenous Fertilizers in India

Sl.No.	*Year*	*Consumption Tonnes/Year*
1.	1951–1952	55000
2.	1961–1962	249800
3.	1971–1972	179800
4.	1981–1982	4068700
5.	1991–1992	8046300
6.	1996–1997	14930000
7.	2000–2001	22383000

Source:

(*i*) Comprehensive industry document Fertilizer Industry, CPCB. New Delhi, COINDS/50/1994–95.

(*ii*) India 1998, Ministry of Information and Broadcasting, New Delhi

Contrary to the belief, agriculture sector is a major source of water pollution. The major factors of importance are reduction in wetland area, monoculture crop production, extensive use of commercial fertilizers and pesticides, irrigation and intensified husbandry. Irrigation, which consumes maximum quantity of water, is known to cause water quality problems such as (*i*) salinization, (*ii*) alkalinization, (*iii*) erosion of irrigated land and (*iv*) increase in sediments in downstream areas like lakes and dams.

There is a tendency in farmers to apply more fertilizers than the optimal in order to reduce a risk of being short of nutrients in good weather years. Excess chemical fertilizers and pesticides are washed off and reach the surface water or groundwater. A survey has shown alarming increase in nitrates of groundwaters of north India.

Livestock breeding and animal husbandry is another source of water pollution from agriculture sector. The major pollution source for animal husbandry are cattle slurry, (dirty water which contains washing from the farm yard and milking parlour), silage effluent (fermented grass used as cattle feed) and animal faeces and urine. Table 98.9 gives the BOD from animal husbandry sources. It is seen that BOD of animal sources is approximately 100 higher than raw domestic sewage. Considering the livestock population of India and BOD load, the estimated figures can be alarming.

There is a major and extensive built-up of water pollution from agriculture. However this is going largely unattended. Popular perceptions as well as regulatory institutions have focused upon urban and industrial pollution. There are no laws governing the pollution from agriculture sector vis-à-vis no extensive studies have been conducted in our country to estimate the water pollution from agriculture sector.

Table 98.9: BOD of Common Animal Husbandry Wastes

Sl.No.	Effluent Type	BOD Range (ppm)
1.	Raw domestic sewage	300–400
2.	Cattle slurry	10000–20000
3.	Poultry slurry	30000–35000
4.	Sillage effluent	30000–80000

Causes for Present State of Affairs

Problems of Wastewater Management in Domestic Sector

Social Awareness

Indians in general are not aware of the environmental consequences of the discharge of the untreated wastewaters. Indian urban centres are administered by municipalities or municipal corporations which are governed by body of elected members. Potable water supply and collection and treatment of wastewater is not an important agenda of municipal elections. Hence wastewater treatment is always a neglected issue and has least priority in policies of the municipalities.

Finance

Municipalities often depend upon the State/Central Government for funds to construct the sewage treatment plants. These funds are either partially available or not available on time.

Bureaucracy

Administrative bureaucratic machinery in the municipalities and in State and Central Governments follow a lengthy and complicated procedure for decision making which cause excessive delay in policy formation and its implementation.

Lack of Trained Manpower

It is often observed that sewage treatment plants are not equipped with qualified and trained manpower to operate them efficiently. The staff is often inexperienced.

Problems of Wastewater Management in Industrial Sector

Technology

Although technology for conventional treatment for water and wastewater is available in the country, various studies in the past have indicated need for specific technologies for specific pollution parameter control. CPCB has assessed various wastewater treatment technologies with respect to know-how, equipment and there availability in India. The details of this study are reported in Table 98.10. Some of these technologies are already developed and demonstrated in the country while some needs development of design criteria and demonstration of applicability and efficiency.

Law Enforcement

The Central Pollution Control Board (CPCB) is the national apex body for assessment, monitoring and control of water pollution. The executive responsibilities for enforcement of the Acts for Prevention and Control of Pollution of Water (1974) and of the Water (Cess) act, 1977 are carried out through the board. Under the Environment (Protection) act, 1986, effluent and emission standards in respect of 61 categories of industries have been notified.

Table 98.10: Assessment of Wastewater Technology Status in India

Sl.No.	Process	Control Parameter	Industries where Required	Present Status in India	Assessment of Import Needs			Notes
					Technology	Haridware		
						Total	Critical Component	
1.	Adsorption	BOD, COD, Toxic Compounds, Colour, Odour	Dyes and intermediates, Textile, Paper, Pharma-ceuticals, Pesticides, Fruits, Canning, Petro-chemicals, Rubber Chemicals, Man-made Fibres	Limited product range	Yes	No	–	Import of know how for technology of activated carbon manufacture and regeneration. Import of technology for synthetic adsorbents. Initial import of special adsorbents including activated carbon.
2.	Biotechnology							
2.1	Conventional	BOD, COD	For all industries– Biodegradable and slowly biodegradable					
2.1.1	Aerobic							
A.	General			Developed	No	No	No	
B.	Rotating Biological Contractors (RBC)			Underdeveloped	Yes	No	Yes	Import of process know how and in Haridware initial import of Disc Segments only.
C.	Diffused Aeration			Limited range	Yes	No	Yes	Import of know how for manufacturing of diffusers and initial import of porous diffusers of various materials.
2.1.2	Anaerobic			Developed	Yes	No	No	Import of know how for the design of reactors including anaerobic activated sludge, egg shaped digesters etc.

Contd...

Table 98.10–Contd...

Sl.No.	Process	Control Parameter	Industries where Required	Present Status in India	Assessment of Import Needs: Technology	Haridware: Total	Haridware: Critical Component	Notes
2.2	Advanced	BOD, COD						
A	Microbial Cultures/ Enzymes			Research stage	Yes	No	Yes	Initial import of know how and cultures/enzymes especially for recovery of single products.
B.	Bioreactor			Underdeveloped	Yes	No	Yes	Import of know how for design of aerobic and anaerobic reactors such as deep shaft, tows, loop, upflow and downflow fixed film and fluidized bed etc. Also initial import of critical comparison, if necessary.
3	Disinfection							
3.1	Conventional (Chlorination etc.)	Pathogens		Developed	No	No	No	
3.2	Ultraviolet	Pathogens		Nil	Yes	No	Yes	Import of know how and critical components
4.	Dissolved Air Floatation	Suspended Solids/ Oil and Grease	Metal processing and flashing, Petroleum refining and petro-chemicals, Dairy, Edible Oil, Textile processing, Dyestuff, Tanneries, Flood and Fruit processing	Limited product range	Yes	No	Yes	Import of know how and critical components
5.	Electro-dialysis and Electro-deposition	Heavy Metals	Metal processing and finishing, Caustic Soda	Nil	Yes	No	Yes	Import of know how and critical components especially membranes

Contd...

Table 98.10–Contd...

Sl.No.	Process	Control Parameter	Industries where Required	Present Status in India	Assessment of Import Needs			Notes
					Technology	Haridware		
						Total	Critical Component	
6.	Evaporation & Crystallization	Toxic Organic and Inorganic Salts	Pulp and Paper, Distillery, Metal Processing and Finishing, Textile, Organic Chemicals and Petrochemicals	Underdeveloped	Yes	No	Yes	Import of know how for various evaporators and critical components
7.	Incineration	BOD, COD, Toxic Organics	Paper, Distillery, Tannery, Pharmaceuticals, Pesticides, Petrochemicals, Petroleum, Dyestuff Intermediates	Underdeveloped	Yes	No	No	Import of know how especially for effective combustion system.
8.	Ion Exchange	Heavy Metals, NH_3, Cyanides, Fluorides	Caustic Soda, Fertilizer, Manmade Fibres, Rayon, Pulp and Paper, Distillery, Pharmaceuticals, Metal processing and finishing, Electronics, Ferrous and Non-ferrous	Limited product range	Yes	No	No	Import of know how for the manufacture of macro-raticular resins and selective ion exchange resins. Initial import of macro-raticular resins and selective ion exchange resins.
9.	Ozonation	BOD, Colour, Odour, Refractory Organics, Toxic Chemicals, Pathogens	Iron and steel, Dyes and Intermediates, Pharmaceuticals, Petrochemicals, Paper, Pesticides	Nil	Yes	No	Yes	Import of know how and initial imports of ozonation
10.	Reserve Osmosis	Dissolved Organics	All industries for reclamation of water	No	Yes	No	Yes	Import of know how and membranes. Import of know how for suitable pump manufacture and critical components

Contd...

Table 98.10–Contd...

Sl.No.	Process	Control Parameter	Industries where Required	Present Status in India	Assessment of Import Needs			Notes
					Technology	Haridware		
						Total	Critical Component	
11	Sedimentation	Suspended Solids, Oil and Grease						
11.1	Conventional			Developed	No	No	No	–
11.2				Underdeveloped	Yes	No	No	Import of know how
12.	Sludge Handling Treatment and Disposal							
12.1	Handling		All industries	Limited product range	Yes	No	No	Import of know how for the manufacture of straw pumps for high consistency sludges and air lift pumps.
12.2	Treatment							
12.2.1	Thickening			Developed	No	No	No	–
12.2.2	Dewatering			Underdeveloped	Yes	No	Yes	Import of know how for manufacture of centrifuge, belt filters etc. Also import of critical components.
12.2.3	Digestion	Organic Sludge		Underdeveloped	Yes			Import of know how for reactor design
12.2.4	Composting	Organic Sludge		Developed	No	No	No	
12.2.5	Thermal Dewatering	Organic and Inorganic Sludge		Underdeveloped	Yes	No	No	Import of know how
12.3	Ultimate Disposal							
12.3.1	Smelting	Toxic Inorganic Sludge		Developed	No	No	No	
12.3.2	Incineration	Organic Sludge		Underdeveloped	Yes	No	No	Import of know how

Contd...

Table 98.10–Contd...

Sl.No.	Process	Control Parameter	Industries where Required	Present Status in India	Assessment of Import Needs			Notes
					Technology	Haridware		
						Total	Critical Component	
12.3.3	Stabilization and Solidification	Toxic and Hazardous Sludges		Nil	Yes	No	No	Import of know how
13.	Ultrafiltrations	COD, Dissolved Polymers and Colloidals	Dairy, Oil Refinery, Wool Industry, Breweries, Paper, Paints	Underdeveloped	Yes	No	Yes	Import of know how and critical components especially membranes.
14.	Control and Monitoring	Various parameters	All industries	Limited range products	Yes	No	Yes	Import of know how for instruments for continuous monitoring and sampling systems, specific ion electrodes, flow measurements and recording. Also import of critical components and initial import of instruments.

State Pollution Control Boards are formed to enforce the laws constituted by Central and State Governments at state level. However these pollution control boards are often deficient of funds, poorly manned and have inadequate infrastructure for proper regulation of the pollution control laws.

Policy

An approach to industrial wastewater treatment should be to consider the problem in its totality and to include both preventive and curative aspects. This means that pollution control begins from the point of generation of wastewater and ends with ultimate disposal point. This concept could be broadly divided into two components, namely inplant control and end of the pipe treatment. In plant control consists of measures, which will reduce the wastewater quantitively and reduce the pollution potential qualitatively. This is achieved through appropriate selection of raw materials, cleaner technologies for production, conservation through recycle and reuse, byproduct recovery and process and equipment modifications. This leads to preventive, internal, systematic, integrated, conservational and low pollution systems, which considerably reduce the load on end of the pipe treatment. End of the pipe treatment, theoretically, deals with treatment and disposal of wastewater, which are residual, after all the inplant measures are applied. In practice, however till recent past, inplant measures were not emphasized and consequently the major pollution control efforts were concentrated on the end of the pipe treatment. But with the scenario of large scale industrial development, the focus of the policy planners and enforcing agencies should shift towards inplant pollution control thus permitting a holistic approach to pollution.

Problems of Environmental Management in Small Scale Industry (SSI)

Management/Economics

No Investment Returns on Pollution Control

The management of SSI's have taken the approach that pollution control is a dead-end investment. In many cases this fear is indeed realised, where the cost of installing pollution does not add to the productivity and to the bottom-line profits of the enterprise. In addition the cost of operating the pollution control equipment cuts deeply into the normal cash-flow of the unit.

Investment for Pollution Control Higher than Initial Investment of Plant

The cost of pollution control has traditionally been compared with the initial investment required to set up the industry. As in the case with many highly polluting small units, *e.g.* glass, lime kiln, and dye-stuffs industry, the initial investment is of the order of 3 to 5 lakhs. Pollution control designed to reduce the pollutants to reasonable levels, would cost about the same amount. This causes the enterpreneurs not to install pollution control devices. However, the cost of pollution control should be compared with the annual turn-over of the industry, because in some of these cases, the turn-over could be many times the initial investment. This fact creates the additional problem, that industry is more willing to enter protracted legal wrangling to buy time, rather than install pollution control equipments. When push comes to shove, they are willing to close down the industry and start over again.

Awareness

The awareness among the management of SSI's is fairly low with respect to technical options, finance availability in terms of subsidies and tax incentives for pollution control, and the legal repercussions of non-compliance with regulations. SSI is typically not aware of national level efforts in development and availability of non-or less- polluting technologies.

Operation of Existing Pollution Control Equipments

Due to the shortage of electricity and water in various states, having large number of small scale industries, *e.g.* Uttar Pradesh and West Bengal, operation of the pollution control devices is severely affected. And when electricity is available, the entrepreneurs choose not to operate the control devices to save on operating costs.

Economies of Scale

Due to the size of SSI sector, cost-effective and technically feasible pollution control becomes exceedingly difficult, if not impossible. Also due to the small size, the units often times operate in an inefficient manner. For example, the small lime kilns operate at low thermal efficiencies losing significant amounts of heat as waste heat. Also, the heat content of the flues are so small that heat recovery is not possible. In another example, hot flue gases at around 600°C from cupolas pose severe problems for removal of particulates from the flue. The option of recovering the waste heat, as steam is infeasible because there is no demand for steam in the foundry or the immediately neighbouring industries.

Finance

Even though the Government provides soft loans for pollution control, with the interest rates in the neighbourhood of 12–14 per cent p.a., it hardly makes the investment in pollution control cost effective. SSI financial institutions and banks providing loans for setting up of SSI's are not aware of the environmental consequences of setting up the industry. As a result, the cost of pollution control has not been included in the feasibility studies, or required in business proposals from the entrepreneurs. If environmental costs are considered, they are done apart from the cost of the project, and not intemalised as a part of the project.

Technology

Know-how for cost-effective control of pollution for the SSI sector is not readily available. A concerted effort in developing pollution control and less polluting production technologies do not exist, and efforts of R&D institutions are sporadic and they do not solve the problems effectively. Reputed consulting engineers have not shown interest in catering to the small scale, where as several unreliable consulting–shops have mushroomed, providing inadequate and at times grossly wrong solutions.

Regulatory/Institutional

No effective regulatory mechanism exists for regulating, controlling and monitoring the pollution status of the SSI. Both the Central Pollution Control Board (CPCB) and the State Pollution Control Board (SPCB) are short of trained manpower and adequate funds to regulate large industries, let alone the small scale sector. Due to their very large numbers the efforts of the PCBs, if any, have been adhoc, and ineffective. Vast differences occur between the levels of effort of the PCBs in various states. This encourages location of SSIs in state providing lax control. This also causes unfair treatment with respect to the environment to the SSIs located in different states, affecting their competitive position.

No database regarding the pollution from the SSIs is available to assist in development of suitable environmental management policies, and regulations. Forthcoming efforts of CPCB and financial institutions towards environmental control have only started scratching the surface.

Problems of Wastewater Management in Agriculture Sector

The successive Governments formed at Centre and at State have always championed the cause of farmers by advocating the policy to increase agriculture production by increasing the chemical fertilizer dosing. These governments have encouraged this policy by providing subsidized chemical fertilizers, which has led to over application of fertilizers. Any attempt to discourage this by increasing the fertilizer price will invite the wrath of powerful farmers lobby, which no government can afford due to its long-term political implications. Similar is the case with livestock breeding. Traditional Indian values encourage to have maximum number of cattle, as it increases ones social standing in rural India. Cows and bull are considered, as holy animals and religion prohibits their slaughtering. These reasons have caused rise in livestock population of the country.

Towards Pragmatic Solutions

To affect and change the burgeoning problem of wastewater treatment in India, pragmatic solution to these problems at the macro level (policy level) as well as at the grass roots levels in terms of the implementation needs to be made. One set of attempts without the other, would as in the past fail to produce the desired levels. Following are some of the possible solutions to tackle this problem.

1. Adoption of low cost wastewater treatment systems for treatment of wastewater from small towns and cities. These technologies include water hyacinth pond, oxidation ponds, and application of wastewater for irrigation and constructed wetlands. These technologies require less capital and operation and maintenance cost but require large quantity of land which is available in plenty in the vicinity of smaller towns. Valuable by-products can also be recovered from these treatment schemes. For example, methane, fertilizer and water for irrigation from water hyacinth pond. It is reported that domestic wastewater (3650×10^6 m^3) can contribute 219×10^3 tonne of nitrogen, 73×10^3 tonne of phosphate, 146×10^3 tonne of potash and 146×10^3 tonne organic matter with an economic value of Rs. 2000 million per annum.
2. By imparting environmental education to the urban population by making them aware about the problems and consequences of discharging untreated wastewater into the environment and their duties towards social hygiene.
3. Providing financial support to the municipalities for wastewater management either from State/Central Governments or from financial organisations at reasonable terms.
4. Reducing the bureaucratic delays simplifying the process of decision making.
5. Providing suitable training to the treatment plant operators for efficient running of the treatment plants. Common effluent treatment plants provide a viable option for a cluster of small industrial units which are not capable of setting up the treatment units individually on their own. Financial incentives in the form of grant should be made available for constructing CETPs.
6. The need to encourage adoption of cleaner technologies (CT) for industrial production is required. Government of India is taking steps in this direction by setting up an Indian center for promotion of cleaner technologies (ICPC) with a large number of agencies operating as a network with access to databases within as well as outside the country. The ICPC, after becoming operational shortly, will be able to service the industry for waste management purposes in the following ways:
   - ☆ Recovering wastes.

- ✰ Treatment and disposal of wastes.
- ✰ Waste exchange to identify users for potential recyclable wastes.

7. It is necessary to underline the fact that mere availability of technology is no guarantee that waste management would be undertaken in a proper manner. Skilled manpower and management package is equally important input for successful completion of waste management programme. Considering the quantum of waste to be handled in the public sector, it is imperative that the constitutional mechanism is suitably strengthened and also made accountable for successfully and timely completing the assigned task. This aspect, however, has not received the requisite attention and needs to vigorously pursued along with the promotion of the research and development efforts in such areas as biotechnology intervention for waste management.
8. Another approach would be to move from CETP to combined effluent treatment plant to jointly treat domestic sewage with industrial wastes. Addition of domestic sewage from surrounding community will not only solve the problem of community waste treatment but will also improve the treatability of industrial wastewater.
9. There are at present several fiscal incentives for installations of pollution control equipment and for shifting industries from congested areas. However, economic instruments should be amended to encourage the shift from curative to preventive measures, internalise the cost of pollution and conserve resources, particularly water. One possible suggestion is effluent should be charged based on the nature and volume of releases to the environment. The charge will be based on the cost of treatment and the flow discharged, thus putting thrust on optimum release of wastewater. The precise choice of economic instruments adopted should be determined by the ease with which releases can be measured, as well as prospective changes in technology and market structure.
10. At present the disincentives include penalties and fines for non compliance under the various acts as well as the water cess. All of these incentives are extremely low in relation to the cost of pollution prevention/control for compliance. This can be overcome by increasing the amounts of penalties and fines.
11. In addition to the above, some market instruments could also be employed. One such instruments is the "Eco-mark" labeling system. This system, through public awareness, could provide market forces to encourage environment friendly production processes.
12. Another marketing instrument could be rating of industries, similar to financial rating (CRISIL) or ISO-9000. A good tool would be BS-7750, the British standard for environmental management system or ISO-14000.
13. Problems associated with SSI could be solved by helping genuine SSI's especially those promoted by young and technically qualified entrepreneurs. Although adequate network of institutional finance exists, yet there is a need for coordinating the flow of both short and long term capital. Government should coordinate making capital available to SSI's, on right time and at appropriate terms. Financial institutions created for assisting SSI's should require environmental impact studies on appropriate scale prior to project financing. Environmental experties should be made available with the financial institutions for this purpose.
14. By providing R&D encouragement; create instruments for R&D related to cleaner technologies of production and pollution control from SSI.

15. Multimedia compliance represents a more holistic approach for achieving a clean environment. This approach examines the big picture emissions, the geographic locations, wind patterns, social impact-in order to come up with a plan that will be both effective and practically reliable. Release to all environmental media (*e.g.* Air emissions, wastewater discharges, and hazardous wastes) are considered simultaneously and in an integrated manner. This means that the environment permit or license for a facility address all environment media, as do compliance inspections and enforcement responses.

The advantages are:

1. It allows the consideration of cross media effects, *e.g.* the effect of wastewater treatment on solid waste and air pollution generation.
2. It allows the facility operation to consider a full range of environmental improvements.

Forward thinking decision makers in several countries have incorporated this new focus into legislation and policy with positive results.

1. By promoting and implementing low cost wastewater treatment schemes for management and treatment of wastewater from agriculture and rural areas. One such example is wetland transformation. The capacity of wetlands to remove nutrients and organic loads from freshwaters has been emphasized. Natural riparian wetlands are believed to play a key role in the removal of nutrients, especially nitrates. Several studies have shown the efficiency of constructed wetlands in groundwater storage and recharge has also been proven in many areas, yet it is often neglected in water management planning.
2. Restoration or creation of buffer strips or resparian ecotone along the water courses can arrest nutrient flow in rivers. Such areas are known to have high nitrogen removal capacity by denitrification and uptake by macrophytes followed by harvesting and storage in the sediments. Wetlands, possibly in the form of wet buffer strips along the water courses, have the same potentials.
3. By implementing the schemes to popularise the use of organic nutrients in agriculture through the use of compost green manures and biofertilisers.
4. By making a national policy of reducing the consumption of chemical fertilisers and pesticides.
5. Controlling the livestock population and improving the quality of breed.
6. By increasing the awareness on environmental sanitation specially in rural areas by providing environmental education.

Conclusion

Water and wastewater management problems in India in different from developed countries as it is linked with gross inadequacy of environmental sanitation, lack of political will, shortage of fund and, no availability of technology and social, religious and cultural values. While taking the stock of the situation, one is left to ponder about how little has been done so far in this regard and how big is the magnitude of the problem ahead.

Index